Foundations of Finitely Supported Structures

Andrei Alexandru • Gabriel Ciobanu

Foundations of Finitely Supported Structures

A Set Theoretical Viewpoint

Springer

Andrei Alexandru
Institute of Computer Science
Romanian Academy
Iaşi, Romania

Gabriel Ciobanu
Institute of Computer Science
Romanian Academy
Iaşi, Romania

ISBN 978-3-030-52964-2 ISBN 978-3-030-52962-8 (eBook)
https://doi.org/10.1007/978-3-030-52962-8

This Springer imprint is published by the registered company Springer Nature Switzerland AG
The registered company address is: Gewerbestrasse 11, 6330 Cham, Switzerland

"The infinite! No other question has ever moved so profoundly the spirit of man; no other idea has so fruitfully stimulated his intellect; yet no other concept stands in greater need of clarification than that of the infinite."

David Hilbert

Contents

Preface

Finitely supported structures have historical roots in the permutation models of Zermelo-Fraenkel set theory with atoms elaborated by Fraenkel and Mostowski in the 1930s in order to prove the independence of the axiom of choice from the other axioms of a set theory with atoms. They are also related to the recent developments of Fraenkel and Mostowski axiomatic set theory and of nominal sets; from the beginning of this century, finitely supported structures appeared in computer science to describe new ways of presenting the syntax of formal systems involving variable-binding operations. Inductively defined finitely supported sets involving the name-abstraction together with Cartesian product and disjoint union can encode syntax modulo renaming of bound variables. In this way, the standard theory of algebraic data types can be extended to include signatures involving binding operators. In particular, there is an associated notion of structural recursion for defining syntax-manipulating functions and a notion of proof by structural induction. Various generalizations of finitely supported sets were used in order to study automata, languages or Turing machines that operate over infinite alphabets; for this a relaxed notion of finiteness called 'orbit finiteness' was defined and means 'having a finite number of orbits under a certain group action'. Finitely supported sets were studied from both a set theoretical perspective (by M.J. Gabbay who introduced the so called axiomatic Fraenkel-Mostowski set theory which is actually Zermelo-Fraenkel set theory with atoms together with a new finite support axiom requiring the existence of a finite support for every hierarchical set theoretical construction) and a categorical perspective (by A. Pitts who defined nominal sets as classical Zermelo-Fraenkel sets equipped with a canonical group action of the group of permutations of a fixed ZF set of basic elements called 'the set of atoms' by analogy with the Fraenkel-Mostowski approach, satisfying additionally a finite support requirement; nominal sets represent an alternative Zermelo-Fraenkel approach to the non-standard axiomatic Fraenkel-Mostowski set theory).

In this book we also equip classical sets with permutation actions. The world of finitely supported sets contains the family of classical (non-atomic) Zermelo-Fraenkel sets (having the property that all of their elements are empty supported), and the family of atomic sets (which contain at least one 'basic element/atom' somewhere in their structure) having finite supports as elements in the powerset of a set equipped with a permutation action. The main goal of this book is to present a set theoretical approach for studying the foundations of finitely supported sets and of related topics. In this sense we analyze the consistency of various forms of choice, as well as the consistency of results regarding cardinality, maximality and infinity, in the framework of finitely supported sets. We also introduce finitely supported algebraic structures as finitely supported sets that are equipped with finitely supported algebraic laws or with finitely supported relations. We present detailed examples of finitely supported partially ordered sets and finitely supported lattices, and we provide new properties of them. Some properties (especially fixed point properties, properties regarding cardinalities order, or results of comparing various forms of infinity) are specific to the theory of finitely supported sets, leading from the finite support requirement. A complete listing of the properties of basic elements (atoms) in the framework of finitely supported sets is also carried out. The notion of infinity is described within finitely supported sets, and several definitions of infinity are compared internally in this new framework. Finally, we present the concepts of freshness and abstractions from a slightly different perspective than in the theory of nominal sets.

The translation of a result from a non-atomic framework into an atomic framework could be quite complicated. For example, in Zermelo-Fraenkel framework it is known that both Kurepa maximal antichain principle and multiple choice principle imply the axiom of choice. Such a result is not preserved in Zermelo-Fraenkel set theory with atoms as proved by Jech. The translation of a result from a non-atomic framework into an atomic framework of sets with finite supports is even much more complicated. We analyze if a classical Zermelo-Fraenkel result (obtained for non-atomic sets) can be adequately reformulated by replacing 'set' with 'finitely supported set (under the canonical permutation action)' in order to remain valid also for atomic sets with finite support. We investigate what results in the classical non-atomic set theory are preserved in the theory of finitely supported atomic sets. We also analyze if there are specific properties of finitely supported sets that do not have a related Zermelo-Fraenkel (non-atomic) correspondent. In this way, infinite structures hierarchically constructed from the related set of basic elements can be characterized in a finitary manner by analyzing their finite supports. A meta-theoretical principle that works within the world of finitely supported sets states that for any finite set S of atoms, anything that is definable (in a higher-order logic) from S-supported structures using S-supported constructions is S-supported. However, the formal application of this method actually consists in a hierarchical step-by-step construction of the support of a certain structure by employing the supports of the substructures of a related structure, and has limitations related to results requiring choice principle or hidden choice.

The book represents a set theoretical development for the (set theoretical) foundations of the theory of finitely supported sets and structures (either originally presented as Fraenkel-Mostowski sets, or later defined as nominal sets). We collect various results on this topic and present them in a uniform manner. More than half of the results presented in this book are original, especially all the results regarding choice principles and their equivalences, results regarding cardinalities (Trichotomy, Cantor-Schröder-Bernstein theorem and its dual, cardinals arithmetic, cardinals ordering, Dedekind infinity, Tarski infinity, Mostowski infinity), results regarding the relationship between various forms of infinity, specific fixed point properties for finitely supported ordered structures, constructions of finitely supported algebraic structures with their specific properties, important properties of atoms (and also of functions on atoms and of higher-order constructions on atoms), and properties of connecting atomic and non-atomic sets. Therefore, this is a pure theoretical book accessible to a broad audience. We do not discuss here computer science applications of finitely supported sets (which are treated in [44] by using nominal sets).

To conclude, we focus on set theoretical foundations and go back to the original Fraenkel and Mostowski approach. Nominal sets are called in this book 'invariant sets' motivated by Tarski's approach on logicality (logical notions are, according to Tarski, those notions which are left invariant under the effect of the one-to-one transformations of the universe of discourse onto itself). We discuss foundations of the finitely supported sets, meaning that we analyze the consistency of various Zermelo-Fraenkel results within the framework of the invariant sets where atomic structures are involved, and also present specific properties of atomic structures. There is no major difference regarding 'Finitely Supported Mathematics' (which is a generic name for the theory of finitely supported algebraic structures) and the 'nominal approach' related to basic definitions, except that the nominal approach (whose value we certainly recognize) is related to computer science applications, while we work on foundations of mathematics and experimental sciences (by studying the validity, the consistency and the inconsistency of various results within the framework of atomic finitely supported sets). Our goal is not to re-brand the nominal framework, but to provide a full collection of set theoretical results regarding the foundations of finitely supported structures.

Acknowledgements

We are grateful to our scientific ancestors and mentors who paved the path to this book, great scientists upon whose shoulders we stand.

Chapter 1
The World of Structures with Finite Supports

Abstract We introduce the theory of atomic finitely supported algebraic structures (that are finitely supported sets equipped with finitely supported internal operations or with finitely supported relations), and describe topics related to this theory such as permutation models of Zermelo-Fraenkel set theory with atoms, Fraenkel-Mostowski set theory, the theory of nominal sets, the theory of orbit-finite sets, and the theory of admissible sets. The motivation for developing such a theory comes from both experimental sciences (by modelling infinite algebraic structures hierarchically defined by involving some basic elements called atoms in a finitary manner, by analyzing their finite supports) and computer science (where finitely supported sets are used in various areas such as semantics, domain theory, automata theory and software verification). We describe the methods of translating the results from the non-atomic framework of Zermelo-Fraenkel sets into the atomic framework of sets with finite supports, focusing on the S-finite support principle and on the constructive method of defining supports. We also emphasize the limits of translating non-atomic results into an atomic set theory by presenting examples of valid Zermelo-Fraenkel results that cannot be formulated using atomic sets.

1.1 A Short Introduction

There does not exist a formal description of infinity in those sciences which are focused on quantitative aspects. Questions such as 'What represents the infinite?', 'How could the infinite be modelled?', or 'Does the infinite really exist or is it just a convention?' naturally appear. In order to provide appropriate answers, we present 'finitely supported mathematics' (FSM) which is a generic name for 'the theory of finitely supported algebraic structures'. FSM is developed by employing the general principle of finite support claiming that any infinite structure hierarchically defined by involving some basic elements called atoms must have a finite support under a canonical permutation action. Informally, in FSM framework we can model infinite structures (defined by involving atoms) by using only a finite number of character-

A. Alexandru, G. Ciobanu, *Foundations of Finitely Supported Structures*,
https://doi.org/10.1007/978-3-030-52962-8_1

istics. More precisely, in FSM we admit the existence of infinite atomic structures, but for such an infinite structure (hierarchically constructed from $\emptyset$ and from the set of atoms) we consider that only a finite family of its elements (i.e., its 'finite support') is 'really important' in order to characterize the related structure, while the other elements are somehow 'similar'. As an intuitive motivation in a λ-calculus interpretation, the finite support of a λ-term modulo α-equivalence is represented by the set of all 'free variables' of the term; these variables are those who are really important in order to characterize the term, while the other variables can be renamed (by choosing new names from an infinite family of names) without affecting the essential properties of the λ-term. This means that we can obtain an infinite family of terms starting from an original one (by renaming its bound variables), but in order to characterize this infinite family of terms it is sufficient to analyze the finite set of free variables of the original term.

Finitely supported mathematics has connections with the Fraenkel-Mostowski permutation model of Zermelo-Fraenkel set theory with atoms [41], with Fraenkel-Mostowski axiomatic set theory [29], with the theory of nominal sets on countable sets of atoms which do not have an internal structure [44], and with the theory of generalized nominal sets on sets of atoms which may have an internal structure [23]. Actually, FSM corresponds to Pitts' nominal sets theory by analyzing nominal sets (or, more generally, finitely supported subsets of nominal sets) endowed with a finitely supported algebraic structure (such as nominal monoids, nominal groups, nominal partially ordered sets, etc) and with the mention that the countability of the set of basic elements is ignored. Nominal sets are called in this book 'invariant sets', motivated by Tarski's approach on logicality; this aspect is explained below.

Intuitively, FSM is the algebraic theory obtained by replacing 'object' with 'finitely supported object (under a canonical permutation action)' in the classical Zermelo-Fraenkel set theory (ZF). The principles of constructing FSM have historical roots both in the definition of logical notions by Alfred Tarski [49] and in the Erlangen Program of Felix Klein for the classification of various geometries according to invariants under suitable groups of transformations [38]. There also exist several similarities between FSM, admissible sets introduced by Barwise [19] and Gandy machines used for describing computability [30]. FSM sets are finitely supported subsets of invariant sets (where invariant sets developed over countable families of atoms are actually nominal sets). We use a slightly different terminology motivated by Tarski's approach regarding logicality (i.e. a logical notion is defined by Tarski as one that is invariant under the permutations of the universe of discourse) and because our results are related to foundations of mathematics, i.e. we study choice principles, results regarding cardinality order, results regarding cardinality arithmetic, results regarding various forms of infinity (Dedekind infinity, Tarski infinity, Mostowski infinity, ascending infinity, etc), results regarding fixed points, results regarding the connections between atomic and non-atomic sets, results regarding finitely supported binary relations and so on. The value of the nominal approach is recognized. However, it is related to computer science applications (which are described well in [44]), while we work on foundations of mathematics by studying the validity, the consistency and the inconsistency of various results in an 'atomic' refinement of ZF set theory. Actually, our work is connected also with the Fraenkel-Mostowski set theory (FM), admissible sets and amorphous sets (that inspired actually the 'nominal approach'). We do not minimize the benefits of nominal approach that has significant applications in areas such as semantics, automata theory and verification, but regarding the foundations of finitely supported structures we consider that 'FM', 'invariant' or 'FSM' are more adequate names. Our book can be seen as an in-depth study on the foundations of FM sets and nominal sets (defined over families of basic elements that are not necessarily countable) accessible even to graduate students. For computer science applications of nominal sets (that are not treated in this book) and for generalizations in this direction (orbit-finite sets, automata over infinite languages etc), we strongly recommend the book [44].

The original motivation for developing finitely supported sets has its roots in the study of the independence of the Axiom of Choice (**AC**) claiming that for any family of nonempty sets $\mathscr{F}$ there exists a set containing exactly a single element from each member of $\mathscr{F}$. Since its first formulation **AC** has conduced to several debates and controversies. The first controversy is about the meaning of the word 'exists' since this term is very 'abstract'. One group of mathematicians (called intuitionists) believes that a set exists only if we are able to provide a method of constructing it. Another controversy is represented by a geometrical consequence of **AC** known as the Banach and Tarski's paradoxical decomposition of the sphere which shows that any solid sphere can be split into finitely many subsets which can themselves be reassembled to form two solid spheres, each of the same size as the original [17]. Questions regarding the independence of **AC** appeared naturally. In 1922, Fraenkel introduced the permutation method to prove the independence of **AC** from a set theory with atoms [26]. In 1935-1940, Gödel proved that **AC** is consistent (it does not induce a contradiction) with the axioms of von Neumann/Bernays/Gödel set theory [31]. In 1963, Cohen proved the independence of **AC** (i.e. the consistency of both **AC** and its negation) from the standard axioms of ZF set theory, using the so called forcing method which is derived from the Fraenkel's original permutation method [37]. The original permutation models of Zermelo-Fraenkel set theory with atoms (ZFA) has been recently rediscovered and extended by Gabbay (in a new axiomatic set theoretical framework) and Pitts (in a ZF alternative categorical framework) [29, 44] in order to solve various problems regarding binding, renamings and fresh names in computer science. The alternative set theory introduced by Gabbay and Pitts was extended by Alexandru and Ciobanu [7] in order to describe those algebraic structures that are are defined with respect to the finite support requirement, and by Bojanczyk, Klin and Lasota [23] by considering the so called 'orbit-finite sets' that replace 'finite sets' in order to solve problems regarding automata, programming languages and Turing machines that operate over infinite alphabets.

1.2 Related Topics

The following topics were developed in the last century regarding finitely supported sets. We just mention them, keeping in mind that we are mainly interested in their links to our approach of the finitely supported structures.

The FM permutation models of ZFA set theory were developed in 1930s by Fraenkel, Lindenbaum and Mostowski in order to prove the independence of **AC** from the other axioms of ZFA set theory, but they have been recently rediscovered by Gabbay and Pitts in order to describe syntax involving binding operations. Several models of ZFA set theory are well-known. We mention permutation Fraenkel basic and second models, and the permutation Mostowski ordered model (defined over countable sets of atoms) [37].

The ZFA universe is described as the von Neumann cumulative hierarchy $\nu(A)$ of sets involving atoms from the set of atoms A (by generalizing the usual hierarchy on ordinals):

- $\nu_0(A) = \emptyset$;
- $\nu_{\alpha+1}(A) = A + \wp(\nu_\alpha(A))$ for every non-limit ordinal α (we require that an ordinal does not have atoms among its elements)
- $\nu_\lambda(A) = \bigcup\limits_{\alpha<\lambda} \nu_\alpha(A)$ (whenever λ a limit ordinal);
- $\nu(A) = \bigcup\limits_{\alpha} \nu_\alpha(A)$,

where $+$ is the disjoint union of sets, and $\wp(X)$ represents the powerset of X. The disjoint union is used to emphasize that elements on $\nu(A)$ are either atoms (objects having no internal structure, i.e. entities containing no other elements) or sets (higher-order constructions on atoms or on the empty set).

FM axiomatic set theory was presented in [29]. It is inspired by FM permutation models of ZFA set theory. However, FM set theory, ZFA set theory and ZF set theory are independent axiomatic set theories. All of these theories are described by axioms, and all of them have specific models. For example, the cumulative hierarchy Fraenkel-Mostowski universe $FM(A)$ presented below is a model of FM set theory, while detailed lists of Cohen models of ZF and Fraenkel-Mostowski permutation models of ZFA can be found in [35]. The axioms of FM set theory are precisely the ZFA axioms over an infinite set of atoms, together with the special axiom of finite support which claims that for each element x in an arbitrary set we can find a finite set supporting x according to a hierarchically constructed group action of the group of all permutations of atoms. Therefore, in the FM universe only finitely supported objects are allowed. The original purpose of axiomatic FM set theory was to provide a mathematical model for variables in a certain syntax. Since they have no internal structure, atoms can be used to represent names. The finite support axiom is motivated by the fact that syntax can only involve finitely many names. Fresh names for the bound variables of a term can always be chosen from the atoms that are outside of the support (outside the set of the free names) of the related term. Binding is modelled by a certain concept of FM abstraction generalizing the notion of abstraction in the λ-calculus; actually FM set theory provides a formal framework for dealing with λ-terms modulo α-conversion. The construction of the universe of all FM sets [29] is inspired by the construction of the universe of all admissible sets over an arbitrary collection of atoms [19]. The FM sets represent a generalization of hereditary finite sets (which are particular admissible sets used to describe 'Gandy machines' [30]); actually, any FM set is an hereditary finitely supported set. Note that the infinite set of atoms in FM set theory does not necessary be countable. The FM set theory is consistent whether the infinite set of atoms is countable or not. In [29] it is used a countable set of atoms in order to define a model of FM set theory for new names in computer science, while in [21] there are described (generalized) FM sets over a set of atoms which do not represent a homogeneous structure. In [24] the authors also use non-countable sets of atoms (such as the set of real numbers) in order to study the minimization of deterministic timed automata.

In a pair $(X,\cdot)$ formed by a ZFA set X and a group action $\cdot$ on X of the group of all (finite) permutations of the set A of atoms, an arbitrary element $x \in X$ is finitely supported if there exists a finite family $S \subseteq A$ such that any permutation of A that fixes S pointwise also leaves x invariant under the group action $\cdot$. An invariant set $(X,\cdot)$ in ZFA framework is actually a classical ZFA set X equipped with an action $\cdot$ on X of the group of all permutations of atoms, having the additional property that any element $x \in X$ is finitely supported. If there exists an action $\cdot$ of the group of permutations of atoms on a set X, then there is an action $\star$ of the group of permutations of atoms on $\wp(X) = \{Y \,|\, Y \subseteq X\}$, defined by $(\pi, Y) \mapsto \pi \star Y := \{\pi \cdot y \,|\, y \in Y\}$ for all permutations of atoms π and all $Y \subseteq X$. In this sense, on $\nu(A)$ we define the action $\cdot$ of the group of all permutation of A recursively by $\pi \cdot a = \pi(a)$ for all $a \in A$, $\pi \cdot X = \{\pi \cdot x \,|\, x \in X\}$ for all elements $X \in \nu(A)$ that are not atoms. A model of axiomatic FM set theory is represented by the von Neumann cumulative hierarchy $FM(A)$ which is a subclass of $\nu(A)$ defined as follows:

- $FM_0(A) = \emptyset$;
- $FM_{\alpha+1}(A) = A + \wp_{fs}(FM_\alpha(A))$ for every non-limit ordinal α;
- $FM_\lambda(A) = \bigcup\limits_{\alpha<\lambda} FM_\alpha(A)$ (whenever λ is a limit ordinal);
- $FM(A) = \bigcup\limits_{\alpha} FM_\alpha(A)$.

where $+$ denotes the disjoint union of sets, and $\wp_{fs}(X)$ represents the family of those finitely supported subsets of X (i.e. the family of those subsets of X which are finitely supported as elements in $\wp(X)$ under the action $\cdot$). It is easy to prove that all $FM_\alpha(A)$ are invariant ZFA sets. The disjoint union is again used to emphasize the difference between 'sets' and 'atoms', i.e. atoms are not sets. A ZFA set X is an FM set (i.e. an element in $FM(A) \setminus A$) if and only if it is finitely supported as an element of $\nu(A)$ under the action $\cdot$ and Y is an FM set or an atom for all $Y \in X$.

Admissible sets and Gandy machines are related to developments of Barwise and Gandy. Gandy proved that any machine satisfying four physical 'principles' is equivalent to some Turing machine. Gandy's four principles define a class of computing machines, namely the 'Gandy machines'. Gandy machines are described by classes of 'states' and 'transition operations between states'. States are represented by hereditary finite sets (which are particular admissible sets) built up from an infinite set U of basic elements (atoms), and transformations are given by restricted operations from states to states. The class HF of all hereditary finite sets over U introduced in Definition 2.1 from [30] is described quite similarly to the von Neumann cumulative hierarchy of FM sets $FM(A)$. The single difference between these approaches is that each HF_{n+1} is defined inductively involving 'finite subsets of $U \cup HF_n$', while each $FM_{\alpha+1}(A)$ is defined inductively by using 'the disjoint union between A and the finitely supported subsets of $FM_\alpha(A)$'; HF is the union of all HF_n (with the remark that the empty set is not used in this construction), and the family of all FM sets is the union of all FM_α from which we exclude the set A of atoms. The support of an element x in HF, obtained according to Definition 2.2(1) from [30], coincides with $supp(x)$ (with notations from Theorem 2.1) if we see x as an FM set. Furthermore, the effect of a permutation π on a structure x described in Definition 2.3 from [30] is defined similarly to the effect of a permutation π on an element $x \in FM(A)$ under the canonical permutation action defined on $FM(A)$. The hierarchical construction of the universe of all FM sets (i.e. sets constructed according to the FM axioms) is therefore inspired by the construction of the universe of all admissible sets over an arbitrary collection of atoms. The FM sets represent a generalization of hereditary finite sets (that are particular admissible sets) because any FM set is an hereditary finitely supported set.

Nominal sets represent a ZF alternative to the non-standard FM set theory. More exactly, nominal sets can be defined both in the ZF framework (in a categorical form) [44] and in the FM framework (as particular FM sets) [29]. In ZF, a nominal set is defined as a usual ZF set endowed with a particular group action of the group of permutations over a certain fixed countable set A satisfying a finite support requirement. The set A is formed by elements whose internal structure is not taken into consideration; it is called either 'the set of atoms' by analogy with the FM framework or the set of 'atomic names' motivated by the nominal aim to provide a formal model for names in syntax. More exactly, a nominal set $(X, \cdot)$ is a classical ZF set X equipped with an action $\cdot$ on X of the group of (finite) permutations of A, having the additional property that any element $x \in X$ is finitely supported. In a pair $(X, \cdot)$ formed by a ZF set X and a group action $\cdot$ on X of the group of all permutations of A, an arbitrary element $x \in X$ is finitely supported if there exists a finite family $S \subseteq A$ such that any permutation of A that fixes S pointwise also leaves x invariant under the group action $\cdot$. Categorically, if $\mathbb{F}$ is the category whose objects are the finite subsets of A and whose morphisms are the one-to-one mappings between them, then the category **Nom** (whose objects are the nominal sets and whose morphisms are the empty supported functions) is a Cartesian closed category that is equivalent to the full subcategory of $\mathbf{Set}^{\mathbb{F}}$ consisting of presheaves that preserve pullbacks.

The definition of nominal sets also makes sense in the FM framework by replacing the fixed ZF set A with the set of atoms in ZFA. In FM framework, a nominal set is a set constructed according to the FM axioms with the additional property of being empty supported (invariant under all permutations). This is because nominal sets need to be closed under the group actions they are equipped with (meaning that the nominality requires empty-supportness at the following order level in a hierarchical construction). This means that the nominal FM sets are precisely the empty supported elements of the large nominal set (nominal class) $FM(A)$. These ways of defining nominal sets lead to similar properties. Intuitively, in a λ-calculus interpretation, we can think of the elements of a nominal set as having a finite set of 'free names'. The action of a permutation on such an element actually represents the renaming of the 'bound names'. Actually, nominal sets represent a categorical mathematical theory of names studying scope, binding, freshness and renaming in formal languages based upon symmetry. They can also be used in domain theory [51], in the theory of abstract interpretation [8], in topology [43], in programming involving binding [48], or in proof theory [52]. Since the principles of structural recursion and induction were proved to be valid in the framework of nominal sets [44], the theory of nominal sets provides a right balance between an informal reasoning and a rigorous formalism.

Similarly to the program of Felix Klein for the classifications of geometries [38], Tarski defined the logical notions as those invariant under all possible one-to-one transformations of the universe of discourse onto itself. In [9] the authors proved that those nominal sets defined in the FM cumulative universe $FM(A)$ are logical in Tarski's view. This is because in the FM framework it can be proved that every one-to-one transformation of A onto itself must be a finite permutation of A (i.e. a permutation that leaves unchanged all but finitely many elements of A), and so the effect of a bijective self-transformation of the universe of discourse on an element x belonging to a nominal (empty supported) set $(X,\cdot)$ from the FM cumulative universe coincides with the effect of a (finite) permutation of atoms on x under the group action $\cdot$. Such an element $x \in X$ is left invariant under the effect of any bijection of atoms if and only if it is equivariant (empty supported) as an element of X. By considering $X = FM(A)$, those equivariant elements of X are precisely the nominal FM sets, which are, therefore, logical notions in Tarski sense.

Generalized nominal set are an extension of nominal sets that provides an important step in the computation with infinite. The theory of nominal sets over a fixed set A of atoms is generalized in [23] to a refined theory of nominal sets over arbitrary (unfixed) sets of data values. This provides the generalized nominal sets. The notion of 'set equipped with a group action of the group of all permutations of the fixed set A' is replaced by the notion of 'set endowed with an action of a subgroup of the symmetric group of $\mathbb{D}$' for an arbitrary set of data values $\mathbb{D}$, and the notion of 'finite set' is replaced by the notion of 'set with a finite number of orbits with respect to the previous group action (orbit-finite set)'. This approach is useful for studying automata on data words, languages over infinite alphabets, or Turing machines that operate over infinite alphabets. Computation in these generalized nominal sets is presented in [22].

Fraenkel-Mostowski generalized set theory was introduced in [28] and generalizes both the size of atoms and the size of support from the FM set theory. More exactly, it is presented a generalization of FM sets by replacing 'finite support' with 'well-orderable (at least countable) support' and by considering an uncountable set of atoms. Notions such as abstraction and freshness quantifier И have also been extended into the new framework. In this sense, for a predicate p, in Fraenkel-Mostowski generalized theory$\text{И}a.p(a)$ means that p holds for all atoms except a well-orderable subset of atoms, while in FM $\text{И}a.p(a)$ means that p holds for all atoms except a finite subset of atoms. This approach allows binding of infinitely many names in syntax instead of only finitely many names.

A framework for modelling bindings using functors on sets was introduced in [20]. It overlaps somehow the nominal sets framework (the fresh induction proof principle introduced in [20] is inspired by the nominal approach), but also provides significant distinctions. In [20], the authors employed functors for modelling the presence of variables instead of sets with atoms. Furthermore, the authors are able to remove the finite support restriction and to accept terms that are infinitely branching, that have infinite depth, or both. Unlike nominal sets theory where atoms can only be manipulated via bijections, the functors described in [20] distinguish between binding variables (managed via bijections) and free variables (managed via possibly non-bijective functions); these functors allow the authors to apply not only swappings or permutations, but also arbitrary substitutions.

Relaxed Fraenkel-Mostowski axiomatic set theory (RFM) represents a refinement of FM set theory obtained by replacing the finite support axiom with a consequence of it which states that any subset of the set A of atoms is either finite or cofinite. Thus, in RFM the finite support axiom is replaced by requiring only an amorphous structure on A. More exactly, the aim of RFM was to replace the requirement 'finite support for all sets (built on a cumulative hierarchy from a family of basic elements)' with 'finite support only for sets of basic elements' in order to obtain similar results as in the FM case. In this sense, although we do not require the existence of a finite support for any hierarchically defined structure, several properties of the group of all bijections of A (such as torsioness or local finiteness) are preserved. RFM was presented first in [7] under the name EFM, and is now described more clearly in Chapter 13 of this book.

1.3 Defining Finitely Supported Structures

In order to describe FSM as a theory of finitely supported algebraic structures, we refer to the theory of nominal sets (with the mention that the requirement regarding the countability of A is irrelevant). We call these sets *invariant sets*, using the motivation of Tarski regarding logicality. The cardinality of the set of atoms *cannot* be internally compared with any other ZF cardinality as proved in Theorem 10.1; thus, we just say that atoms form an infinite set without any specifications regarding its cardinality. FSM is actually represented by finitely supported subsets of invariant sets together with finitely supported internal algebraic operations or with finitely supported relations (that should be finitely supported as subsets in the Cartesian product of two invariant sets). Our goal is to provide an fundamental study of these structures and to manage those structures hierarchically defined from an infinite set of basic elements, by emphasizing their finite supports. The goal of nominal approach was to provide a framework for manipulating syntax involving binding, renaming and fresh names. All the result in this book are obviously valid for nominal sets. We use the terminology 'FSM' instead of 'nominal' because we provide a set theoretical approach for finitely supported algebraic structures, with applications in algebra and logic, that is closer to Fraenkel and Mostowski approach and to Tarski's concept of logicality.

Adjoin to ZF a fixed infinite (not necessarily countable) set A formed by elements whose internal structure is irrelevant, only their identity being taken into consideration. In this sense we refer to A as being a fixed (distinctly emphasized) infinite ZF set which is called 'the set of atoms' by analogy with ZFA approach. The set theoretical (hierarchical) constructions that involve atoms are called 'atomic' (elements, sets, structures, etc) and be analyzed separately from the ordinary ZF constructions (which are non-atomic). An 'atomic' construction can also contain non-atomic elements, but it is required that at least one atom was involved somewhere (even at an earlier stage) in its construction. An invariant set $(X,\cdot)$ is actually a classical ZF set X equipped with an action $\cdot$ on X of the group of (finite) permutations of A (see Definition 2.1 and the comments below it regarding permutations), having the additional property that any element $x \in X$ is finitely supported. In a pair $(X,\cdot)$ formed by a ZF set X and a group action $\cdot$ on X of the group of all permutations of A, an arbitrary element $x \in X$ is finitely supported if there exists a finite family $S \subseteq A$ such that any permutation of A that fixes S pointwise also leaves x invariant under the group action $\cdot$. An empty supported element $x \in X$ is called equivariant. 'Invariant set' is the correspondent of 'nominal set' if A is countable. If there exists an action $\cdot$ of the group of permutations of A on a set X, then there is an action $\star$ of the group of permutations of A on $\wp(X) = \{Y \mid Y \subseteq X\}$, defined by $(\pi, Y) \mapsto \pi \star Y := \{\pi \cdot y \mid y \in Y\}$ for all permutations π of A and all $Y \subseteq X$. A subset of X is called finitely supported if it is finitely supported as an element in $\wp(X)$ with respect to the action $\star$. The Cartesian product of two invariant sets $(X,\cdot)$ and $(Y,\diamond)$ is an invariant set with the action $(\pi,(x,y)) \mapsto (\pi \cdot x, \pi \diamond y)$. Generally, an FSM set is either an invariant set or a finitely supported subset of an invariant set. A relation (or, particularly, a function) between two FSM sets is finitely supported/equivariant if it is finitely supported/equivariant as a subset of the Cartesian product of those two FSM sets. Particularly, a function between two FSM sets $(X,\cdot)$ and $(Y,\diamond)$ is supported by a finite set S if and only if $f(\pi \cdot x) = \pi \diamond f(x)$, $\pi \cdot x \in X$ and $\pi \diamond f(x) \in Y$ for all $x \in X$ and all permutations π that fixes S pointwise. Whenever an element $a \in A$ appears in (the construction of) an FSM set $(X,\cdot)$, the effect of a permutation of A π on a under $\cdot$ is $\pi(a)$. Ordinary ZF sets defined without involving atoms (i.e. without involving elements of A) are trivial invariant sets. Generally, an algebraic structure is invariant (or finitely supported) if it can be represented as an invariant set (or as a finitely supported subset of an invariant set) endowed with an equivariant algebraic law (or with an algebraic law which is finitely supported as a subset of an invariant set). Detailed formal definitions can be found in Chapter 2. The theory of finitely supported sets allows one to define finitely supported algebraic structures (as finitely supported sets endowed with finitely supported algebraic laws or with finitely supported relations) which are the core of FSM.

Concretely, FSM represents a reformulation of the ZF algebra obtained by replacing 'structure' with 'invariant/finitely supported structure', that is, all the structures defined in FSM (either atomic or non-atomic ones) must be finitely supported according to canonical hierarchically defined permutation actions (the non-atomic ones are trivial, while the atomic ones are equipped with canonical actions described in Proposition 2.2). FSM is defined on the ZF axioms (by taking A as a fixed ZF set and by considering only finitely supported structures), but it can be adequately reformulated according to ZFA axioms (if the fixed ZF set A considered in its construction is replaced by the set of atoms in ZFA). Thus, FSM is not defined as a model of ZF (or ZFA) set theory, but as an independent set theory whose axioms are the ZF axioms where we adjoin an infinite ZF set A (formed by basic elements) and we impose the additional axiom of finite support for all constructions that involve elements of A. Thus, ZF is strictly contained in FSM. Alternatively, A can be considered the set of atoms in ZFA (obtained by modifying the axiom of extensionality) and FSM corresponds to FM axiomatic set theory (without any restriction regarding countability of atoms).

Actually, FSM contains both the family of 'non-atomic' (i.e., ordinary) ZF sets which are proved to be trivial FSM sets (i.e., their elements are left unchanged under the effect of the canonical permutation action), and the family of 'atomic' sets (i.e., sets that contain at least an atom somewhere in their structure) with finite supports (hierarchically constructed from the empty set and the fixed ZF set A of atoms). The main task is to analyze whether a classical ZF result (obtained in the framework of non-atomic sets) can be adequately reformulated by replacing 'non-atomic ZF element/set/structure' with 'atomic and finitely supported element/set/structure' in order to be valid also for atomic sets with finite supports.

Note that the FSM sets are not closed under ZF subset constructions, meaning that there exist subsets of FSM sets that fail to be finitely supported (for example the simultaneously ZF infinite and coinfinite subsets of the set A). Thus, for proving results in FSM we cannot use related results from the ZF framework without reformulating them with respect to the finite support requirement. This means that in order to reformulate a general ZF result into FSM, the proof of the related FSM result should not break the principle that any construction has to be finitely supported. More exactly, non-finitely supported structures are not allowed in this theory (they simply do not exist), that is they can appear nowhere, nor even in an intermediate step of a proof. Informally, we cannot use a result (or a construction) outside FSM in order to prove an FSM result.

To conclude, FSM contains the entire ZF universe and additionally, the family of atomic finitely supported sets. Thus, results in ZF may lose their validity when transferred into atomic FSM, and so they cannot be directly applied to prove FSM properties. This is exactly what we need to verify in this book: whether a certain ZF result is preserved when rephrased into FSM. Actually, we prove in this book that some consistent ZF results remain consistent in FSM by moving the framework from 'non-atomic' to 'atomic with finite support', while other results that are consistent with the ZF axioms lose their validity when transferred into FSM.

For studying the validity of an FSM result, one must prove that all the structures/constructions involved in the statement or in the proof of the related result are finitely supported. We present three general methods of proving that a certain structure is finitely supported.

- The first method is a 'hierarchically constructive' one for supports.
- The second method is represented by a general equivariance/finite support principle which is defined by using the higher-order logic [44].
- The third method is given by a refinement of the equivariance/finite support principle allowing to prove boundedness properties for supports.

The hierarchically constructive method for supports represents an hierarchical (step-by-step) construction of the support of a certain structure by employing the supports of the sub-structures of a related structure. It means to anticipate a possible (intuitive) candidate for the support of an object (this candidate could be, for example, the empty set, the union of the supports of the sub-structures of the related object when this union is finite etc), and then to prove that this candidate is indeed a support by using the classical properties of finitely supported sets.

Equivariance principle states that any function or relation that is defined from equivariant functions and subsets using classical higher-order logic is itself equivariant, with requirement that we restrict any quantification over functions or subsets to range over ones that are equivariant. However, this method has some limitations because, in the form it is carried out in [44], it does not emphasize directly relationship properties between the supports of the structures involved in a higher-order construction (such relationship properties are mandatory, for example when we want to prove that a higher-order structure is uniformly supported, i.e. all its elements are supported by the same set of atoms). This issue was solved by describing the next method which is a refinement of the equivariance principle inspired by the constructive method for supports.

The S-finite support principle states that for any finite set $S \subseteq A$, anything that is definable (in higher-order logic) from S-supported structures using S-supported constructions is S-supported. However, the formal application of this method generally overlaps on the constructive method for supports. Furthermore, the related S-finite support reasoning has limitations related to results requiring choice principle. There does not exist a full study regarding the validity of all choice forms in FSM. Although the non-validity of most forms of choice in FSM is proved in this book, there are still many open problems in this topic (such as hidden choice in the proofs of various results).

1.4 Limits of Reformulating ZF Results in FSM

Although FSM has historical origins in the construction of permutation models for ZFA set theory, it is independent of them (it is actually related to the theory of nominal sets) having its own axioms. Moreover, we mention that the set of atoms in FSM should not necessarily be countable. The formal connections between FSM and the axiomatic theory of FM sets will be presented in Chapter 2. It is worth noting that FSM can be adequately presented in ZFA, if the fixed ZF set A (also called the set of atoms by analogy with ZFA) used in its construction is replaced by the set of atoms in ZFA (described as objects containing no elements, by modifying the axiom of extensionality from ZF); this is because we did not require an internal structure for the elements of A. Since we can identify the FM sets with (hereditary) finitely supported subsets of certain invariant sets, the results in FSM can be adequately rephrased also in the world of FM sets. Summarizing, FSM is a theory of algebraic structures obtain by replacing the term '(non-atomic) ZF structure' by '(atomic) finitely supported structure (under the canonical group action of the group of all permutations of A)', which can be as well reformulated over ZFA. However, there are many limitations when reformulating the results from ZF (or ZFA) into FSM, i.e. when reformulating a result from a non-atomic framework to an atomic framework. We emphasize various examples of non-transferable results between related frameworks such as ZF, ZFA and FSM.

Jech-Sochor theorem (Theorem 6.1 in [37]) states that permutation models of ZFA can be embedded into symmetric models of ZF. As a consequence, a statement which holds in a given permutation model of ZFA and whose validity depend only on a certain fragment of that model, also holds in some well-founded model of ZF. Therefore, the proof of consistency in ZF for a certain existential statement reduces to the construction of a permutation model of ZFA. For example, the existence of sets that cannot be well-ordered and of sets that cannot be totally ordered is consistent with ZF because such sets exist in (models of) ZFA. Similarly, by constructing adequate permutation models of ZFA and by involving Jech-Sochor embedding theorem, we can prove that the following statements are consistent with ZF: 'There exists a vector space which has no basis' and 'There exists a free group such that not every subgroup of it is free'. However, there does not exist a full equivalence between the consistency of results in ZF and the consistency of the appropriate results in ZFA.

The ZF relationship results do not necessarily remain valid when reformulating them in ZFA. There exist statements that are equivalent to axiom of choice in ZF, but are weaker than axiom of choice in ZFA. For example, multiple choice principle (Definition 2.5 in [34]) and Kurepa antichain principle (claiming that each partially ordered set has a maximal subset of pairwise incomparable elements) are such statements. According to Theorem 5.4 in [32], multiple choice principle and Kurepa antichain principle are both equivalent to the axiom of choice in ZF. However, in Theorem 9.2 of [37] it is proved that multiple choice principle is valid in the Second Fraenkel Model (model N2 from [35]), while the axiom of choice fails in this model. Furthermore, Kurepa maximal antichain principle is valid in the Basic

Fraenkel Model (model N1 from [35]), while multiple choice principle fails in this model. This means that the following two statements (that are valid in ZF) '*Kurepa principle implies axiom of choice*' and '*Multiple choice principle implies axiom of choice*' fail in ZFA. Thus, there exist valid ZF theorems that lose their validity when transferred into ZFA set theory. However, the goal of this book is not to study the transfer of ZF results into ZFA, but the transferability of ZF results into FSM. Although Jech-Sochor theorem establish a form of transferability of the consistency of ZFA results into ZF, there does not exist a general rule for translating ZF results into ZFA (this is not an obvious task).

The construction of FSM makes sense on both ZF and ZFA frameworks. However, the ZF and ZFA results are not necessary valid in FSM. This is because the family of sets in FSM (formed by invariant sets or finitely supported subsets of invariant sets) is not closed under subsets construction. Thus, for reformulating a ZF or ZFA result into FSM we must analyze if there exists a valid result obtained by replacing 'object' with 'finitely supported object' in the ZF result, or in the ZFA result respectively. Therefore, for proving results in FSM we cannot use related results from ZF (or ZFA) set theory without priory reformulating them with respect to the finite support requirement. We always need to verify if the proof of such results employs only finitely supported constructions under canonical atomic actions.

The ZF results are not necessarily valid when we reformulate them in FSM. Various choice principles are proved to be independent of the ZF axioms. For example, the ordering principle claiming that every set can be totally ordered is satisfied in Cohen's First Model (model M1 from [35]). However, this principle is not satisfied in Shelah's Second Model (model M38 from [35]), meaning that the ordering principle is independent of the ZF axioms. The axiom of dependent choice (Definition 2.11 in [34]) is valid in Shelah's Second Model but fails in Cohen's First Model, meaning that it also independent of the axioms of ZF set theory. The axiom of countable choice (Definition 2.5 in [34]) is valid in Shelah's Second Model but fails in Cohen's First Model, meaning that this principle is also independent of the axioms of ZF set theory. The prime ideal theorem (Definition 2.15 in [34]) is independent of the axioms of ZF set theory because it is valid in Cohen's First Model, but it fails in Shelah's Second Model. Although the previously mentioned choice principles are *independent* of the ZF axioms, in Chapter 3 we prove that all of them are *inconsistent* with (their negations can be proved in) FSM.

Similarly, the ZFA results are not necessarily valid in FSM when reformulating them in terms of finitely supported objects. Various choice principles are proved to be independent of the ZFA axioms. For example, the ordering principle is satisfied in Mostowski Ordered Model (model N3 in [35]), but it fails in Fraenkel's Second Model, which means the ordering principle is independent of the axioms of ZFA [37]. The prime ideal theorem is satisfied in Mostowski Ordered Model (Theorem 7.16 in [32]), but it fails in Fraenkel's Second Model [37], which means the prime ideal theorem is independent of the axioms of ZFA. The countable choice principle is also independent of the axioms of ZFA because it is satisfied in Howard-Rubin's First Model (model N38 from [35]), but it fails in Fraenkel's Second Model [34]. Although the previously mentioned choice principles are *independent* of the ZFA axioms, they are all *inconsistent* with FSM, when FSM is obtained by replacing the fixed ZF set A with the set of atoms in ZFA (see Chapter 3).

The above paragraph just states that a ZFA result involving arbitrary atomic sets cannot be directly reformulated by employing only finitely supported atomic sets (when FSM is developed over ZFA). For example, a ZFA infinite and coinfinite subset of atoms may exist (such a construction is not in contradiction with ZFA axioms, and so it is consistent within ZFA), but an FSM infinite and coinfinite subset of atoms may not exist. However, it is obvious that a general property that is valid for all atomic sets also holds for atomic sets with finite support. The translation of ZFA results into FSM is not treated in this book where we actually study the translation of ZF results into FSM by adding to the non-atomic sets an atomic structure. Sometimes it is more difficult to translate a result from ZF into ZFA than from ZF into FSM (we proved above that there are ZF results that do not remain valid in ZFA). In fact, we present the translation of results from ZF (not ZFA) into FSM, by moving from the 'non-atomic' framework to the 'atomic with finite support' framework. We do not know whether the results we study in this book (and which are consistent with ZF axioms) are also consistent with ZFA axioms. Above we mentioned two examples of valid ZF results (namely 'Kurepa antichain principle implies axiom of choice', and 'multiple choice principle implies axiom of choice') that lose their validity in ZFA. We remind that FSM is not a model of ZF. It may be associated somehow to a permutation model of ZFA when A is the set of atoms in ZFA (with the mention that FM set theory is however developed by M.Gabbay and A.Pitts as an independent set theory with its own axioms and its own models), but, even in this case, we do not know yet if the ZF results we discuss in this book are valid in ZFA in order to try translating them in FSM by using results in model theory. Thus, we study the validity of ZF results when replacing '(infinite) ordinary (non-atomic) set' with '(infinite) FSM set, i.e. with (infinite) atomic finitely supported subset of an invariant set' in their statement by using the S-finite support principle or the hierarchically constructive method for defining supports. Since FM sets are particular FSM sets (i.e. they are *hereditary* finitely supported subsets of invariant sets) the results obtained in FSM lead immediately to valid FM properties.

Chapter 2
Finitely Supported Sets: Formal Results

Abstract We formally describe finitely supported sets as classical Zermelo-Fraenkel sets equipped with canonical permutation actions, satisfying a certain finite support requirement. We provide higher-order constructions of atomic sets starting from some basic atomic sets. We present basic properties of finitely supported sets and of mappings between finitely supported sets. We also prove that mappings defined on some specific atomic sets have surprising (fixed points) properties. Particularly, finitely supported self-mappings defined on the finite powerset of atoms have infinitely many fixed points if they satisfy some particular properties (such as strict monotony, injectivity or surjectivity). Finally, we describe the Fraenkel-Mostowski axiomatic set theory that is connected with the theory of finitely supported algebraic structures.

2.1 Basic Properties

Adjoin to ZF an infinite set A (formed by elements whose internal structure is irrelevant, the only relevant attribute being their identity) which is called 'the set of atoms (basic elements)' and is emphasized distinctly from the ordinary ZF sets. We refer to A as a fixed ZF set, while noting that the internal structure of the elements of A is not taken into consideration. We refer to A simply as an 'infinite ZF set' without discussing its ZF cardinality. Actually, we will prove that there does not exist a finitely supported bijection between A and an ordinary (non-atomic) ZF set, and so we cannot associate to A a certain ZF cardinality (such as countable, uncountable, etc). This is a distinction between FSM approach and permutation models of set theory with atoms.

Ordinary ZF structures are those structures that are classically defined as in ZF (meaning that they are defined without involving atoms), while atomic ZF structures are those containing a hierarchically defined sub-structure over A. In FSM every atomic structure should have a finite support and should be hierarchically defined according to the rules in Proposition 2.2. The following results remain valid in the

A. Alexandru, G. Ciobanu, *Foundations of Finitely Supported Structures*,
https://doi.org/10.1007/978-3-030-52962-8_2

ZFA framework by replacing the fixed ZF set A with the set of atoms (elements with no internal structure) from ZFA, obtained by modifying/weakening the ZF axiom of extensionality, and by replacing 'ZF' with 'ZFA' in their statements.

The equivalence of various definitions of finiteness is a consequence of **AC** which will be proved to be inconsistent with FSM. To avoid any doubt, where the word 'finite' appears in this book without other specification, it means 'it corresponds one-to-one and onto to a finite ordinal', i.e. 'it can be represented as $\{x_1, \ldots, x_n\}$ for some $n \in \mathbb{N}$'. Consequently, 'infinite' without other specification means 'which is not finite', i.e. the 'FSM classical infinite' according to Definition 9.1.

Definition 2.1
- Let X, Y be ZF sets. A mapping $f : X \to Y$ is *one-to-one (injective)* if, whenever $x \in X$, $y \in Y$ with $x \neq y$, we have $f(x) \neq f(y)$.
- Let X, Y be ZF sets. A mapping $f : X \to Y$ is *onto (surjective)* if its image $f(X)$ coincides with Y, i.e. if for any $y \in Y$ there exists $x \in X$ such that $f(x) = y$.
- Let X, Y be ZF sets. A mapping $f : X \to Y$ is *bijective* if it is one-to-one and onto.
- Let X, Y be ZF sets. A mapping $f : X \to Y$ is *invertible* if there is $g : Y \to X$ such that $(f \circ g)(y) = y$ for all $y \in Y$ and $(g \circ f)(x) = x$ for all $x \in X$. The mapping g is called *the inverse* of f, and it is denoted by f^{-1}.
- A *transposition of atoms (transposition of A)* is a function $(a\,b) : A \to A$ defined by $(a\,b)(a) = b$, $(a\,b)(b) = a$ and $(a\,b)(c) = c$ for $c \neq a, b$.
- A *permutation of atoms (permutation of A)* is a one-to-one and onto function on A which permutes at most finitely many elements.

A mapping is invertible if and only if it is bijective. Each transposition is bijective, and the inverse of a transposition coincide with itself. A permutation of A is actually a bijection of A which leaves unchanged all but finitely many atoms. Let S_A be the group of all permutations of A. In fact, S_A is the set of those bijections of A which can be expressed by composing finitely many transpositions. We will prove in this book that an arbitrary bijection of A is finitely supported if and only if it is a permutation of A in the sense of Definition 2.1 (see Proposition 2.11); this is why we considered only (finite) permutations of A instead of arbitrary one-to-one transformations of A onto itself. Thus, the notions 'bijection of A' and 'permutation of A' coincide in FSM. This means that when we use these terms we actually refer to a 'finite/finitary permutation of A' with respect to the general terminology in the theory of symmetric groups; we have previously used the term '(finite) permutation of A' in order to emphasize that in FSM permutations of atoms are actually finite compositions of transpositions of atoms (meaning that they permute at most finitely many atoms) - see also Remark 2.2.

Definition 2.2 1. Let X be a ZF set. An *S_A-action (permutation action)* on X is a function $\cdot : S_A \times X \to X$ having the properties that $Id \cdot x = x$ and $\pi \cdot (\pi' \cdot x) = (\pi \circ \pi') \cdot x$ for all $\pi, \pi' \in S_A$ and $x \in X$, where $Id : A \to A$ (also denoted by 1_A) is the identity mapping on A defined by $Id(a) = a$ for all $a \in A$.

2. An *S_A-set* is a pair $(X, \cdot)$ where X is a ZF set, and $\cdot : S_A \times X \to X$ is an S_A-action on X. We simply use X whenever no confusion arises.

3. Let $(X, \cdot)$ be an S_A-set. We say that $S \subset A$ *supports* x (or x is supported by S) whenever for each $\pi \in Fix(S)$ we have $\pi \cdot x = x$, where $Fix(S) = \{\pi \in S_A \mid \pi(a) = a \text{ for all } a \in S\}$. An element which is supported by a finite subset of atoms is called *finitely supported.*
4. Let $(X, \cdot)$ be an S_A-set. We say that X is an *invariant set* if for each $x \in X$ there exists a finite set $S_x \subset A$ which supports x.

Lemma 2.1 *Let $\pi \in S_A$, $\pi \neq Id$. Then π can be expressed as a non-empty composition of finitely many transpositions of form (ab) with $\pi(a) \neq a$ and $\pi(b) \neq b$.*

Proof We prove the result by using induction on the size of the finite non-empty set $\{a \in A \mid \pi(a) \neq a\}$. The related set cannot have size 1. This is because whenever $\pi(a) = y \neq a$, we cannot have $\pi(y) = y$ due to the injectivity of π, and so we also have $\pi(y) \neq y$. If the related set has size 2, there should exist $c, d \in A$ such that $\pi(c) \neq c$, $\pi(d) \neq d$ and $\pi(x) = x$ for all atoms $x \neq c, d$. Due to the injectivity of π we should have $\pi(c) = d$ and $\pi(d) = c$; otherwise, if $\pi(c) = e \neq c, d$ then we could not have $\pi(e) = e$. Thus, π coincides with the transposition (cd).

Let us assume now that $S = \{a \in A \mid \pi(a) \neq a\}$ has the size greater than 2. Let $b \in S$ and $\sigma = \pi \circ (b\pi^{-1}(b))$. Clearly, $\sigma(b) = \pi((b\pi^{-1}(b))(b)) = \pi(\pi^{-1}(b)) = b$. For any $a \neq b$ we have that: $\pi(a) = a$ implies $\sigma(a) = \pi((b\pi^{-1}(b))(a)) = \pi(a) = a$. Therefore, $\{a \in A \mid \sigma(a) \neq a\} \subseteq \{a \in A \mid \pi(a) \neq a\} \setminus \{b\}$ (claim 1). Thus, the size of the set $\{a \in A \mid \sigma(a) \neq a\}$ is less than the size of the set $\{a \in A \mid \pi(a) \neq a\}$; according to the inductive hypothesis, this means that σ can be expressed as a finite composition of transpositions (uv) having the property $\sigma(u) \neq u$ and $\sigma(v) \neq v$. Due to (claim 1), for these transpositions (uv) we also have $\pi(u) \neq u$ and $\pi(v) \neq v$ (claim2). Furthermore, $\pi = \pi \circ ((b\pi^{-1}(b)) \circ (b\pi^{-1}(b))) = (\pi \circ (b\pi^{-1}(b))) \circ (b\pi^{-1}(b)) = \sigma \circ (b\pi^{-1}(b))$. Since we have $\pi(b) \neq b$ (because $b \in S$) and $\pi(\pi^{-1}(b)) = b \neq \pi^{-1}(b)$, the result follows from (claim2). □

Theorem 2.1 *Let X be an S_A-set. For each $x \in X$ we define $\mathcal{F}_x = \{S \subset A \mid S \text{ finite}, S \text{ supports } x\}$. If $\mathcal{F}_x$ is non-empty, then it has a least element; this element is* the support of x, *and we denote it by $supp(x)$.*

Proof We define $supp(x) = \cap\{S \subset A \mid S \text{ finite}, S \text{ supports } x\} = \bigcap_{S \in \mathcal{F}_x} S$. Let π be a permutation from $Fix(supp(x))$. We prove that $\pi \cdot x = x$. If $\pi = Id$, the claim follows obviously from the definition of a permutation action. If $\pi \neq Id$, according to Lemma 2.1, π can be expressed as a non-empty composition of finitely many transpositions of form (ab) with $\pi(a) \neq a$ and $\pi(b) \neq b$. Since π fixes each element in $supp(x)$, a transposition (ab) should satisfy that $a, b \notin supp(x)$. Thus, π can be expressed as a non-empty composition of finitely many transpositions of form (ab) with $a, b \notin supp(x)$.

Let us fix two arbitrary atoms $a, b \notin supp(x)$. Since $supp(x) = \bigcap_{S \in \mathcal{F}_x} S$, there exist $S_1, S_2 \in \mathcal{F}_x$ such that $a \notin S_1$ and $b \notin S_2$. If $S_1 = S_2$, then (because S_1 supports x and $a, b \notin S_1$, i.e. $(ab) \in Fix(S_1)$) we have $(ab) \cdot x = x$, according to Definition 2.2. Now let us suppose $S_1 \neq S_2$. Since $S_1 \cup S_2$ is finite and A is infinite, we can find $c \in A \setminus$

$(S_1 \cup S_2)$ and $a \neq c \neq b$. Since $a, c \notin S_1$ and S_1 supports x, it follows that $(ca) \cdot x = x$. Since $c, b \notin S_2$ and S_2 supports x, we have $(cb) \cdot x = x$. It follows that $(ab) \cdot x = (ab) \cdot ((cb) \cdot x) = ((ab) \circ (cb)) \cdot x = ((cb) \circ (ac)) \cdot x = (cb) \cdot ((ac) \cdot x) = x$. According to the definition of a permutation action, we should also have $\pi \cdot x = x$. Therefore, $supp(x)$ supports x, and $supp(x)$ is minimal among the finite sets supporting x. □

Corollary 2.1 *Let X be an invariant set, and for each $x \in X$ we define $\mathscr{F}_x = \{S \subset A \mid S \text{ finite}, S \text{ supports } x\}$. Then $\mathscr{F}_x$ has a least element which also supports x; we call this element* the support of x, *and denote it by $supp(x)$.*

Definition 2.3 Let $(X, \cdot)$ be an invariant set. An element $x \in X$ is called *equivariant* if it has an empty support, i.e. $\pi \cdot x = x$ for each $\pi \in S_A$.

Proposition 2.1 *Let $(X, \cdot)$ be an S_A-set and let $\pi \in S_A$ be an arbitrary permutation of atoms. Then for each $x \in X$ which is finitely supported we have that $\pi \cdot x$ is finitely supported and $supp(\pi \cdot x) = \pi(supp(x))$.*

Proof Let $\pi \in S_A$ be an arbitrary permutation of atoms, and $x \in X$ a finitely supported element. Firstly, we show that $\pi(supp(x))$ supports $\pi \cdot x$. Let us consider $\sigma \in Fix(\pi(supp(x)))$. This means $\sigma(\pi(a)) = \pi(a)$ for all $a \in supp(x)$. Therefore, $\pi^{-1}(\sigma(\pi(a))) = \pi^{-1}(\pi(a)) = a$ for all $a \in supp(x)$. So we get $\pi^{-1} \circ \sigma \circ \pi \in Fix(supp(x))$. However $supp(x)$ supports x (by Theorem 2.1). According to Definition 2.2, we have $(\pi^{-1} \circ \sigma \circ \pi) \cdot x = x$. Since $\cdot$ is a group action and the composition of permutations of atoms is associative, the last equality is equivalent to $\sigma \cdot (\pi \cdot x) = \pi \cdot x$. Hence, whenever $x \in X$ is finitely supported, we have that $\pi \cdot x$ is finitely supported. Moreover, $supp(\pi \cdot x) \subseteq \pi(supp(x))$ for each $x \in X$ which is finitely supported and each $\pi \in S_A$ (1). We now apply (1) for elements $\pi^{-1} \in S_A$ and $\pi \cdot x \in X$ (we already know the latter is finitely supported). We get $supp(\pi^{-1} \cdot \pi \cdot x) \subseteq \pi^{-1}(supp(\pi \cdot x))$. Composing with π in the last relation, we obtain $\pi(supp(x)) \subseteq supp(\pi \cdot x)$. □

2.2 Constructions of Finitely Supported Sets

The following result presents the rules for a hierarchical construction of FSM sets starting from the basic set of atoms. Proof is straightforward (and left to the reader).

Proposition 2.2

1. *The set A of atoms is an S_A-set with the* canonical S_A-action $\cdot : S_A \times A \to A$ *defined by $\pi \cdot a := \pi(a)$ for all $\pi \in S_A$ and $a \in A$. $(A, \cdot)$ is an invariant set because for each $a \in A$ it follows that $supp(a) = \{a\}$.*
2. *The set S_A is an S_A-set with the S_A-action $\cdot : S_A \times S_A \to S_A$ defined by $\pi \cdot \sigma := \pi \circ \sigma \circ \pi^{-1}$ for all $\pi, \sigma \in S_A$. $(S_A, \cdot)$ is an invariant set because for each $\sigma \in S_A$ it follows that $supp(\sigma) = \{a \in A \mid \sigma(a) \neq a\}$.*

3. *Any ordinary (non-atomic) ZF set X (such as* $\mathbb{N}, \mathbb{Z}, \mathbb{Q}$ *or* $\mathbb{R}$ *for example) is an* S_A*-set with the (single possible)* S_A*-action* $\cdot : S_A \times X \to X$ *defined by* $\pi \cdot x := x$ *for all* $\pi \in S_A$ *and* $x \in X$*. Moreover, X is also an invariant set because for each* $x \in X$ *it follows that* $supp(x) = \emptyset$*.*
4. *If* $(X, \cdot)$ *is an* S_A*-set, then the powerset* $\wp(X) = \{Y \mid Y \subseteq X\}$ *is also an* S_A*-set with the* S_A*-action* $\star : S_A \times \wp(X) \to \wp(X)$ *defined by* $\pi \star Y = \{\pi \cdot y \mid y \in Y\}$ *for all* $\pi \in S_A$*, and all* $Y \subseteq X$*. For each* S_A*-set* $(X, \cdot)$*, we denote by* $\wp_{fs}(X)$ *the set formed from those subsets of X which are finitely supported as elements of* $\wp(X)$ *under the action* $\star$ *in the sense of Definition 2.2. According to Proposition 2.1,* $(\wp_{fs}(X), \star|_{\wp_{fs}(X)})$ *is an invariant set, where* $\star|_{\wp_{fs}(X)} : S_A \times \wp_{fs}(X) \to \wp_{fs}(X)$ *is defined by* $\pi \star|_{\wp_{fs}(X)} Y := \pi \star Y$ *for all* $\pi \in S_A$ *and* $Y \in \wp_{fs}(X)$*.*
5. *Let* $(X, \cdot)$ *and* $(Y, \diamond)$ *be* S_A*-sets. The Cartesian product* $X \times Y$ *is also an* S_A*-set with the* S_A*-action* $\otimes : S_A \times (X \times Y) \to (X \times Y)$ *defined by* $\pi \otimes (x, y) = (\pi \cdot x, \pi \diamond y)$ *for all* $\pi \in S_A$ *and all* $x \in X$*,* $y \in Y$*. If* $(X, \cdot)$ *and* $(Y, \diamond)$ *are invariant sets, then* $(X \times Y, \otimes)$ *is also an invariant set.*
6. *Let* $(X, \cdot)$ *and* $(Y, \diamond)$ *be* S_A*-sets. We define the disjoint union of X and Y by* $X + Y = \{(0,x) \mid x \in X\} \cup \{(1,y) \mid y \in Y\}$*.* $X + Y$ *is an* S_A*-set with the* S_A*-action* $\star : S_A \times (X + Y) \to (X + Y)$ *defined by* $\pi \star z = (0, \pi \cdot x)$ *if* $z = (0, x)$ *and* $\pi \star z = (1, \pi \diamond y)$ *if* $z = (1, y)$*. If* $(X, \cdot)$ *and* $(Y, \diamond)$ *are invariant sets, then* $(X + Y, \star)$ *is also an invariant set: each* $z \in X + Y$ *is either of the form* $(0, x)$ *and supported by the finite set supporting x in X, or of the form* $(1, y)$ *and supported by the finite set supporting y in Y.*

Definition 2.4

1. Let $(X, \cdot)$ be an S_A-set. A subset Z of X is called *finitely supported* if and only if $Z \in \wp_{fs}(X)$ with the notations from Proposition 2.2, i.e. if and only if Z is finitely supported (in the sense of Definition 2.2) as an element of the S_A-set $\wp(X)$ with respect to the action $\star$ defined on $\wp(X)$ as in Proposition 2.2(4). A subset Z of X is *uniformly supported* if all the elements of Z are supported by the same set S (and so Z is itself supported by S as an element of $\wp_{fs}(X)$). Generally, an *FSM set* is a finitely supported subset (possibly equivariant) of an invariant set. An FSM set is also called *finitely supported set*, meaning that it is finitely supported as a subset of an invariant set.
2. Let $(X, \cdot)$ be a finitely supported subset of an S_A-set $(Y, \cdot)$. A subset Z of Y is called *finitely supported subset of X* (and we denote this by $Z \in \wp_{fs}(X)$) if and only if $Z \in \wp_{fs}(Y)$ and $Z \subseteq X$. Similarly, we say that a uniformly supported subset of Y contained in X is a *uniformly supported subset of X*.
3. An action of S_A an a ZF set X is called *canonical permutation action* (canonical S_A-action) if it is considered with respect to the definitions of the S_A-actions provided in Proposition 2.2 for atoms, subsets, Cartesian products and disjoint unions. In this book we use only canonical permutation actions.

Proposition 2.3 *Let* π *be a permutation of A, and X a subset of an* S_A*-set Y. We have* $\pi \star X = X$ *if and only if* $\pi \star X \subseteq X$*, where* $\star$ *is defined on* $\wp(Y)$ *as in Proposition 2.2(4).*

Proof According to Proposition 2.2(4), we have that X is an element of the S_A-set $\wp(Y)$. The direct implication is obvious. For the converse, let us assume by contradiction that $\pi\star X \subsetneq X$. Since $\star$ is a group action and π has an inverse (and so, for two subsets U, V of Y we have $U = V$ if and only if $\pi\star U = \pi\star V$), we get (by induction) $\pi^n\star X \subsetneq X$ for all $n\in\mathbb{N}$ (the trivial remark that $U\subseteq V\subseteq Y$ implies $\pi\star U\subseteq \pi\star V$ is also involved). However, since π is of finite order (being a finite composition of transpositions), i.e. $\pi^m = Id$ for some $m\in\mathbb{N}$, we obtain $X\subsetneq X$; a contradiction. □

From Definition 2.4 and Proposition 2.3, a subset Z of an invariant set $(X,\cdot)$ is finitely supported by a set $S\subseteq A$ if and only if $\pi\star Z\subseteq Z$ for all $\pi\in Fix(S)$, i.e. if and only if $\pi\cdot z\in Z$ for all $z\in Z$ and all $\pi\in Fix(S)$.

In the view of Definition 2.4(3), in FSM we consider that A is the invariant set equipped with the canonical S_A-action defined on Proposition 2.2(1). Any invariant set should be equipped with the canonical action hierarchically constructed as in Proposition 2.2 (for powersets, Cartesian products, disjoint unions and higher-order constructions). Whenever an atom a appears in (the construction of) an invariant set $(X,\cdot)$, the effect of a permutation of atoms π on a under $\cdot$ is $\pi(a)$. However, for ordinary ZF sets (defined without involving any atoms) the trivial actions are the only possible S_A-actions on these sets, and so these sets are invariant. Intuitively, if we consider the set of all positive integers $\mathbb{N}$, we know that $\mathbb{N}$ is defined by starting with $\emptyset$, and then by forming the sets $\{\emptyset\}$, $\{\emptyset,\{\emptyset\}\}$ and so on. Since $\emptyset$ is not defined by employing atoms, it cannot be changed under the effect of a certain permutation of atoms.

Example 2.1 [29]

1. If X' is the set of λ-terms t, then action $\star$ of S_A on X' is inductively defined by:
   - *variable*: $\pi\star a = \pi(a)$ whenever a is a variable (corresponding to atoms), and π is a permutation of atoms;
   - *application*: $\pi\star(tt') = (\pi\star t)(\pi\star t')$ for all λ-terms t and t' and all $\pi\in S_A$;
   - *abstraction*: $\pi\star(\lambda a.t) = \lambda(\pi(a)).(\pi\star t)$ for all variables a, all λ-terms t and all $\pi\in S_A$.

   $(X',\star)$ is an invariant set, and the support of a λ-term t is the finite set of atoms occurring in t, whether as free, bound or binding occurrences.
2. Let X be the set of α-equivalence classes of the λ-calculus terms t. We can define an action $\cdot$ of S_A on X by $\pi\cdot[t]_\alpha = [\pi\star t]_\alpha$ for all λ-terms t and all $\pi\in S_A$ (where $[t]_\alpha$ represents the α-equivalence class of the λ-term t). If two λ-terms t and t' are α-equivalent, it is clear that $\pi\star t =_\alpha \pi\star t'$, and so the action $\cdot$ is well-defined. $(X,\cdot)$ is an invariant set. If t is chosen to be a representative of its α-equivalence class, then $supp(t)$ coincides with $fn(t)$, where $fn(t)$ is the set of free variables of t defined by the λ-calculus rules.

According to Example 2.1, an α-equivalence class of terms does not contain bound names. We cannot define a function $bn : X \to \wp_{fin}(A)$ which would be able to extract exactly the bound names for a term t [29]. Thus, α-equivalent terms are identified in FSM since two α-equivalent terms have the same set of free variables.

Proposition 2.4 *Let X be a finitely supported subset of an invariant set $(U,\cdot)$. Then $U \setminus X$ is finitely supported and $supp(U \setminus X) = supp(X)$.*

Proof We claim that $supp(X)$ supports $U \setminus X$. Indeed, let $\pi \in Fix(supp(X))$. We also have $\pi^{-1} \in Fix(supp(X))$. Let $y \in U \setminus X$. If $\pi \cdot y \in X$, then $y = \pi^{-1} \cdot (\pi \cdot y) \in \pi^{-1} \star X = X$, which is a contradiction. Therefore, $\pi \cdot y \in U \setminus X$, which means $supp(X)$ supports $U \setminus X$, and so $supp(U \setminus X) \subseteq supp(X)$. Now, let $\sigma \in Fix(supp(U \setminus X))$, and so $\sigma^{-1} \in Fix(supp(U \setminus X))$. Let $z \in X$. If $\sigma \cdot z \in U \setminus X$, then $z \in \sigma^{-1} \star (U \setminus X) = U \setminus X$, which is a contradiction. Therefore, $\sigma \cdot z \in X$, which means that $supp(U \setminus X)$ supports X, and so $supp(X) \subseteq supp(U \setminus X)$. □

Proposition 2.5

1. Let X be a uniformly supported subset of an invariant set $(U,\cdot)$. Then X is finitely supported and $supp(X) = \cup\{supp(x) \,|\, x \in X\}$.
2. Let X be a finite subset of an invariant set $(U,\cdot)$. Then X is finitely supported and $supp(X) = \cup\{supp(x) \,|\, x \in X\}$.

Proof 1. Since X is uniformly supported, there is a finite subset of atoms T such that T supports every $x \in X$, i.e. $supp(x) \subseteq T$ for all $x \in X$. Therefore, $\cup\{supp(x) \,|\, x \in X\} \subseteq T$. Clearly, $supp(X) \subseteq \cup\{supp(x) \,|\, x \in X\}$. Conversely, let $a \in \cup\{supp(x) \,|\, x \in X\}$. Thus, there exists $x_0 \in X$ such that $a \in supp(x_0)$. Let b be an atom such that $b \notin supp(X)$ and $b \notin T$. Such an atom exists because A is infinite, while $supp(X)$ and T are both finite. We prove by contradiction that $(b\ a) \cdot x_0 \notin X$. Indeed, suppose that $(b\ a) \cdot x_0 = y \in X$. According to Proposition 2.1, we have $supp(y) = (b\ a)(supp(x_0))$. Since $a \in supp(x_0)$, we have $b = (b\ a)(a) \in (b\ a)(supp(x_0)) = supp((b\ a) \cdot x_0) = supp(y)$. Since $supp(y) \subseteq T$, we get $b \in T$, a contradiction. Thus, $(b\ a) \star X \neq X$, where $\star$ is the standard S_A-action on $\wp(U)$ defined in Proposition 2.2.

Since $b \notin supp(X)$, we prove by contradiction that $a \in supp(X)$. Indeed, suppose that $a \notin supp(X)$. It follows that the transposition $(b\ a)$ fixes each element from $supp(X)$, i.e. $(b\ a) \in Fix(supp(X))$. Since $supp(X)$ supports X, by Definition 2.2, it follows that $(b\ a) \star X = X$, which is a contradiction. Thus, $a \in supp(X)$.

2. Any finite set $X = \{x_1, \dots, x_k\}$ is uniformly supported by $S = supp(x_1) \cup \dots \cup supp(x_k)$, and so the result follows from the above item. □

Corollary 2.2 *Let X be a uniformly supported subset of an invariant set. Then X is uniformly supported by $supp(X)$.*

Proof Since $supp(X) = \cup\{supp(x) \,|\, x \in X\}$, we have $supp(x) \subseteq supp(X)$ for all $x \in X$ which means $supp(X)$ supports every $x \in X$. □

Corollary 2.3 *Let X be a cofinite subset of an invariant set $(U,\cdot)$.*
Then X is finitely supported and $supp(X) = \cup\{supp(x)\,|\,x \in U \setminus X\}$.

Proof Since X is cofinite, we have that $U \setminus X$ is finite, and so $U \setminus X$ is finitely supported. According to Proposition 2.4, we have that $X = U \setminus (U \setminus X)$ is finitely supported and $supp(X) = supp(U \setminus X)$. Since $U \setminus X$ is finite, from Proposition 2.5 we have that $supp(U \setminus X) = \cup\{supp(x)\,|\,x \in U \setminus X\}$. □

Corollary 2.4 *Let X be a finite invariant set. Then X is necessarily a trivial invariant set, i.e. there exists only one possible S_A-action on X, $\cdot : S_A \times X \to X$ defined by $\pi \cdot x = x$ for all $x \in X$.*

Proof Let X be a finite invariant set. We can equivalently say that X is a finite subset of the invariant set $(X,\cdot)$. Let $X = \{x_1,\dots,x_k\}$. According to Proposition 2.5, we have $supp(X) = supp(x_1) \cup \dots \cup supp(x_k)$. However, since X is itself equivariant, we have that $supp(X) = \emptyset$. We obtain that $supp(x_1) \cup \dots \cup supp(x_k) = \emptyset$, and so $supp(x_i) = \emptyset$ for all $i \in \{1,\dots,k\}$. Thus, $\pi \cdot x_i = x_i$, for all $x_i \in X$ and all $\pi \in S_A$, and so X is a trivial invariant set. □

Corollary 2.5 *Let $(X,\cdot)$ be an S_A-set.*

1. *If X is an invariant set, then the finite powerset of X, namely $\wp_{fin}(X) = \{Y \subseteq X\,|\,Y \text{ finite}\}$ is an equivariant subset of $\wp_{fs}(X)$. Similarly, the cofinite powerset of X, namely $\wp_{cofin}(X) = \{Y \subseteq X\,|\,X \setminus Y \text{ finite}\}$ is an equivariant subset of $\wp_{fs}(X)$.*
2. *$T_{fin}(X) = \{(x_1,\dots,x_m) \subseteq (X \times \dots \times X)\,|\,m \geq 0 \text{ and } x_i \neq x_j \text{ for all } i,j=1,\dots,m\}$ containing all the finite injective tuples of elements from X is an S_A-set with the S_A-action $\star : S_A \times T_{fin}(X) \to T_{fin}(X)$ defined by the pointwise operation $\pi \star (x_1,\dots,x_m) = (\pi \cdot x_1,\dots,\pi \cdot x_m)$ for all $(x_1,\dots,x_m) \in T_{fin}(X)$. The family $T^*_{fin}(X)$ of all non-empty tuples of elements from X is also an S_A set. If X is an invariant set, then both families $T_{fin}(X)$ and $T^*_{fin}(X)$ are invariant sets.*

Proof The first item follows from the trivial remark that, because permutations of atoms are one-to-one mappings, the image of a finite subset of an invariant set under such a permutation is a finite subset of the related invariant set. Similarly, the image of a cofinite subset of an invariant set under a permutation of atoms is a cofinite subset of the related invariant set. The first part of the second item is a direct consequence of the fact that permutations of atoms are injective, and so finite injective tuples are transformed into finite injective tuples. Moreover, if X is invariant we have that each injective tuple $(x_1,\dots,x_m)$ is finitely supported with $supp(x_1,\dots,x_m) = supp(x_1) \cup \dots \cup supp(x_m)$; this follows from Proposition 2.6. □

It is worth noting that not any ZF subset of an invariant set is finitely supported. This becomes clear by analyzing the following result.

Theorem 2.2 *In FSM we have $\wp_{fs}(A) = \wp_{fin}(A) \cup \wp_{cofin}(A)$.*

Proof If $B \subset A$ and B is finite, then B is finitely supported with $supp(B) = B$ (see Proposition 2.5). If $C \subseteq A$ and C is cofinite (i.e. its complement is finite), then C is finitely supported with $supp(C) = A \setminus C$ (see Corollary 2.3). However, if $D \subseteq A$ is neither finite nor cofinite, then D is not finitely supported. Indeed, assume by contradiction that there exists a finite set of atoms S supporting D. Since S is finite and both D and its complement C_D are infinite, we can take $a \in D \setminus S$ and $b \in C_D \setminus S$. Then the transposition (ab) fixes S pointwise, but $(ab) \star D \neq D$ because $(ab)(a) = b \notin D$; this contradicts the assertion that S supports D. Therefore, $\wp_{fs}(A) = \wp_{fin}(A) \cup \wp_{cofin}(A)$. □

We conclude that $\wp(A)$ exists in ZF (and in ZFA, under the axiom of powerset, if the fixed set A from the construction of FSM, formed by elements whose internal structure is not taken into consideration, is replaced by the set of atoms in ZFA obtained by weakening the ZF axiom of extensionality), but only the elements in $\wp_{fin}(A) \cup \wp_{cofin}(A) \subsetneq \wp(A)$ belong to FSM.

Proposition 2.6 *Let $(X_1, \cdot), \ldots, (X_n, \cdot)$ be invariant sets (equipped with possibly different S_A-actions generically denoted by $\cdot$). Then $supp((x_1, \ldots, x_n)) = supp(x_1) \cup \ldots \cup supp(x_n)$ for all $x_i \in X_i$, $i \in \{1, \ldots, n\}$.*

Proof Let $U = (x_1, \ldots, x_n)$ and $S = supp(x_1) \cup \ldots \cup supp(x_n)$. Obviously, S supports U. Indeed, let us consider $\pi \in Fix(S)$. We have that $\pi \in Fix(supp(x_i))$ for all $i \in \{1, \ldots, n\}$. Therefore, $\pi \cdot x_i = x_i$ for all $i \in \{1, \ldots, n\}$, and so $\pi \otimes (x_1, \ldots, x_n) = (\pi \cdot x_1, \ldots, \pi \cdot x_n) = (x_1, \ldots, x_n)$, where $\otimes$ represents the S_A-action on $X_1 \times \ldots \times X_n$ described in Proposition 2.2(5). Thus, $supp(U) \subseteq S$. It remains to prove that $S \subseteq supp(U)$. Fix $\pi \in Fix(supp(U))$. Since $supp(U)$ supports U, we have $\pi \otimes (x_1, \ldots, x_n) = (x_1, \ldots, x_n)$, and so $(\pi \cdot x_1, \ldots, \pi \cdot x_n) = (x_1, \ldots, x_n)$, from which we get $\pi \cdot x_i = x_i$ for all $i \in \{1, \ldots, n\}$. Thus, $supp(U)$ supports x_i for all $i \in \{1, \ldots, n\}$, and so $supp(x_i) \subseteq supp(U)$ for all $i \in \{1, \ldots, n\}$. Therefore, $S = supp(x_1) \cup \ldots \cup supp(x_n) \subseteq supp(U)$. □

Due to Proposition 2.1, if X is a finitely supported subset of an invariant set Y, the uniform powerset of X denoted by $\wp_{us}(X) = \{Z \subseteq X \,|\, Z \text{ uniformly supported}\}$ is a subset of $\wp_{fs}(Y)$ supported by $supp(X)$. This is because, whenever $Z \subseteq X$ is uniformly supported by S and $\pi \in Fix(supp(X))$, we have $\pi \star Z \subseteq \pi \star X = X$ and $\pi \star Z$ is uniformly supported by $\pi(S)$. Furthermore, $\wp_{fin}(X)$ is a subset of $\wp_{us}(X)$ because any finite subset Z of X is uniformly supported by $supp(Z) = \cup\{supp(z) \,|\, z \in Z\}$. We consider that $\emptyset$ belongs to $\wp_{us}(X)$, being a finite subset of X.

2.3 Functions Between Finitely Supported Sets

Recall that a function $f : X \to Y$ is a particular relation on (a subset of) $X \times Y$.

Definition 2.5 Let X and Y be invariant sets.

A function $f : X \to Y$ is *finitely supported* if $f \in \wp_{fs}(X \times Y)$.

Let $Y^X = \{f \subseteq X \times Y \mid f$ is a function from the underlying set of X to the underlying set of $Y\}$.

Proposition 2.7 *Let $(X, \cdot)$ and $(Y, \diamond)$ be invariant sets. Then Y^X is an S_A-set with the S_A-action $\star : S_A \times Y^X \to Y^X$ defined by $(\pi \star f)(x) = \pi \diamond (f(\pi^{-1} \cdot x))$ for all $\pi \in S_A$, $f \in Y^X$ and $x \in X$. A function $f : X \to Y$ is finitely supported in the sense of Definition 2.5 if and only if it is finitely supported with respect to the permutation action $\star$.*

Proof We already know that functions from X to Y are subsets of the Cartesian product $X \times Y$, which is an invariant set. Furthermore, $\wp(X \times Y)$ is an S_A-set and $\pi \star f = \{(\pi \cdot x, \pi \diamond y) \mid (x, y) \in f\}$. Thus, $\pi \star f$ is a function with the domain $\pi \cdot X = X$. Moreover, $(\pi \star f)(\pi \cdot x) = \pi \diamond f(x)$. Let $x' = \pi \cdot x$, and so $x = \pi^{-1} \cdot x'$. We obtain $(\pi \star f)(x') = \pi \diamond (f(\pi^{-1} \cdot x'))$. The mapping $x \mapsto x'$ is bijective, and it follows that $(\pi \star f)(x) = \pi \diamond (f(\pi^{-1} \cdot x))$ for all $\pi \in S_A$, $f \in Y^X$ and $x \in X$. Therefore, each function $f : X \to Y$ is finitely supported with respect to the permutation action $\star$ if and only if $f \in \wp_{fs}(X \times Y)$.

Proposition 2.8 *Let $(X, \cdot)$ and $(Y, \diamond)$ be invariant sets. Let $f \in Y^X$ and $\sigma \in S_A$ be arbitrary elements. Let $\star : S_A \times Y^X \to Y^X$ be the S_A-action on Y^X, defined by: $(\pi \star f)(x) = \pi \diamond (f(\pi^{-1} \cdot x))$ for all $\pi \in S_A$, $f \in Y^X$ and $x \in X$. Then $\sigma \star f = f$ if and only if for all $x \in X$ we have $f(\sigma \cdot x) = \sigma \diamond f(x)$.*

Proof Let $\sigma \in S_A$ be a permutation of atoms such that $\sigma \star f = f$. From Proposition 2.7 we know that for each $x \in X$ we have $(\sigma \star f)(\sigma \cdot x) = \sigma \diamond f(x)$. Since $\sigma \star f = f$, it follows that $f(\sigma \cdot x) = \sigma \diamond f(x)$ for all $x \in X$. Conversely, let us suppose that for $\sigma \in S_A$ and all $x \in X$ we have $f(\sigma \cdot x) = \sigma \diamond f(x)$. For each $x \in X$ we have $(\sigma \star f)(x) = \sigma \diamond (f(\sigma^{-1} \cdot x)) = f(\sigma \cdot (\sigma^{-1} \cdot x)) = f(x)$. □

Definition 2.5 can be generalized in the following way.

Definition 2.6 Let X and Y be invariant sets, Z a finitely supported subset of X, and T a finitely supported subset of Y. A function $f : Z \to T$ is *finitely supported* if $f \in \wp_{fs}(X \times Y)$. The family of the finitely supported functions between Z and T is denoted by T^Z_{fs}.

Proposition 2.9 *Let $(X, \cdot)$ and $(Y, \diamond)$ be invariant sets, and let Z be a finitely supported subset of X and T be a finitely supported subset of Y. Let $f : Z \to T$ be a function. The function f is finitely supported in the sense of Definition 2.6 by a finite set S of atoms if and only if for all $x \in Z$ and all $\pi \in Fix(S)$ we have $\pi \cdot x \in Z$, $\pi \diamond f(x) \in T$ and $f(\pi \cdot x) = \pi \diamond f(x)$.*

Proof We assume that f is finitely supported in the sense of Definition 2.6 by S. Then $\pi \star f = f$ for all $\pi \in Fix(S)$, where $\star$ represents the S_A-action on $\wp(X \times Y)$ defined as in Proposition 2.2(4). Let $x \in Z$ and $\pi \in Fix(S)$ be arbitrary elements. Then there exists a unique $y \in T$ such that $(x, y) \in f$. Since $\pi \star f = f$, we have $(\pi \cdot x, \pi \diamond y) \in f \subseteq (Z \times T)$. Thus, $\pi \cdot x \in Z$, $\pi \diamond f(x) \in T$ and $f(\pi \cdot x) = \pi \diamond y =$

$\pi \diamond f(x)$. Conversely, we assume that there exists a finite set S of atoms such that for all $x \in Z$ and all $\pi \in Fix(S)$ we have $\pi \cdot x \in Z$, $\pi \diamond f(x) \in T$ and $f(\pi \cdot x) = \pi \diamond f(x)$. We claim that $\pi \star f = f$ for all $\pi \in Fix(S)$. Let us consider such a π. Fix some $\sigma \in Fix(S)$, and consider (x,y) an arbitrary element in f. We have $f(x) = y$, and so $(\sigma \cdot x, \sigma \diamond y) \in f$. Thus, $\sigma \otimes (x,y) = (\sigma \cdot x, \sigma \diamond y) \in f$, where $\otimes$ represents the S_A-action on $X \times Y$ defined as in Proposition 2.2(5). We actually proved that $\sigma \star f \subseteq f$ whenever σ fixes S pointwise (1). If σ fixes S pointwise, then σ^{-1} also fixes S pointwise. By applying (1), firstly for $\sigma = \pi$, and secondly for $\sigma = \pi^{-1}$, we get $\pi \star f \subseteq f$ and $\pi^{-1} \star f \subseteq f$, and because $\star$ is a group action, we also get $f \subseteq \pi \star f$, and so $\pi \star f = f$. □

Corollary 2.6 *Let $(X, \cdot)$ be an invariant set.*
The function $x \mapsto supp(x)$ from X to $\wp_{fin}(A)$ is equivariant. .

Proof The condition required in Proposition 2.9 (namely $supp(\pi \cdot x) = \pi \star supp(x) = \pi(supp(x))$ for all $x \in X$) is satisfied according to Proposition 2.1. □

Not every function between two nominal sets is finitely supported.

Proposition 2.10 *Let X be an infinite ordinary ZF set, and $f : A \to X$ a function.*

1. *If f is finitely supported, then there is $x \in X$ such that $\{y \in A \mid f(y) \neq x\}$ is finite.*
2. *If there is $x \in X$ such that $\{y \in A \mid f(y) \neq x\}$ is finite, then f is finitely supported, and $supp(f) = \{y \in A \mid f(y) \neq x\}$.*

Proof Firstly, we remind that an ordinary ZF set $(X, \diamond)$ (i.e. a set X which is constructed without involving atoms) is necessarily trivial, i.e. $\pi \diamond z = z$ for all $z \in X$.

1. Let $f : A \to X$ be a finitely supported function. Let us fix an element $b \in A$ with $b \notin supp(f)$. Such an atom b exists because A is infinite, while $supp(f)$ is finite. Let $c \neq b$ be another arbitrary element from $A \setminus supp(f)$. Since $b \notin supp(f)$, we have that (bc) fixes every element from $supp(f)$, i.e. $(bc) \in Fix(supp(f))$. However, $supp(f)$ supports f. According to Proposition 2.8, we have $f((bc)(a)) = (bc) \diamond f(a) = f(a)$ for all $a \in A$. In particular, $f(c) = f((bc)(b)) = f(b)$. Since c has been chosen arbitrarily from $A \setminus supp(f)$, it follows that $f(c) = f(b)$, for all $c \in A \setminus supp(f)$. If $supp(f) = \{a_1, \dots, a_n\}$, then $Im(f) = \{f(a_1)\} \cup \dots \cup \{f(a_n)\} \cup \{f(b)\}$. We take $x = f(b)$ because at most $f(a_1), \dots, f(a_n)$ can be different from x, while $f(a) = x$ for all $a \in A \setminus \{a_1, \dots, a_n\}$.

2. We claim that $S = \{y \in A \,|\, f(y) \neq x\}$ supports f. Let $\pi \in Fix(S)$. According to Proposition 2.8, we have to prove that $f(\pi(a)) = f(a)$ for all $a \in A$. Let $a \in A$ be an arbitrary element. If $f(a) \neq x$, we have $a \in S$, and $\pi(a) = a$ because π fixes S pointwise. Therefore, $f(\pi(a)) = f(a)$. Now let $a \in A$ with $f(a) = x$. Let us assume by contradiction that $f(\pi(a)) \neq x$. It follows that $\pi(a) \in S$ and $\pi(\pi(a)) = \pi(a)$. Since π is bijective, we get $\pi(a) = a$ and $f(\pi(a)) = f(a) = x$ which contradicts our assumption that $f(\pi(a)) \neq x$. Therefore, $f(\pi(a)) = x = f(a)$. We proved that S supports f. It remains to prove that S is minimal between the finite sets supporting f. Let V be a finite set supporting f. We claim $S \subseteq V$. Let $b \in S$, i.e. $f(b) \neq x$. Assume by contradiction that $b \notin V$. Let c be an arbitrary element from $A \setminus (V \cup \{b\})$. Since

$b,c \notin V$, we have $(bc) \in Fix(V)$. However, V supports f. This means $f((bc)(a)) = f(a)$ for all $a \in A$. In particular, $f(c) = f((bc)(c)) = f(b) \neq x$. Therefore, $c \in S$, and $A \setminus (V \cup \{b\}) \subseteq S$. Since $A \setminus (V \cup \{b\})$ is infinite, we obtain that S is cofinite which contradicts the assumption that S is finite. Therefore, $b \in V$, and $S \subseteq V$. Thus, S is the least finite set supporting f, i.e. S is the support of f. □

Remark 2.1 If X is an infinite ordinary ZF set and $f : A \to X$ a finitely supported function, then $Im(f)$ is obviously finite. However, the converse of this assertion is not true. There may exist an ordinary ZF set X and a ZF function $f : A \to X$ such that $Im(f)$ is finite, but f is not finitely supported. Indeed, let X be an ordinary ZF set with at least two elements and let us fix two elements $u, v \in X$. Let P be a ZF subset of A which is simultaneously infinite and coinfinite. Such a subset of A exists in ZF. We define $g : A \to X$ by $g(x) = \begin{cases} u & \text{for} x \in P \\ v & \text{for} x \in A \setminus P \end{cases}$. For any permutation of atoms π, we have that $g(\pi(x)) = g(x)$ for all $x \in A$ if and only if P is fixed by π (i.e. $\pi(x) \in P$ whenever $x \in P$). Therefore, because P is not finitely supported, it follows that g is not finitely supported.

Proposition 2.11 *Let $f : A \to A$ be a finitely supported injection on A.*
Then $\{a \in A \,|\, f(a) \neq a\}$ is finite, and $supp(f) = \{a \in A \,|\, f(a) \neq a\}$.

Proof Firstly, we prove that for each $a \in A$ we have that $a \notin supp(f)$ implies $f(a) = a$. Let $a \notin supp(f)$. Assume that $f(a) \neq a$. Let us consider two atoms $b, c \notin supp(f)$ such that a, b, c, are all different (such atoms exist because $supp(f)$ is finite, while A is infinite). Since $supp(f)$ supports f and $(ab) \in Fix(supp(f))$, then $(ab) \star f = f$, where $\star$ is the S_A-action on A^A presented in Proposition 2.7. Similarly, $(ac) \star f = f$. According to Proposition 2.8, we have $f(b) = f((ab)(a)) = (ab)(f(a))$. However, $f(a) \neq a$. Since f is an injection, it follows that $f(a) = b$ (otherwise, we would have $f(b) = f(a)$ with $b \neq a$). However, from $f((ac)(a)) = (ac)(f(a))$, it follows that $f(c) = (ac)(b) = b = f(a)$, which contradicts the injectivity of f. Thus, $f(a) = a$. This means $S = \{a \in A \,|\, f(a) \neq a\} \subseteq supp(f)$. Since $supp(f)$ is finite, we have that S is finite.

Now we prove that the finite set S supports f. Indeed, let us consider $\pi \in Fix(S)$, i.e. $\pi(a) = a$ whenever $f(a) \neq a$. We claim that $f(\pi(x)) = \pi(f(x))$ for all $x \in A$. Indeed, let us fix an arbitrary element $x \in A$. If $f(x) \neq x$, then $\pi(x) = x$ and $f(\pi(x)) = f(x)$. However, since f is injective, we also have $f(f(x)) \neq f(x)$, and so $\pi(f(x)) = f(x)$. Thus, $f(\pi(x)) = \pi(f(x))$. On the other hand, if $f(x) = x$, then $\pi(f(x)) = \pi(x)$. Suppose that $f(\pi(x)) \neq \pi(x)$. This means $\pi(\pi(x)) = \pi(x)$, and so $\pi(x) = x$. Then $f(\pi(x)) = f(x) = x = \pi(x)$, which contradicts the assumption that $f(\pi(x)) \neq \pi(x)$. We obtain that $f(\pi(x)) = \pi(x)$, and so $f(\pi(x)) = \pi(f(x))$. According to Proposition 2.8, we have that S supports f. Since $supp(f)$ is minimal between the finite sets supporting f, it follows that $S = supp(f)$. □

Remark 2.2 Note that in FSM every construction (in particular, every function) should be finitely supported. If we considered the entire group of bijections over A instead of S_A to define the permutation actions and finitely supported elements, i.e.

if we defined that *an element x is finitely supported by a finite set S of atoms if and only if for every bijection of atoms f that fixes S pointwise we have that x is invariant under the action of f*, this definition would finally be equivalent to Definition 2.2(3). This is because we would obtain (according to the new definition) that a bijection of atoms f is finitely supported if and only if $f \in S_A$. Indeed, if f is a bijection of A supported by a finite set of atoms T (under the new definition written above with italic fonts), one can prove that whenever $a \notin T$ we have $f(a) = a$. This assertion follows as in the proof of Proposition 2.11 by replacing $supp(f)$ with T and by using the definition of a finitely supported function that leads from the definition of a finitely supported subset of a set (reformulated in the sense of the new italic definition 'of being finitely supported' presented above); for proving Proposition 2.9 involved in the proof of Proposition 2.11 we used only the fact that permutations of atoms are bijective without involving their finiteness, and so Proposition 2.9 holds as well under the new italic definition described above. Conversely, an element g from S_A is finitely supported under the above new definition by the finite set of atoms changed by g (this follows exactly as in the last part of the proof of Proposition 2.11). When finite support requirement is involved, the set of all one-to-one transformations of A onto itself (i.e. the set of all bijections of A) coincides with S_A. Therefore, in FSM it is sufficient to consider only the permutations of A (i.e. only the bijections of A that can be expressed as finite compositions of transpositions) instead of arbitrary bijections of A, and to define finitely supported elements by employing Definition 2.2(3) instead of the alternative definition described (with italic fonts) in this remark.

In the classical ZF set theory it is known that for any set X we have $|\wp(X)| = 2^{|X|}$. This results is preserved in FSM (as well as other properties of cardinalities).

Theorem 2.3 *Let $(X, \cdot)$ be a finitely supported subset of an invariant set $(Z, \cdot)$. There exists a one-to-one mapping from $\wp_{fs}(X)$ onto $\{0,1\}^X_{fs}$ which is finitely supported by $supp(X)$, where $\wp_{fs}(X)$ is considered the family of those finitely supported subsets of Z contained in X.*

Proof Let Y be a finitely supported subset of Z contained in X, and φ_Y be the characteristic function on Y, i.e. $\varphi_Y : X \to \{0,1\}$ is defined by

$$\varphi_Y(x) \stackrel{def}{=} \begin{cases} 1 \text{ for} x \in Y \\ 0 \text{ for} x \in X \setminus Y \end{cases}.$$

We prove that φ_Y is a finitely supported function from X to $\{0,1\}$ (according to Proposition 2.2, $\{0,1\}$ is a trivial invariant set), and the mapping $Y \mapsto \varphi_Y$ defined on $\wp_{fs}(X)$ is also finitely supported in the sense of Definition 2.6.

Firstly, we prove that φ_Y is supported by $supp(Y) \cup supp(X)$. Let us take $\pi \in Fix(supp(Y) \cup supp(X))$. Thus, $\pi \star Y = Y$ (where $\star$ represents the canonical permutation action on $\wp(Z)$), and so $\pi \cdot x \in Y$ if and only if $x \in Y$. Since we additionally have $\pi \star X = X$, we obtain $\pi \cdot x \in X \setminus Y$ if and only if $x \in X \setminus Y$. Thus, $\varphi_Y(\pi \cdot x) = \varphi_Y(x)$ for all $x \in X$. Furthermore, because π fixes $supp(X)$ pointwise we have $\pi \cdot x \in X$ for all $x \in X$, and from Proposition 2.9 we get that φ_Y is supported by $supp(Y) \cup supp(X)$.

We remark that $\{0,1\}^X_{fs}$ is a finitely supported subset of the set $(\wp_{fs}(Z \times \{0,1\}), \widetilde{\star})$. Let $\pi \in Fix(supp(X))$ and $f: X \to \{0,1\}$ finitely supported. We have $\pi \widetilde{\star} f = \{(\pi \cdot x, \pi \diamond y) \mid (x,y) \in f\} = \{(\pi \cdot x, y) \mid (x,y) \in f\}$ because $\diamond$ is the trivial action on $\{0,1\}$. Thus, $\pi \widetilde{\star} f$ is a function with the domain $\pi \star X = X$ which is finitely supported as an element of $(\wp(Z \times \{0,1\}), \widetilde{\star})$ according to Proposition 2.1. Moreover, $(\pi \widetilde{\star} f)(\pi \cdot x) = f(x)$ for all $x \in X$ (1).

According to Proposition 2.9, to prove that the function $g := Y \mapsto \varphi_Y$ defined on $\wp_{fs}(X)$ (with the codomain contained in $\{0,1\}^X_{fs}$) is supported by $supp(X)$, we have to prove that $\pi \widetilde{\star} g(Y) = g(\pi \star Y)$ for all $\pi \in Fix(supp(X))$ and all $Y \in \wp_{fs}(X)$ (where $\widetilde{\star}$ symbolizes the induced S_A-action on $\{0,1\}^X_{fs}$). This means that we need to verify the relation $\pi \widetilde{\star} \varphi_Y = \varphi_{\pi \star Y}$ for all $\pi \in Fix(supp(X))$ and all $Y \in \wp_{fs}(X)$. Let us consider $\pi \in Fix(supp(X))$ (which means $\pi \cdot x \in X$ for all $x \in X$) and $Y \in \wp_{fs}(X)$. For any $x \in X$, we know that $x \in \pi \star Y$ if and only if $\pi^{-1} \cdot x \in Y$. Thus, $\varphi_Y(\pi^{-1} \cdot x) = \varphi_{\pi \star Y}(x)$ for all $x \in X$, and so $(\pi \widetilde{\star} \varphi_Y)(x) \overset{(1)}{=} \varphi_Y(\pi^{-1} \cdot x) = \varphi_{\pi \star Y}(x)$ for all $x \in X$. Moreover, from Proposition 2.1, $\pi \star Y$ is a finitely supported subset of Z contained in $\pi \star X = X$, and $\{0,1\}^X_{fs}$ can be represented as a finitely supported subset of $\wp_{fs}(Z \times \{0,1\})$ (supported by $supp(X)$). According to Proposition 2.9 we have that g is a finitely supported function from $\wp_{fs}(X)$ to $\{0,1\}^X_{fs}$.

Obviously, g is one-to-one. Now we prove that g is onto. Let us consider an arbitrary finitely supported function $f: X \to \{0,1\}$. Let $Y_f \overset{def}{=} \{x \in X \mid f(x) = 1\}$. We claim that $Y_f \in \wp_{fs}(X)$. Let $\pi \in Fix(supp(f))$. According to Proposition 2.9 we have $\pi \cdot x \in X$ and $f(\pi \cdot x) = f(x)$ for all $x \in X$. Thus, for each $x \in Y_f$, we have $\pi \cdot x \in Y_f$. Therefore $\pi \star Y_f = Y_f$, and so Y_f is finitely supported by $supp(f)$ as a subset of Z, and it is contained in X. A simple calculation show us that $g(Y_f) = f$, and so g is onto. □

Corollary 2.7 *Let $(X, \cdot)$ be a non-trivial invariant set.*

There exists an equivariant one-to-one mapping from $\wp_{fs}(X)$ onto $\{0,1\}^X_{fs}$.

Proof If X is a non-trivial invariant set, then X is an empty supported subset of itself. The results follows from Theorem 2.3, because $supp(X) = \emptyset$. □

Theorem 2.4 *Let X, Y, Z be finitely supported subsets of invariant sets. There exists a bijection between $Z^{X \times Y}_{fs}$ and $(Z^Y_{fs})^X_{fs}$, finitely supported by $S = supp(X) \cup supp(Y) \cup supp(Z)$.*

Proof We generically denote the (possibly different) actions of the invariant sets containing X, Y, Z by $\cdot$, the actions on Cartesian products by $\otimes$ and the actions of function spaces by $\star$. Let us define $\varphi : Z^{X \times Y}_{fs} \to (Z^Y_{fs})^X_{fs}$ in the following way. For each finitely supported mapping $f: X \times Y \to Z$ and each $x \in X$ we consider $\varphi(f): X \to Z^Y_{fs}$ to be the function defined by $(\varphi(f)(x))(y) = f(x,y)$ for all $y \in Y$. Let us prove that φ is well-defined. For a fixed $x \in X$ we firstly prove that $\varphi(f)(x)$ is a finitely supported mapping from Y to Z. Indeed, according to Proposition 2.9 (since π fixes $supp(f)$ pointwise and $supp(f)$ supports f), for $\pi \in Fix(supp(x) \cup supp(f) \cup S)$ we have $(\varphi(f)(x))(\pi \cdot y) = f(x, \pi \cdot y) = f(\pi \cdot$

$x, \pi \cdot y) = f(\pi \otimes (x,y)) = \pi \cdot f(x,y) = \pi \cdot (\varphi(f)(x))(y)$ for all $y \in Y$, and by using again Proposition 2.9, we obtain that $\varphi(f)(x)$ is a finitely supported function. Now we prove that $\varphi(f) : X \to Z_{fs}^{Y}$ is finitely supported by $supp(f) \cup S$. Let $\pi \in Fix(supp(f) \cup S)$. In the view of Proposition 2.9 we have to prove that $\varphi(f)(\pi \cdot x) = \pi \star \varphi(f)(x)$ for all $x \in X$. Fix $x \in X$ and consider an arbitrary $y \in Y$. We have $(\varphi(f)(\pi \cdot x))(y) = f(\pi \cdot x, y)$. According to Proposition 2.9, we also have $(\pi \star \varphi(f)(x))(y) = \pi \cdot (\varphi(f)(x))(\pi^{1} \cdot y) = \pi \cdot f(x, \pi^{-1} \cdot y) = f(\pi \otimes (x, \pi^{-1} \cdot y)) = f(\pi \cdot x, y)$. Thus, $\varphi(f) : X \to Z_{fs}^{Y}$ is finitely supported. Now we claim that φ is finitely supported by S. Let $\pi \in Fix(S)$. In the view of Proposition 2.9 we have to prove that $\varphi(\pi \star f) = \pi \star \varphi(f)$ for all $f : X \times Y \to Z$. Fix $f : X \times Y \to Z$. We have to prove that $\varphi(\pi \star f)(x) = (\pi \star \varphi(f))(x)$ for all $x \in X$. Fix some $x \in X$ and consider an arbitrary $y \in Y$. We have $(\varphi(\pi \star f)(x))(y) = (\pi \star f)(x,y) = \pi \cdot f(\pi^{-1} \otimes (x,y)) = \pi \cdot f(\pi^{-1} \cdot x, \pi^{-1} \cdot y)$. Furthermore, $((\pi \star \varphi(f))(x))(y) = (\pi \star \varphi(f)(\pi^{-1} \cdot x))(y) = \pi \cdot (\varphi(f)(\pi^{-1} \cdot x))(\pi^{-1} \cdot y) = \pi \cdot f(\pi^{-1} \cdot x, \pi^{-1} \cdot y)$, and so our claim follows.

Similarly, we define $\psi : (Z_{fs}^{Y})_{fs}^{X} \to Z_{fs}^{X \times Y}$ in the following way. For any finitely supported function $g : X \to Z_{fs}^{Y}$, define $\psi(g) : X \times Y \to Z$ by $\psi(g)(x,y) = (g(x))(y)$ for all $x \in X$ and $y \in Y$. Firstly we prove that $\psi(g)$ is well-defined. Let $\pi \in Fix(supp(g))$. According to Proposition 2.9, we have $\psi(g)(\pi \otimes (x,y)) = \psi(g)(\pi \cdot x, \pi \cdot y) = (g(\pi \cdot x))(\pi \cdot y) = (\pi \star g(x))(\pi \cdot y) = \pi \cdot (g(x))(\pi^{-1} \cdot (\pi \cdot y)) = \pi \cdot (g(x))(y) = \pi \cdot \psi(x,y)$ for all $(x,y) \in X \times Y$. Thus, in the view of Proposition 2.9, we conclude that $\psi(g)$ is supported by $supp(g)$. Let us prove now that ψ is finitely supported by S. We have to prove that for $\pi \in Fix(S)$ we get $\psi(\pi \star g) = \pi \star \psi(g)$ for any finitely supported function $g : X \to Z_{fs}^{Y}$. Fix such a g, and consider arbitrary $x \in X$ and $y \in Y$. We have $\psi(\pi \star g)(x,y) = ((\pi \star g)(x))(y) = (\pi \star g(\pi^{-1} \cdot x))(y) = \pi \cdot (g(\pi^{-1} \cdot x))(\pi^{-1} \cdot y) = \pi \cdot \psi(g)(\pi^{-1} \cdot x, \pi^{-1} \cdot y) = \pi \cdot \psi(g)(\pi^{-1} \otimes (x,y)) = (\pi \star \psi(g))(x,y)$, i.e. the desired result. It is routine to prove that $\psi \circ \varphi = 1|_{Z_{fs}^{X \times Y}}$ and $\varphi \circ \psi = 1|_{(Z_{fs}^{Y})_{fs}^{X}}$, and so ψ and φ are bijective, one being the inverse of the other. □

Corollary 2.8 *Let X, Y, Z be invariant sets.*
There exists an equivariant bijection between $Z_{fs}^{X \times Y}$ and $(Z_{fs}^{Y})_{fs}^{X}$.

Theorem 2.5 *Let X, Y, Z be finitely supported subsets of invariant sets. There exists a bijection between Z_{fs}^{X+Y} and $Z_{fs}^{X} \times Z_{fs}^{Y}$, finitely supported by $S = supp(X) \cup supp(Y) \cup supp(Z)$.*

Proof We generically denote the (possibly different) actions of the invariant sets containing X, Y, Z by $\cdot$, the actions on Cartesian products by $\otimes$, the actions of function spaces by $\star$, and the actions on disjoint unions by $\diamond$. We define $\varphi : Z_{fs}^{X+Y} \to Z_{fs}^{X} \times Z_{fs}^{Y}$ as follows: if $f : X + Y \to Z$ is a finitely supported mapping, then $\varphi(f) = (f_1, f_2)$ where $f_1 : X \to Z$, $f_1(x) = f((0,x))$ for all $x \in X$, and $f_2 : Y \to Z$, $f_2(y) = f((1,y))$ for all $y \in Y$. Clearly, φ is well-defined because f_1 and f_2 are both supported by $supp(f)$. Furthermore, φ is bijective. It remains to prove that φ is supported by S. Let $\pi \in Fix(S)$ and consider an arbitrary $f : X + Y \to Z$. We have $\varphi(\pi \star f) = (g_1, g_2)$ where $g_1(x) = (\pi \star f)((0,x)) = \pi \cdot f(\pi^{-1} \diamond (0,x)) = \pi \cdot f((0, \pi^{-1} \cdot x)) = \pi \cdot f_1(\pi^{-1} \cdot x) = (\pi \star f_1)(x)$ for all $x \in X$, and similarly, $g_2(y) =$

$(\pi\star f)((1,y)) = \pi\cdot f(\pi^{-1}\diamond(1,y)) = \pi\cdot f((1,\pi^{-1}\cdot y)) = \pi\cdot f_2(\pi^{-1}\cdot y) = (\pi\star f_2)(y)$ for all $y\in Y$. Thus, $\varphi(\pi\star f) = (g_1,g_2) = (\pi\star f_1,\pi\star f_2) = \pi\otimes(f_1,f_2) = \pi\otimes\varphi(f)$. According to Proposition 2.9, we have that φ is supported by S. □

Corollary 2.9 *Let X,Y,Z be invariant sets.*
There exists an equivariant bijection between Z_{fs}^{X+Y} and $Z_{fs}^{X}\times Z_{fs}^{Y}$.

Theorem 2.6 *Let X,Y,Z be finitely supported subsets of invariant sets. There exists a bijection between $(X\times Y)_{fs}^{Z}$ and $X_{fs}^{Z}\times Y_{fs}^{Z}$, finitely supported by $S = supp(X)\cup supp(Y)\cup supp(Z)$.*

Proof We generically denote the (possibly different) actions of the invariant sets containing X,Y,Z by $\cdot$, the actions on Cartesian products by $\otimes$ and the actions of function spaces by $\star$. We define $\varphi : X_{fs}^{Z}\times Y_{fs}^{Z}\to (X\times Y)_{fs}^{Z}$ by $\varphi(f_1,f_2)(z) = (f_1(z),f_2(z))$ for all $f_1\in X_{fs}^{Z}$, all $f_2\in Y_{fs}^{Z}$ and all $z\in Z$. Fix some finitely supported mappings $f_1 : Z\to X$ and $f_2 : Z\to Y$. For $\pi\in Fix(supp(f_1)\cup supp(f_2))$, according to Proposition 2.9, we have $\varphi(f_1,f_2)(\pi\cdot z) = (f_1(\pi\cdot z),f_2(\pi\cdot z)) = (\pi\cdot f_1(z),\pi\cdot f_2(z)) = \pi\otimes(f_1(z),f_2(z)) = \pi\otimes\varphi(f_1,f_2)(z)$ for all $z\in Z$. Thus, $\varphi(f_1,f_2)$ is a finitely supported mapping, and so φ is well-defined. Furthermore, φ is bijective. Let us prove now that φ is finitely supported by S. Let $\pi\in Fix(S)$. Fix some arbitrary $f_1\in X_{fs}^{Z}$, $f_2\in Y_{fs}^{Z}$ and $z\in Z$. We have $\varphi(\pi\otimes(f_1,f_2))(z) = \varphi(\pi\star f_1,\pi\star f_2)(z) = ((\pi\star f_1)(z),(\pi\star f_2)(z)) = (\pi\cdot f_1(\pi^{-1}\cdot z),\pi\cdot f_2(\pi^{-1}\cdot z)) = \pi\otimes(f_1(\pi^{-1}\cdot z),f_2(\pi^{-1}\cdot z)) = \pi\otimes\varphi(f_1,f_2)(\pi^{-1}\cdot z) = (\pi\star\varphi(f_1,f_2))(z)$. According to Proposition 2.9, φ is finitely supported. □

Corollary 2.10 *Let X,Y,Z be invariant sets.*
There exists an equivariant bijection between $(X\times Y)_{fs}^{Z}$ and $X_{fs}^{Z}\times Y_{fs}^{Z}$.

Proposition 2.12 *Let $(X,\cdot)$ be an invariant set containing no infinite uniformly supported subset. Let $f : X\to X$ be a finitely supported injective function. Then for each $x\in X$ there exists $n\in\mathbb{N}$ such that $f^n(x) = x$, where $f^n : X\to X$ represents the n-composition of f with itself.*

Proof Let us fix an arbitrary element $x_0\in X$. We consider the sequence $(f^n(x_0))_{n\in\mathbb{N}}$ which has the first term x_0 and the general term $x_{n+1} = f(x_n)$ for all $n\in\mathbb{N}$. We prove that x_{n+1} is supported by $supp(f)\cup supp(x_n)$ for all $n\in\mathbb{N}$. Indeed, let us fix some $n\in\mathbb{N}$. Let $\pi\in Fix(supp(f)\cup supp(x_n))$. Thus, $\pi\cdot x_n = x_n$. According to Proposition 2.9, because π fixes $supp(f)$ pointwise and $supp(f)$ supports f, we get $\pi\cdot x_{n+1} = \pi\cdot f(x_n) = f(\pi\cdot x_n) = f(x_n) = x_{n+1}$. Since $supp(x_{n+1})$ is the least set supporting x_{n+1}, we obtain $supp(x_{n+1})\subseteq supp(f)\cup supp(x_n)$. Since n was arbitrarily chosen, by finite recursion, we have $supp(x_n)\subseteq supp(f)\cup supp(x_0)$ for all $n\in\mathbb{N}$. Thus, $(x_n)_{n\in\mathbb{N}}$ is a uniformly supported subset of X, and so $(x_n)_{n\in\mathbb{N}}$ must be finite. Therefore, there should exist two distinct elements $n,m\in\mathbb{N}$ such that $x_n = x_m$, i.e. $f^n(x_0) = f^m(x_0)$. Suppose $n < m$. Since f is injective, we get $x_0 = f^{m-n}(x_0)$, and so the result follows. □

Theorem 2.7 (Infinite Pigeonhole Principle) *Suppose that there exists an FSM colouring on the elements of an infinite invariant set $(X,\cdot)$ with finitely many colours belonging to a certain invariant set $(Y,\diamond)$. Then there is an infinite set $M \subseteq X$ which is finitely supported, such that the elements of M are FSM coloured with the same colour.*

Proof We have that the set of colours $\{c_1,\ldots,c_m\}$ forms a subset of Y which is finitely supported by $S = supp(c_1) \cup \ldots \cup supp(c_m)$. By hypothesis, there exists an FSM colouring on X, which means there exists a finitely supported function $f : X \to \{c_1,\ldots,c_m\}$. We denote by $X_i = f^{-1}(c_i) = \{x \in X \,|\, f(x) = c_i\}$ for all $i \in \{1,\ldots,n\}$. We claim that $X_i \in \wp_{fs}(X)$ for all $i \in \{1,\ldots,n\}$. Indeed, we prove that each X_i is supported by $supp(f) \cup supp(c_i)$. Let $\pi \in Fix(supp(f) \cup supp(c_i))$, and $y \in X_i = f^{-1}(c_i)$. This means $f(y) = c_i$. Thus, because π fixes $supp(f)$ pointwise and $supp(f)$ supports f, we have $f(\pi \cdot y) = \pi \diamond f(y) = \pi \diamond c_i$, However, since π fixes $supp(c_i)$ pointwise and $supp(c_i)$ supports c_i, we have $\pi \diamond c_i = c_i$, and so $\pi \cdot y \in f^{-1}(c_i)$. Therefore, X_i is finitely supported for all $i \in \{1,\ldots,n\}$. Since f is a function, it follows that $X = \underset{i}{\cup} X_i$. Since X is infinite, at least one X_i should be infinite. Such an X_i is the required subset M of X. Furthermore, the mapping $g = f|_M : M \to \{c_i\}$ is supported by $supp(f) \cup supp(M) = supp(f) \cup supp(c_i)$. □

2.4 Particular Functions on the Finite Powerset of Atoms

Finitely supported self-mappings defined on the finite powerset of atoms have some interesting properties.

Proposition 2.13 *Let $f : \wp_{fin}(A) \to \wp_{fin}(A)$ be finitely supported and injective. For each $X \in \wp_{fin}(A)$ we have $X \setminus supp(f) \neq \emptyset$ if and only if $f(X) \setminus supp(f) \neq \emptyset$. Moreover, $X \setminus supp(f) = f(X) \setminus supp(f)$. Furthermore, if f is monotone (i.e. order preserving), then $X \setminus supp(f) = f(X \setminus supp(f))$ for all $X \in \wp_{fin}(A)$, and $f(supp(f)) = supp(f)$.*

Proof Let $Y \in \wp_{fin}(A)$. Then we have $supp(Y) = Y$. According to Proposition 2.9, for any permutation $\pi \in Fix(supp(f) \cup supp(Y)) = Fix(supp(f) \cup Y)$ we have $\pi \star f(Y) = f(\pi \star Y) = f(Y)$ which means $supp(f) \cup Y$ supports $f(Y)$, that is $f(Y) = supp(f(Y)) \subseteq supp(f) \cup Y$ (1). If $Y \subseteq supp(f)$, we get $f(Y) \subseteq supp(f)$ (2). Let $X \in \wp_{fin}(X)$ with $X \subseteq supp(f)$. From (2) we get $f(X) \subseteq supp(f)$. Conversely, assume $f(X) \subseteq supp(f)$. Applying (2), by induction we get $f^n(X) \subseteq supp(f)$ for all $n \in \mathbb{N}^*$ (3). Since $supp(f)$ is finite, there should exist $m,k \in \mathbb{N}^*$ with $m \neq k$ such that $f^m(X) = f^k(X)$. Assume $m > k$. Due to the injectivity of f we get $f^{m-k}(X) = X$ which by (3) leads to $X \subseteq supp(f)$. Therefore, $X \subseteq supp(f)$ if and only if $f(X) \subseteq supp(f)$, and so $X \setminus supp(f) \neq \emptyset$ if and only if $f(X) \setminus supp(f) \neq \emptyset$.

Let $Z \in \wp_{fin}(A)$ such that $f(Z) \setminus supp(f) \neq \emptyset$, or equivalently $Z \setminus supp(f) \neq \emptyset$. Thus, Z has the form $Z = \{a_1,\ldots,a_n,b_1,\ldots,b_m\}$ with $a_1,\ldots,a_n \in supp(f)$ and $b_1,\ldots,b_m \in A \setminus supp(f)$, $m \geq 1$, or the form $Z = \{b_1,\ldots,b_m\}$ with $b_1,\ldots,b_m \in$

$A \setminus supp(f), m \geq 1$. According to (1), $f(Z)$ should be $f(Z) = \{c_1, \dots, c_k, b_{i_1}, \dots, b_{i_l}\}$ with $c_1, \dots, c_k \in supp(f)$ and $b_{i_1}, \dots, b_{i_l} \in A \setminus supp(f)$, or $f(Z) = \{b_{i_1}, \dots, b_{i_l}\}$ with $b_{i_1}, \dots, b_{i_l} \in A \setminus supp(f)$, having in any case the property that $\{b_{i_1}, \dots, b_{i_l}\}$ is non-empty (i.e. it should contain at least one element, say b_{i_1}) and $\{b_{i_1}, \dots, b_{i_l}\} \subseteq \{b_1, \dots, b_m\}$. If $m = 1$, then $l = 1$, $b_{i_1} = b_1$, and we are done, so let $m > 1$. Assume by contradiction that there exists $j \in \{1, \dots, m\}$ such that $b_j \notin \{b_{i_1}, \dots, b_{i_l}\}$. Then $(b_{i_1}\ b_j) \star Z = Z$ since both $b_{i_1}, b_j \in Z$ and Z is a finite subset of atoms (b_{i_1} and b_j are interchanged in Z under the effect of the transposition $(b_{i_1}\ b_j)$, but the whole Z is left invariant). Furthermore, since $b_{i_1}, b_j \notin supp(f)$ we have $(b_{i_1}\ b_j) \in Fix(supp(f))$, and by Proposition 2.9 we get $f(Z) = f((b_{i_1}\ b_j) \star Z) = (b_{i_1}\ b_j) \star f(Z)$ which is a contradiction because $b_{i_1} \in f(Z)$ while $b_j \notin f(Z)$. Thus, $\{b_{i_1}, \dots, b_{i_l}\} = \{b_1, \dots, b_m\}$, and so $Z \setminus supp(f) = f(Z) \setminus supp(f)$. The case $supp(f) = \emptyset$ is included in the above analysis and leads to $f(\emptyset) = \emptyset$ and $f(X) = X$ for all $X \in \wp_{fin}(A)$.

Assume now that f is monotone. Let us fix $X \in \wp_{fin}(A)$, and consider the case $X \setminus supp(f) \neq \emptyset$, that is $X = \{a_1, \dots, a_n, b_1, \dots, b_m\}$ with $a_1, \dots, a_n \in supp(f)$ and $b_1, \dots, b_m \in A \setminus supp(f)$, $m \geq 1$, or $X = \{b_1, \dots, b_m\}$ with $b_1, \dots, b_m \in A \setminus supp(f)$, $m \geq 1$. Therefore we get $X \setminus supp(f) = \{b_1, \dots, b_m\}$, and by involving the above arguments, we should have $f(X \setminus supp(f)) = \{u_1, \dots, u_i, b_1, \dots, b_m\}$ with $u_1, \dots, u_i \in supp(f)$ or $f(X \setminus supp(f)) = \{b_1, \dots, b_m\}$. In either case we have $X \setminus supp(f) \subseteq f(X \setminus supp(f))$, and since f is monotone we construct an ascending chain $X \setminus supp(f) \subseteq f(X \setminus supp(f)) \subseteq \dots \subseteq f^n(X \setminus supp(f)) \subseteq \dots$. Since for any $n \in \mathbb{N}$ we have that $f^n(X \setminus supp(f))$ is supported by $supp(f) \cup supp(X \setminus supp(f)) = supp(f) \cup supp(X)$ and $\wp_{fin}(A)$ does not contain an infinite uniformly supported subset (the elements of $\wp_{fin}(A)$ supported by $supp(f) \cup supp(X)$ are precisely the subsets of $supp(f) \cup supp(X)$), the related chain should be stationary, that there exists $m \in \mathbb{N}$ such that $f^m(X \setminus supp(f)) = f^{m+1}(X \setminus supp(f))$, which, due to the injectivity of f, leads to $X \setminus supp(f) = f(X \setminus supp(f))$.

The remaining case is $X \subseteq supp(f)$. Then $X \setminus supp(f) = \emptyset$, and $f(\emptyset) \subseteq supp(f)$. In the finite set $supp(f)$ we can define the chain $\emptyset \subseteq f(\emptyset) \subseteq f^2(\emptyset) \subseteq \dots \subseteq f^n(\emptyset) \subseteq \dots$ which is uniformly supported by $supp(f)$. Therefore the related chain should be stationary, which means that there exists $m \in \mathbb{N}$ such that $f^m(\emptyset) = f^{m+1}(\emptyset)$. According to the injectivity of f, we get $X \setminus supp(f) = \emptyset = f(\emptyset) = f(X \setminus supp(f))$.

According to (2) we have $f(supp(f)) \subseteq supp(f)$, and since f is preserves the inclusion relation, we construct in $supp(f)$ the chain $\dots \subseteq f^n(supp(f)) \subseteq \dots \subseteq f(supp(f)) \subseteq supp(f)$. Since $supp(f)$ is finite, the chain should be stationary, and so $f^{m+1}(supp(f)) = f^m(supp(f))$ for some positive integer m, which, because f is injective, leads to $f(supp(f)) = supp(f)$. □

Remark 2.3 From the proof of Proposition 2.13, if $f : \wp_{fin}(A) \to \wp_{fin}(A)$ is finitely supported (even if it is not injective) with $X \subseteq supp(f)$, we have $f(X) \subseteq supp(f)$. If $f(X) \setminus supp(f) \neq \emptyset$, then $X \setminus supp(f) = f(X) \setminus supp(f)$.

Corollary 2.11 *Let $f : \wp_{fin}(A) \to \wp_{fin}(A)$ be finitely supported and surjective. Then for each $X \in \wp_{fin}(A)$ we have $X \setminus supp(f) \neq \emptyset$ if and only if $f(X) \setminus supp(f) \neq \emptyset$. In either of these cases $X \setminus supp(f) = f(X) \setminus supp(f)$. If, furthermore, f is monotone, then $X \setminus supp(f) = f(X \setminus supp(f))$ for all $X \in \wp_{fin}(A)$, and $f(supp(f)) = supp(f)$.*

Proof From the proof of (the second part of) Proposition 9.4, a finitely supported surjective mapping $f : \wp_{fin}(A) \to \wp_{fin}(A)$ should be injective. Now the result follows from Proposition 2.13. □

Theorem 2.8 *Let $f : \wp_{fin}(A) \to \wp_{fin}(A)$ be finitely supported and strictly monotone (i.e. f has the property that $X \subsetneq Y$ implies $f(X) \subsetneq f(Y)$). Then we have $X \setminus supp(f) = f(X \setminus supp(f))$ for all $X \in \wp_{fin}(A)$.*

Proof Let $X \in \wp_{fin}(A)$. Since the support of a finite subset of atoms coincides with the related subset, then $supp(X) = X$ and $supp(f(X)) = f(X)$. According to Proposition 2.9, for any permutation $\pi \in Fix(supp(f) \cup supp(X))$=$Fix(supp(f) \cup X)$ we have $\pi \star f(X) = f(\pi \star X) = f(X)$ which means $supp(f) \cup X$ supports $f(X)$, that is $f(X) = supp(f(X)) \subseteq supp(f) \cup X$ (claim 1).

If $supp(f) = \emptyset$, then we have $f(X) \subseteq X$ for all $X \in \wp_{fin}(A)$. If there exists $Y \in \wp_{fin}(A)$ with $f(Y) \subsetneq Y$, then we can construct the sequence $\ldots \subsetneq f^j(Y) \subsetneq \ldots \subsetneq f^2(Y) \subsetneq f(Y) \subsetneq Y$ which is infinite and uniformly supported by $supp(Y) \cup supp(f)$ (using an induction on j and Proposition 2.9). This is a contradiction, because the finite set Y cannot have infinitely many subsets, and so $f(X) = X$ for all $X \in \wp_{fin}(A)$.

Assume now that $supp(f)$ is non-empty. If $X \subseteq supp(f)$, then $f(X \setminus supp(f)) = f(\emptyset) = \emptyset = X \setminus supp(f)$. The second identity follows because f is strictly monotone; otherwise we could construct an infinite strictly ascending chain in $\wp_{fin}(A)$, uniformly supported by $supp(f)$, namely $\emptyset \subsetneq f(\emptyset) \subsetneq \ldots \subsetneq f^k(\emptyset) \subsetneq \ldots$, contradicting the fact that $\wp_{fin}(A)$ does not contain an infinite uniformly supported subset.

We prove the following intermediate result. Let us consider an arbitrary set $Z = \{b_1, \ldots, b_n\}$ such that $b_1, \ldots, b_n \in A \setminus supp(f)$, $n \geq 1$ and $f(Z) \setminus supp(f) \neq \emptyset$. We prove that $f(Z) = Z$ (claim 2). According to (claim 1), $f(Z)$ should be $f(Z) = \{c_1, \ldots, c_k, b_{i_1}, \ldots, b_{i_l}\}$ with $c_1, \ldots, c_k \in supp(f)$ and $b_{i_1}, \ldots, b_{i_l} \in A \setminus supp(f)$, or $f(Z) = \{b_{i_1}, \ldots, b_{i_l}\}$ with $b_{i_1}, \ldots, b_{i_l} \in A \setminus supp(f)$. In both cases we have the property that $\{b_{i_1}, \ldots, b_{i_l}\}$ is *non-empty* (i.e. it should contain at least one element, say b_{i_1}, because we assumed that $f(Z)$ contains at least one element outside $supp(f)$) and $\{b_{i_1}, \ldots, b_{i_l}\} \subseteq \{b_1, \ldots, b_n\}$. If $n = 1$, then $l = 1$ and $b_{i_1} = b_1$. Now let us consider $n > 1$. Assume by contradiction that there exists $j \in \{1, \ldots, n\}$ such that $b_j \notin \{b_{i_1}, \ldots, b_{i_l}\}$. Then $(b_{i_1}\ b_j) \star Z = Z$ since both $b_{i_1}, b_j \in Z$ and Z is a finite subset of atoms (b_{i_1} and b_j are interchanged in Z under the effect of the transposition $(b_{i_1}\ b_j)$, while the other atoms belonging to Z are left unchanged, which means the entire Z is left invariant under the effect of the related transposition). Furthermore, since $b_{i_1}, b_j \notin supp(f)$ we have $(b_{i_1}\ b_j) \in Fix(supp(f))$, and by Proposition 2.9 we get $f(Z) = f((b_{i_1}\ b_j) \star Z) = (b_{i_1}\ b_j) \star f(Z)$ which is a contradiction because $b_{i_1} \in f(Z)$ while $b_j \notin f(Z)$. Thus, $\{b_{i_1}, \ldots, b_{i_l}\} = \{b_1, \ldots, b_n\}$. Now we prove that $f(Z) = Z$. Assume by contradiction that we are in the case $f(Z) = \{c_1, \ldots, c_k, b_1, \ldots, b_n\}$ with $c_1, \ldots, c_k \in supp(f)$. Then $Z \subsetneq f(Z)$, and since f is strictly monotone we can construct a strictly ascending chain $Z \subsetneq f(Z) \subsetneq \ldots \subsetneq f^i(Z) \subsetneq \ldots$. Since for any $i \in \mathbb{N}$ we have that $f^i(Z)$ is supported by $supp(f) \cup supp(Z)$ (this follows by induction on i using Proposition 2.9) and $\wp_{fin}(A)$ does not contain an infinite uniformly supported subset (the elements of $\wp_{fin}(A)$ supported by $supp(f) \cup supp(Z)$ are precisely the subsets of $supp(f) \cup supp(Z)$), we lead to a contradiction. Thus, $f(Z) = Z$.

We return to the proof of the theorem and we consider the remaining case $X \setminus supp(f) \neq \emptyset$. We should have that $X = \{a_1, \dots, a_t, d_1, \dots, d_m\}$ with $a_1, \dots, a_t \in supp(f)$ and $d_1, \dots, d_m \in A \setminus supp(f)$, $m \geq 1$, or $X = \{d_1, \dots, d_m\}$ with $d_1, \dots, d_m \in A \setminus supp(f)$, $m \geq 1$. We have $X \setminus supp(f) = \{d_1, \dots, d_m\}$. Denote by $V = X \setminus supp(f)$. If $f(V) \setminus supp(f) \neq \emptyset$, then $f(V) = V$ according to (claim 2). Assume by contradiction that $f(V) \setminus supp(f) = \emptyset$, that is, $f(V) = \{x_1, \dots, x_i\}$ with $x_1, \dots, x_i \in supp(f)$, $i \geq 1$ (we cannot have $f(V) = \emptyset$ because f is strictly monotone $f(\emptyset) = \emptyset$ and $\emptyset \subsetneq V$). Since $supp(f)$ has only finitely many subsets, A is infinite and f is *strictly* monotone, there should exist $W \in \wp_{fin}(A)$, $W \subsetneq A \setminus supp(f)$ such that $V \subsetneq W$ and $f(W)$ *contains at least one element outside* $supp(f)$; for example, we can choose finitely many distinct atoms $d_{m+1}, \dots, d_{m+2^{|supp(f)|}+1} \in A \setminus (supp(f) \cup \{d_1, \dots, d_m\})$, and consider $W = \{d_1, \dots, d_m, d_{m+1}, \dots, d_{m+2^{|supp(f)|}+1}\}$; since $\{d_1, \dots, d_m\} \subsetneq \{d_1, \dots, d_m, d_{m+1}\} \subsetneq \dots \subsetneq \{d_1, \dots, d_m, \dots, d_{m+2^{|supp(f)|}+1}\}$ and f is strictly monotone, we get that $f(W)$ should contain at least one element outside the finite set $supp(f)$. But in this case $f(W) = W$ according to (claim 2), and since $f(V) \subsetneq f(W) = W$, we get $\{x_1, \dots, x_i\} \subseteq W$, i.e. $x_1, \dots, x_i$ are outside $supp(f)$, a contradiction. Thus, we necessarily have $f(V) \setminus supp(f) \neq \emptyset$, and so $f(V) = V$, that is $X \setminus supp(f) = f(X \setminus supp(f))$ for all $X \in \wp_{fin}(A)$. □

Proposition 2.14 *There do not exist neither a finitely supported injective mapping* $f : \wp_{fin}(A) \to T_{fin}(A)$*, nor a finitely supported injective mapping* $f : T_{fin}(A) \to \wp_{fin}(A)$.

Proof For the first part, let us assume by contradiction that $f : \wp_{fin}(A) \to T_{fin}(A)$ is finitely supported and injective. Let $X \in \wp_{fin}(A)$. Since the support of a finite subset of atoms coincides with the related subset, and the support of a finite tuple of atoms is represented by the set of atoms forming the related tuple, according to Proposition 2.9, for any permutation $\pi \in Fix(supp(f) \cup supp(X)) = Fix(supp(f) \cup X)$ we have $\pi \star f(X) = f(\pi \star X) = f(X)$. This means $supp(f) \cup X$ supports $f(X)$, that is $supp(f(X)) \subseteq supp(f) \cup X$, and so the atoms forming $f(X)$ are contained in $supp(f) \cup X$ (claim 1). Since f is injective and $supp(f)$ is finite, there exist two distinct atoms $b_1, b_2 \in A \setminus supp(f)$ such that $f(\{b_1, b_2\})$ contains at least one atom outside $supp(f)$. Connecting this assertion with (claim 1), we have that at least b_1 or b_2 are in the tuple $f(\{b_1, b_2\})$. Say b_1 is in the tuple $f(\{b_1, b_2\})$. Since $(b_1\, b_2) \in Fix(supp(f))$, from Proposition 2.9 we get $f(\{b_1, b_2\}) = f((b_1\, b_2) \star \{b_1, b_2\}) = (b_1\, b_2) \star f(\{b_1, b_2\}) = (b_1\, b_2)(f(\{b_1, b_2\}))$, which is a contradiction because b_2 replaces b_1 in the injective tuple $f(\{b_1, b_2\})$ by applying the transposition $(b_1\, b_2)$.

For the second part of the result, let us assume by contradiction that $f : T_{fin}(A) \to \wp_{fin}(A)$ is finitely supported and injective. Let $X \in T_{fin}(A)$. Since the support of a finite subset of atoms coincides with the related subset, and the support of a finite tuple of atoms is represented by the set of atoms forming the related tuple, according to Proposition 2.9, for any permutation $\pi \in Fix(supp(f) \cup supp(X))$ we have $\pi \star f(X) = f(\pi \star X) = f(X)$. This means $supp(f) \cup supp(X)$ supports $f(X)$, that is $f(X) = supp(f(X)) \subseteq supp(f) \cup supp(X)$, and so $f(X)$ is contained in the union between $supp(f)$ and the set of atoms forming X (claim 1). Since f is injective and $supp(f)$ is finite, there exist two distinct atoms $b_1, b_2 \in A \setminus supp(f)$ such that

$f((b_1,b_2))$ contains at least one atom outside $supp(f)$. Connecting this assertion with (claim 1), we have that at least b_1 or b_2 belong to $f((b_1,b_2))$. We distinguish three cases.

Case 1. $b_1,b_2 \in f((b_1,b_2))$. This means the transposition $(b_1\, b_2)$ interchanges b_1 and b_2 in $f((b_1,b_2))$, but leaves the set $f((b_1,b_2))$ unchanged, namely $(b_1\, b_2) \star f((b_1,b_2)) = f((b_1,b_2))$. Then, since $(b_1\, b_2) \in Fix(supp(f))$, according to Proposition 2.9, we obtain $f((b_2,b_1)) = f((b_1\, b_2) \star (b_1,b_2)) = (b_1\, b_2) \star f((b_1,b_2)) = f((b_1,b_2))$, contradicting the injectivity of f.

Case 2. $b_1 \in f((b_1,b_2))$ and $b_2 \notin f((b_1,b_2))$. According to (claim 1), all the other elements in $f((b_1,b_2))$ (if they exist) belong to $supp(f)$ (claim 2). Let $c_1 \in A \setminus supp(f)$ distinct from b_1,b_2. According to (claim 2) $c_1 \notin f((b_1,b_2))$, and so the transposition $(b_2\, c_1)$ fixes $f((b_1,b_2))$ pointwise. Since $(b_2\, c_1) \in Fix(supp(f))$ according to Proposition 2.9, we obtain $f((b_1,c_1)) = f((b_2\, c_1) \star (b_1,b_2)) = (b_2\, c_1) \star f((b_1,b_2)) = f((b_1,b_2))$, contradicting the injectivity of f.

Case 3. $b_2 \in f((b_1,b_2))$ and $b_1 \notin f((b_1,b_2))$. It is similar to Case 2. □

Some other fixed point properties of finitely supported self-mappings defined on $\wp_{fin}(A)$ are presented in Proposition 5.5 and Corollary 5.6(1).

2.5 Fraenkel-Mostowski Axiomatic Set Theory

The previous results are valid if A is an arbitrary fixed infinite ZF set. Since we did not require a certain internal structure of the elements of A, these results can be reformulated in the ZFA framework by replacing the fixed ZF set A with the set of atoms in ZFA (also denoted by A). If A is a set of atoms in ZFA (elements with no internal structure), as in [29] we can slightly modify the usual von Neumann hierarchy on ordinals and define a ZFA class (i.e. a 'large' set) $FM(A)$ equipped with an S_A-action, in which all the elements have the finite support property.

Recall the usual von Neumann cumulative hierarchy of sets:

- $v_0 = \emptyset$;
- $v_{\alpha+1} = \wp(v_\alpha)$;
- $v_\lambda = \bigcup_{\alpha<\lambda} v_\alpha$ (λ a limit ordinal).

A model v of ZF set theory is represented by the union of all v_α. More generally, given a set A (of atoms) we can similarly define a cumulative hierarchy of sets involving atoms from A [29]:

- $v_0(A) = \emptyset$;
- $v_{\alpha+1}(A) = A + \wp(v_\alpha(A))$;
- $v_\lambda(A) = \bigcup_{\alpha<\lambda} v_\alpha(A)$ (λ a limit ordinal),

where $+$ denotes the disjoint union of sets. Let $v(A)$ be the union of all $v_\alpha(A)$. Using the names *atm* and *set* for the functions $x \mapsto (0,x)$ and $x \mapsto (1,x)$ (the notations are those used in Proposition 2.2(6)), we have that every element x of $v(A)$ is either of the form $atm(a)$ with $a \in A$, or of the form $set(X)$ where X is a set formed at an earlier ordinal stage than x. We call ZFA sets the elements of the form $set(X)$, and call atoms the elements of the form $atm(a)$.

On $\nu(A)$ we can recursively define an S_A-action $\cdot$ as follows:

$$\pi \cdot atm(a) = atm(\pi(a)), \quad \pi \cdot set(X) = set(\{\pi \cdot x \,|\, x \in X\}).$$

Let us consider:

- $FM_0(A) = \emptyset$;
- $FM_{\alpha+1}(A) = A + \wp_{fs}(FM_\alpha(A))$ (whenever the ordinal α has a successor);
- $FM_\lambda(A) = \bigcup\limits_{\alpha<\lambda} FM_\alpha(A)$ (if λ is a limit ordinal),

where $+$ represents again the disjoint union of sets defined in Proposition 2.2(6). From Proposition 2.2, each $FM_\alpha(A)$ is an invariant set. The union of all $FM_\alpha(A)$ is called the *Fraenkel-Mostowski universe* and is denoted by $FM(A)$. We mention that $FM(A)$ is a ZFA class; we refer to it as a (large) ZFA set because we actually consider only the (properties of the) group action it is equipped with. Furthermore, the algebraic properties of sets equipped with permutation actions (which are of interest in this book) can be translated into algebraic properties of classes equipped with permutation actions.

Every element x of $FM(A)$ is either of the form $atm(a)$ with $a \in A$, or of the form $set(X)$, where X is a finitely supported set formed at an earlier ordinal stage than x. We call FM sets the elements of the form $set(X)$, and atoms the elements of the form $atm(a)$. The FM universe $FM(A)$ is a subset (more exactly, a subclass) of $\nu(A)$. According to Proposition 2.1, the S_A-action $\cdot$ on $\nu(A)$ leads to an S_A-action on $FM(A)$. Thus, $FM(A)$ is an S_A-class together with the S_A-action $\cdot$ (it is actually a 'large' S_A-set having the same properties as an S_A-set). Furthermore, $FM(A)$ is an invariant class because it contains only finitely supported elements with respect to $\cdot$.

An element $x \in \nu(A)$ that is not an atom is an FM set (i.e. $x \in FM(A) \setminus A$) if and only if the following conditions are satisfied:

- y is an FM set or an atom for all $y \in x$, and
- x has finite support with respect to the action $\cdot$.

An FM set is actually an hereditary finitely supported subset of the invariant class $FM(A)$. An FM set X is not itself closed under the S_A-action $\cdot$ defined on $FM(A)$, unless $supp(X) = \emptyset$. Thus, an FM set is not necessarily equivariant in $FM(A)$ in the sense of Definition 2.2. This means that the restriction on a certain FM set X of the S_A-action $\cdot$ on $FM(A)$ does not necessarily lead to a new group action of S_A on X (since the codomain of the function $\cdot|_X$ is not necessarily X). Only an FM set with empty support is itself closed under the restriction of the S_A-action $\cdot$ on it. According to these remarks, and because invariant sets need to be closed under the actions they are equipped (meaning that 'being an invariant set' means 'being an equivariant element at the following order level in the hierarchical construction'), the invariant (or nominal) sets in the FM cumulative hierarchy are defined as those equivariant (i.e. empty supported) elements of $FM(A)$. This means that an FM set X is invariant if and only if the restriction $\cdot|_X$ of $\cdot$ on X is itself an S_A-action on X in the sense of Definition 2.2(1).

We provide an axiomatic presentation of FM set theory. The axioms are the ZFA axioms [37] together with the additional axiom of finite support (axiom 11).

Definition 2.7 The following axioms define FM set theory:

1. $\forall x.(\exists y.y \in x) \Rightarrow x \notin A$ (only non-atoms can have elements)
2. $\forall x, y.(x \notin A \, and \; y \notin A \; and \; \forall z.(z \in x \Leftrightarrow z \in y)) \Rightarrow x = y$ (modified ZF axiom of extensionality to allow atoms)
3. $\forall x, y.\exists z.z = \{x, y\}$ (axiom of pairing)
4. $\forall x.\exists y.y = \{z \,|\, z \subseteq x\}$ (axiom of powerset)
5. $\forall x.\exists y.y \notin A \, and \, y = \{z \,|\, \exists w.(z \in w \, and \, w \in x)\}$ (axiom of union)
6. $\forall x.\exists y.(y \notin A \, and \, y = \{f(z) \,|\, z \in x\})$, *for each functional formula* $f(z)$ (axiom of replacement)
7. $\forall x.\exists y.(y \notin A \, and \, y = \{z \,|\, z \in x \, and \, p(z)\})$, *for each formula* $p(z)$ (axiom of separation)
8. $(\forall x.(\forall y \in x.p(y)) \Rightarrow p(x)) \Rightarrow \forall x.p(x)$ (induction principle)
9. $\exists x.(\emptyset \in x \, and \, (\forall y.y \in x \Rightarrow y \cup \{y\} \in x))$ (axiom of infinity)
10. *A is not finite.*
11. $\forall x.\exists S \subset A.$ *S is finite, and S supports x.* (finite support axiom)

It is easy to see that $\nu(A)$ is a model of ZFA set theory. Furthermore, $FM(A)$ is a model of FM set theory; a complete proof can be found in Remark 9.1.7 and Section 11.5 from [27]. This proves relative consistency of FM set theory with respect to ZFA set theory.

Regarding the FM axiomatic set theory [27], we can point out the following:

- It is an independent set theory (not a model of ZFA) with its own axioms and with a specific model $FM(A)$.
- The finite support requirement is an axiom, and so non-finitely supported elements do not exist, which means they cannot be involved nor even in an intermediate step of a proof.
- There exist elements in ZFA that are not correctly defined in FM set theory (such as simultaneously infinite and coinfinite subsets of atoms), but every element in FM set theory is correctly defined in ZFA set theory.
- Every result/proof in FM set theory should involve only finitely supported elements. Thus, the validity of the ZFA results should be verified with respect to the finite support requirement. Only those results that can be reformulated using **only** finitely supported constructions (in every step of the proofs) are valid in FM set theory.

Chapter 3
Choice Principles for Finitely Supported Structures

Abstract The validity and the non-validity of choice principles in various models of Zermelo-Fraenkel set theory and of Zermelo-Fraenkel set theory with atoms (including the symmetric models and the permutation models) was investigated in the last century. Actually, choice principles are proved to be independent of the set of axioms for Zermelo-Fraenkel and Zermelo-Fraenkel with atoms, respectively. Since the theory of finitely supported algebraic structures is connected to the related permutation models, it became an open problem to study the consistency of choice principles with this new framework. We prove that many choice principles are inconsistent within finitely supported atomic structures, and so some paradoxes (such as Banach-Tarski paradoxical decomposition of a sphere) are eliminated in the new atomic context. However, no non-atomic result is weakened, meaning that the independence of the choice principles in the non-atomic frameworks of Zermelo-Fraenkel sets is not affected. Proving the inconsistency of choice principles in FSM (i.e. the non-validity of their atomic FSM formulations) is not an easy task because the Zermelo-Fraenkel results between choice principles are not necessarily preserved into this new framework, unless we reprove them with respect to the finite support requirement.

3.1 Choice Principles in ZF

Several ZF choice principles, independent of the axioms of ZF set theory, are presented in [34] and [37]. We note them below.

- **Hausdorff maximal principle (HP)**: Any poset P contains a maximal totally ordered subset;
- **Zorn lemma (ZL)**: Any non-empty inductive poset P (i.e. any poset P for which every totally ordered subset of P has an upper bound in P) has a maximal element;

A. Alexandru, G. Ciobanu, *Foundations of Finitely Supported Structures*,
https://doi.org/10.1007/978-3-030-52962-8_3

- **Axiom of dependent choice (DC)**: let R be a non-empty relation on a set X with the property that for each $x \in X$ there exists $y \in X$ with xRy. Then there exists a function $f : \mathbb{N} \to X$ such that $f(n)Rf(n+1)$ for all $n \in \mathbb{N}$;
- **Axiom of countable choice (CC)**: given any countable family (sequence) of non-empty sets $\mathscr{F}$, it is possible to select a single element from each member of $\mathscr{F}$;
- **Axiom of countable choice over finite sets (CC(fin))**: for any countable family (sequence) of non-empty finite sets $\mathscr{F}$, it is possible to select a single element from each member of $\mathscr{F}$;
- **Axiom of choice over finite sets (AC(fin))**: given any family of non-empty finite sets $\mathscr{F}$, it is possible to select a single element from each member of $\mathscr{F}$;
- **Axiom of partial countable choice (PCC)**: for any countable family (sequence) of non-empty sets $\mathscr{F} = (X_n)_n$, there exists an infinite subset M of $\mathbb{N}$ such that it is possible to select a single element from each member of the family $(X_m)_{m \in M}$;
- **Boolean prime ideal theorem (PIT)**: every Boolean algebra with $0 \neq 1$ has a maximal ideal (and hence a prime ideal);
- **Boolean ultrafilter theorem (UFT)**: in a Boolean algebra, every filter can be enlarged to a maximal one;
- **Kinna-Wagner selection principle (KW)**: given any family $\mathscr{F}$ of sets of cardinality at least 2, there exists a function f on $\mathscr{F}$ such that $f(X)$ is a non-empty proper subset of X for each $X \in \mathscr{F}$;
- **Ordering principle (OP)**: every set can be totally ordered;
- **Order extension principle (OEP)**: every partial order relation on a set can be enlarged to a total order relation;
- **Axiom of Dedekind infiniteness (Fin)**: every infinite set X allows an injection $i : \mathbb{N} \to X$.
- **Surjection inverse principle (SIP)**: any surjective mapping has a right inverse.
- **Finite powerset equipollence principle (FPE)**: for every infinite set X there exists a bijection between X and $\wp_{fin}(X)$.
- **The generalized continuum hypothesis (GCH)**: if an infinite set Y is placed between an infinite set X and the powerset of X, then Y either has the same cardinality as X or the same cardinality as the powerset of X.

According to [34], the following implications are valid in ZF set theory:

1. **GCH** $\Rightarrow$ **AC** $\Leftrightarrow$ **SIP** $\Leftrightarrow$ **ZL** $\Leftrightarrow$ **HP** $\Leftrightarrow$ **FPE** $\Rightarrow$ **DC** $\Rightarrow$ **CC** $\Leftrightarrow$ **PCC** $\Rightarrow$ **CC(fin)**;
2. **PIT** $\Leftrightarrow$ **UFT**;
3. **AC** $\Rightarrow$ **UFT** $\Rightarrow$ **OEP** $\Rightarrow$ **OP** $\Rightarrow$ **AC(fin)** $\Rightarrow$ **CC(fin)**;
4. **CC** $\Rightarrow$ **Fin** $\Leftrightarrow$ **CC(fin)**;
5. **KW** $\Rightarrow$ **OP**.

Furthermore, all the previously mentioned choice principles are independent of the standard axioms of the ZF set theory (meaning that their non-atomic formulations are valid in some models of ZF, while the negations of these non-atomic formulations are valid in other models of ZF). However, we should remark that the ZF results are not necessarily valid in FSM (nor even in ZFA), as we remarked in Section 1.4. When we work in FSM, in order to prove that a certain choice principle is consistent or not, we cannot use a ZF theorem regarding the relationship between choice principles in ZF in order to prove that a certain choice principle is consistent (i.e. it does not lead to a contradiction) or not (i.e. its negation is provable).

This is because in FSM all relationships between various choice principles should be independently reproved according to the finite support requirement in order to remain valid also when finitely supported atomic structures are involved. In [7] we presented an FSM preliminary relationships between FSM choice principles in order to prove their inconsistency. Based on our previous work [3], here we provide an independent proof for the inconsistency of each of the above choice principles in FSM by constructing finitely supported atomic sets for which each of the related choice principles fail.

3.2 A Deep Study of Choice Principles in FSM

The choice principles presented above can be formulated in FSM by requiring that all the constructions which appear in their statement are finitely supported (in order to be consistent with the FSM principle of finite support).

- **AC** has the form "Given any invariant set X, and any finitely supported family $\mathscr{F}$ of non-empty, finitely supported subsets of X, there exists a finitely supported choice function on $\mathscr{F}$".
- **ZL** has the form "Let P be a finitely supported subset of a non-empty invariant partially ordered set (i.e. a finitely supported subset of a non-empty invariant set equipped with an equivariant order relation) with the property that every finitely supported totally ordered subset of P has an upper bound in P. Then P has a maximal element".
- **HP** has the form "Let P be a finitely supported subset of a non-empty invariant partially ordered set (i.e. a finitely supported subset of a non-empty invariant set equipped with an equivariant order relation). Then there exists a maximal finitely supported totally ordered subset of P".
- **DC** has the form "Let R be a non-empty, finitely supported relation on a finitely supported subset Y of an invariant set X having the property that for each $x \in Y$ there exists $y \in Y$ with xRy. Then there exists a finitely supported function $f : \mathbb{N} \to Y$ such that $f(n)Rf(n+1)$ for all $n \in \mathbb{N}$".
- **CC** has the form "Given any invariant set X, and any countable family $\mathscr{F} = (X_n)_n$ of subsets of X such that the mapping $n \mapsto X_n$ is finitely supported, there exists a finitely supported choice function on $\mathscr{F}$".
- **CC(fin)** has the form "Given any invariant set X, and any countable family $\mathscr{F} = (X_n)_n$ of finite subsets of X such that the mapping $n \mapsto X_n$ is finitely supported, there exists a finitely supported choice function on $\mathscr{F}$".
- **AC(fin)** has the form "Given any invariant set X, and any finitely supported family $\mathscr{F}$ of non-empty finite subsets of X, there exists a finitely supported choice function on $\mathscr{F}$".

- **PCC** has the form "Given any invariant set X, and any countable family $\mathscr{F} = (X_n)_n$ of subsets of X such that the mapping $n \mapsto X_n$ is finitely supported, there exists an infinite subset M of $\mathbb{N}$ with the property that there is a finitely supported choice function on $(X_m)_{m \in M}$".
- **PIT** has the form "Every finitely supported Boolean subalgebra of an invariant Boolean algebra (i.e. every finitely supported subset L' of an invariant set $(L, \cdot)$ endowed with an equivariant lattice order $\sqsubseteq$ on L and with the additional condition that L and L' are distributive and uniquely complemented) with $0 \neq 1$ has a maximal finitely supported ideal".
- **UFT** has the form "Every finitely supported filter of a finitely supported Boolean subalgebra of a certain invariant Boolean algebra can be extended to a finitely supported ultrafilter".
- **KW** has the form "Given any invariant set X, and any finitely supported family $\mathscr{F}$ of non-empty, finitely supported subsets of X of cardinality at least 2, there exists a finitely supported function f on $\mathscr{F}$ such that $f(Y)$ is a proper non-empty subset of Y for each $Y \in \mathscr{F}$".
- **OP** has the form "For every finitely supported subset X of an invariant set there exists a finitely supported total order relation on X".
- **OEP** has the form "Every finitely supported partial order relation on a finitely supported subset of an invariant set can be enlarged to a finitely supported total order relation".
- **Fin** has the form "Given any infinite, finitely supported subset X of an invariant set, there exists a finitely supported injection from $\mathbb{N}$ to X".
- **SIP** has the form "Given a finitely supported subset X of an invariant set, a finitely supported subset Y of an invariant set, and a finitely supported surjection from X onto Y, there exists a finitely supported right inverse of f".
- **FPE** has the form "For every infinite, finitely supported subset X of an invariant set there exists a finitely supported bijection between X and $\wp_{fin}(X)$".
- **GCH** has the from "Let X be finitely supported subset of an invariant set. If Y is an infinite, finitely supported subset of an invariant set having the property that there is a finitely supported injection from X to Y and a finitely supported injection from Y to $\wp_{fs}(X)$, then there exist either a finitely supported bijection from X to Y or a finitely supported bijection from Y to $\wp_{fs}(X)$".

We say that a result presented in terms of finitely supported objects is provable in FSM when it can be derived from the axioms describing FSM. We say that a result (or, particularly, a choice principle) rephrased in terms of finitely supported objects is consistent with respect to FSM axioms if it does not introduce a contradiction, i.e. if it is possible to be true in FSM. A related result (or, particularly, a choice principle) is inconsistent (i.e. not consistent) with respect to FSM axioms if it is not true in FSM, i.e. if its negation can be proved by using FSM axioms.

Theorem 3.1 *None of the choice principles* ***AC, ZL, HP, DC, CC, PCC, AC(fin), Fin, PIT, UFT, OP, KW, OEP, SIP, FPE,*** *and* ***GCH*** *(expressed with respect to the finite support requirement as above) is consistent with Finitely Supported Mathe-*

matics, meaning that the negation of each of the above principles is a logical consequence of FSM.

Proof Firstly, we remark that the ZF relationships between choice principles do not hold in FSM unless we are able to reprove them in terms of finitely supported objects, and so each choice principle must be analyzed separately in FSM. Thus, we do not use ZF relationships between choice principles in order to prove this theorem.

By contradiction, we prove that each of the above choice principles is inconsistent with FSM, meaning that the negation of each of them is provable in FSM when the finite support requirement is involved. Thus, for each choice principle, by presenting relevant counterexamples, we prove that if we assume that the related choice principle is a true statement in FSM, then this would lead to a contradiction.

- Suppose **KW** is true in FSM. Consider the invariant set $(A,\cdot)$ of atoms equipped with the canonical action of the group of permutations of atoms $(\pi,a)\mapsto \pi\cdot a \stackrel{def}{=} \pi(a)$. Each permutation of atoms acts on a subset X of A by considering the pointwise action of that permutation on each atom from X. Let $\mathscr{F}:=\{X \mid X\subset A, X \mathit{finite}, |X|=2\}$ be the family of (finitely supported) 2-element sets of atoms. $\mathscr{F}$ together with the action of the group of permutations of atoms $(\pi,\{x,y\})\mapsto \pi\star\{x,y\} \stackrel{def}{=} \{\pi(x),\pi(y)\}$ is itself an invariant set (more exactly, it is an equivariant subset of the invariant set of all finitely supported subsets of A). Let f be a finitely supported Kinna-Wagner selection function on $\mathscr{F}$. Let S be a finite set (of atoms) supporting f. We may select a pair $Y:=\{a,b\}$ from $\mathscr{F}$ such that a and b do not belong to S. This is because A is infinite, while $S\subseteq A$ is finite. Let π be a permutation of atoms which fixes S pointwise, and interchanges a and b. Since f satisfies the property that $f(X)$ is a non-empty proper (one-sized) subset of X for all $X\in\mathscr{F}$, we have $f(Y)=\{a\}$ or $f(Y)=\{b\}$. Since π interchanges a and b, we have $\pi\star f(Y)=\pi(f(Y))\neq f(Y)$ (where $\star$ represents the canonical action on $\wp_{fs}(A)$). However, $\pi\star Y=\{\pi(a),\pi(b)\}=\{b,a\}=Y$. Since π fixes S pointwise and S supports f, according to Proposition 2.8 we have $\pi(f(Y))=\pi\star f(Y)=f(\pi\star Y)=f(Y)$, a contradiction. Thus, **KW** is inconsistent with FSM. Similarly, we remark that **AC(2)** (the axiom of choice for infinite families of 2-sized subsets of an invariant set) is inconsistent with FSM (because the family of one-sized subsets of A, emphasized in the proof of the inconsistency of **KW** as containing the codomain of the KW selection function f, can be identified with A), and so neither **AC(fin)** is true in FSM. This means that the negations of **KW** and **AC(fin)** are logical consequences of the finite support requirement in FSM.
- Suppose **OP** is true in FSM, and so there exists a finitely supported total order relation $<$ on the invariant set A. Let S be a finite set supporting $<$, and $a,b,c\in A\setminus S$ with $a<b$. Since (ac) fixes S pointwise and $<$ is a subset of the Cartesian product $A\times A$ supported by S, we have $(ac)(a)<(ac)(b)$, that is $c<b$. However, we also have that (ab) and (bc) fixes S pointwise, and so $((ab)\circ(bc))(a)<((ab)\circ(bc))(b)$, that is $b<c$. A contradiction, and so **OP** is not true in FSM.

- Suppose that **OEP** holds in FSM. Then the equivariant partial order relation $\leq$ on the invariant set A, defined by $x \leq y$ if and only if $x = y$ can be enlarged to a finitely supported total order relation on A. However, form the above item there does not exist a finitely supported total order relation on A.
- We prove now that the negation of **ZL** holds in FSM. More exactly, we prove that the family of all finite subsets of A, $(\wp_{fin}(A), \subseteq, \star)$ is an invariant set having the property that any finitely supported totally ordered subset of $\wp_{fin}(A)$ has an upper bound in $\wp_{fin}(A)$, but $\wp_{fin}(A)$ does not have a maximal element. Firstly, we remark that $\wp_{fin}(A)$ is an invariant set together with the permutation action $(\pi, X) \mapsto \pi \star X \stackrel{def}{=} \{\pi \cdot a \,|\, a \in X\} = \{\pi(a) \,|\, a \in X\}$. The relation $\subseteq$ on $\wp_{fin}(A)$ is equivariant because whenever $X \subseteq Y$ we have $\pi \star X \subseteq \pi \star Y$ for all $\pi \in S_A$.
  Let $\mathscr{F}$ be a finitely supported totally ordered subset of the invariant set $\wp_{fin}(A)$. Since $\mathscr{F}$ is totally ordered with respect to the inclusion relation on $\wp_{fin}(A)$, then there do not exist two different finite subsets of A of the same cardinality belonging to $\mathscr{F}$. Since $\mathscr{F}$ is finitely supported, then there exists a finite set $S \subseteq A$ such that $\pi \star Y \in \mathscr{F}$ for each $Y \in \mathscr{F}$ and each π that fixes S pointwise. However, since permutations of atoms are bijective, for each $Y \in \mathscr{F}$ and each permutation of atoms π we have that the cardinality of $\pi \star Y$ coincides with the cardinality of Y. Since there do not exist two distinct elements in $\mathscr{F}$ having the same cardinality, we conclude that $\pi \star Y = Y$ for all $Y \in \mathscr{F}$ and all π fixing S pointwise. Thus, $\mathscr{F}$ is uniformly supported by S (i.e. all the elements of $\mathscr{F}$ are supported by the same set S). However, there could exist only finitely many finite subsets of A (i.e. only finitely many elements in $\wp_{fin}(A)$, and hence only finitely many elements in $\mathscr{F}$) supported by S, namely the subsets of S. Thus, $\mathscr{F}$ must be finite, and so there exists the (finite) union of the members of $\mathscr{F}$ which is an elements of $\wp_{fin}(A)$ (expressed as a finite union of finite sets) and an upper bound for $\mathscr{F}$. Suppose by contradiction that there exists a maximal element X_0 of $\wp_{fin}(A)$. Then $X_0 = \{a_1, \ldots a_n\}$, $a_1, \ldots, a_n \in A$ for some $n \in \mathbb{N}$. Since A is infinite, there exists $b \in A \setminus X_0$. However, $\{a_1, \ldots a_n\} \subsetneq \{a_1, \ldots a_n, b\}$ with $\{a_1, \ldots a_n, b\} \in \wp_{fin}(A)$,which contradicts the maximality of X_0.
- We prove that the negation of **HP** is a logical consequence of FSM. For this, we prove that $(\wp_{fin}(A), \subseteq, \star)$ is an invariant partially ordered set, but $\wp_{fin}(A)$ does not have a maximal totally ordered subset. Suppose by contradiction that $\mathscr{F}$ is a maximal finitely supported totally ordered subset of $\wp_{fin}(A)$. As in the above item, since $\mathscr{F}$ is finitely supported and total, it has to uniformly supported, and so it must be finite (because $\wp_{fin}(A)$ does not contain an infinite uniformly supported subset). Thus, there exists $X_0 = \underset{X \in \mathscr{F}}{\cup} X \in \wp_{fin}(A)$ which is defined as a finite union of finite sets. Since X_0 is a finite subset of A, there exists $Y \in \wp_{fin}(A)$ (which is finite and so it is finitely supported) such that $X_0 \subsetneq Y$. Furthermore, Y is a strict superset for all the members of $\mathscr{F}$, and so $\mathscr{F} \cup \{Y\}$ is totally ordered. Moreover, $\mathscr{F} \cup \{Y\}$ is supported by $supp(\mathscr{F}) \cup supp(Y) = supp(\mathscr{F}) \cup Y$. Since $\mathscr{F} \subsetneq \mathscr{F} \cup \{Y\}$ this contradicts the maximality of $\mathscr{F}$.
- Let us assume that **Fin** is true in FSM. Thus, we can find a finitely supported injection $f : \mathbb{N} \to A$. Let us consider $m, n \in \mathbb{N}$ such that $m \neq n$ and $f(m), f(n) \notin$

$supp(f)$. These atoms exist because $f(\mathbb{N}) \subseteq A$ is infinite, while $supp(f) \subseteq A$ is finite. Hence $(f(m)\, f(n)) \star f = f$, where $\star$ is the canonical action on $A^{\mathbb{N}}$ defined as in Proposition 2.7. Let us denote $(f(m)\, f(n))$ by π. Since the single possible S_A-action $\diamond$ on $\mathbb{N}$ is the trivial one (defined in Proposition 2.2(3)), according to Proposition 2.7, we have $f(m) = (\pi \star f)(m) = \pi(f(\pi^{-1} \diamond m)) = \pi(f(m)) = f(n)$ which contradicts the injectivity of f. Thus, **Fin** is in contradiction with the finite support principle from FSM.

- From Proposition 5.2.2 in [43] (whose proof remains valid even when the set of atoms is not countable), there exists an invariant Boolean algebra having a finitely supported filter that cannot be extended to a finitely supported ultrafilter. Therefore, **UFT** fails in the framework of invariant sets.
  Alternatively, the proof of Theorem 4.39 from [34] can be reformulated in FSM because, with the notations in [34], each finite subset F of X is obviously finitely supported, and the mappings $F \mapsto X_F$ and $(E,F) \mapsto A_{(E,F)}$ are finitely supported by $supp(F) \cup supp(R)$ and $supp(E) \cup supp(F) \cup supp(R)$, respectively; it follows that **UFT** $\Rightarrow$ **OEP** holds in FSM. Since the negation of **OEP** is provable in FSM, we have that the negation of **UFT** is provable in FSM.
- By contradiction, assume that the choice principle **PIT** is true in FSM. Thus, every invariant Boolean algebra with $0 \neq 1$ has a maximal finitely supported ideal, and hence a maximal finitely supported filter. We prove that any equivariant filter of an arbitrary invariant Boolean algebra can be extended to a finitely supported maximal filter. Indeed, consider an invariant Boolean algebra $(B, \wedge, \vee, \cdot)$ and let $\mathscr{F}$ be an equivariant filter in B. Thus, $\mathscr{F}' = \{x' \mid x \in \mathscr{F}\}$ (where x' represents the complement of x) is an equivariant ideal in B. We define the relation $\sim_{\mathscr{F}'}$ on B by $x \sim_{\mathscr{F}'} y$ if and only if $(x \wedge y') \vee (y \wedge x') \in \mathscr{F}'$. Since the operations $\wedge$, $\vee$ and *complement* are all equivariant functions (according to the equivariance principle in FSM), and because $\mathscr{F}'$ is an equivariant subset of B, it follows that $\sim_{\mathscr{F}'}$ is also an equivariant subset of $B \times B$. Moreover, the quotient lattice $B/\mathscr{F} \stackrel{def}{=} B/\sim_{\mathscr{F}'}$ is an invariant set (with the S_A-action $\star$ defined by $\pi \star [x]_{\sim_{\mathscr{F}'}} = [\pi \cdot x]_{\sim_{\mathscr{F}'}}$ for all $\pi \in S_A, x \in B$). Thus, because $(B/\mathscr{F}, \bar{\wedge}, \bar{\vee})$ is also a Boolean algebra (according to the general theory of Boolean algebras), from Proposition 2.8, it follows that $(B/\mathscr{F}, \bar{\wedge}, \bar{\vee})$ is an invariant Boolean algebra, where the equivariant operations $\bar{\wedge}, \bar{\vee}$ on $B/\mathscr{F}$ are defined by $[x]_{\sim_{\mathscr{F}'}} \bar{\wedge} [y]_{\sim_{\mathscr{F}'}} = [x \wedge y]_{\sim_{\mathscr{F}'}}$ and $[x]_{\sim_{\mathscr{F}'}} \bar{\vee} [y]_{\sim_{\mathscr{F}'}} = [x \vee y]_{\sim_{\mathscr{F}'}}$ for all $[x]_{\sim_{\mathscr{F}'}}, [y]_{\sim_{\mathscr{F}'}} \in B/\mathscr{F}$. According to **PIT**, there exists a finitely supported maximal filter G in $B/\mathscr{F}$. Consider the natural mapping from B onto the corresponding quotient space, $f : B \to B/\mathscr{F}$ defined by $f(x) = [x]_{\sim_{\mathscr{F}'}}$, for all $x \in B$. By its definition, we have that f is equivariant. According to Proposition 2.8, we have that $f^{-1}(G)$ is supported by $supp(f) \cup supp(G) = \emptyset \cup supp(G) = supp(G)$. Moreover, $f^{-1}(G)$ is a maximal filter in B such that $\mathscr{F} \subseteq f^{-1}(G)$. Thus, $f^{-1}(G)$ is a finitely supported maximal filter in B that enlarges $\mathscr{F}$. However, as it is proved in Proposition 5.2.2 of [43], there exists an invariant Boolean algebra having an equivariant filter that cannot be extended to a finitely supported ultrafilter. The related filter is the filter f de-

fined on page 151 in [43]. We obtain a contradiction, and so **PIT** fails in FSM.

- The choice principles **CC** and **PCC** fail when the finite support principle is involved. We consider the countable family $(X_n)_n$ where X_n is the set of all injective n-tuples from A. Since A is infinite, it follows that each X_n is non-empty. In FSM, each X_n is equivariant because A is an invariant set and each permutation of atoms is a bijective function; more exactly, the image an injective n-tuple of atoms under an arbitrary permutation of atoms $\pi \in S_A$ is another injective n-tuple of atoms. Therefore, the family $(X_n)_n$ is equivariant and the mapping $n \mapsto X_n$ is also equivariant.
If we assume that **CC** is true, then according to the formulation of **CC** in FSM, there exists a finitely supported choice function f on $(X_n)_n$. Let $f(X_n) = y_n$ with each $y_n \in X_n$. Let $\pi \in Fix(supp(f))$. According to Proposition 2.9, and because each element X_n is equivariant according to its definition, we obtain that $\pi \cdot y_n = \pi \cdot f(X_n) = f(\pi \star X_n) = f(X_n) = y_n$, where by $\star$ we denoted the S_A-action on $(X_n)_n$ and by $\cdot$ we denoted the S_A-action on $\underset{n}{\cup} X_n$. Therefore, each element y_n is supported by $supp(f)$. However, since each y_n is a finite tuple of atoms, we have $supp(y_n) = y_n$ for all $n \in \mathbb{N}$. Since $supp(y_n) \subseteq supp(f)$ for all $n \in \mathbb{N}$, we obtain $y_n \subseteq supp(f)$ for all $n \in \mathbb{N}$. Since each y_n has exactly n elements, this contradicts the finiteness of $supp(f)$.
If we assume that **PCC** is true, then according to the formulation of **PCC** in FSM, there exists an infinite subset M of $\mathbb{N}$ and a finitely supported choice function g on $(X_m)_{m \in M}$. Let $g(X_m) = y_m$ with each $y_m \in X_m$. As in the paragraph above we obtain $y_m \subseteq supp(g)$ for all $m \in M$. Since y_m has exactly m elements for each $m \in M$, and since M is infinite, we contradict the finiteness of $supp(g)$.
- We prove that the choice principle **DC** fails in FSM. Assume by contradiction that **DC** holds in FSM. Let us consider the invariant set $(\wp_{fin}(A), \star)$ The relation $R \overset{def}{=} \subsetneq$ defined on $\wp_{fin}(A)$ is equivariant because whenever $X \subsetneq Y$ we have $\pi \star X = \{\pi(a) \,|\, a \in X\} \subsetneq \{\pi(a) \,|\, a \in Y\} = \pi \star Y$ for all $\pi \in S_A$. Let $X \in \wp_{fin}(A)$. Then $X = \{a_1, \dots a_n\}, a_1, \dots, a_n \in A$ for some $n \in \mathbb{N}$. Since A is infinite, there exists $b \in A \setminus X$, and so $X = \{a_1, \dots a_n\} \subsetneq \{a_1, \dots a_n, b\} \overset{def}{=} Y$ with $Y \in \wp_{fin}(A)$. Thus, for all $X \in \wp_{fin}(A)$ there is $Y \in \wp_{fin}(A)$ such that XRY. Then, from **DC**, there exists a finitely supported function $f : \mathbb{N} \to \wp_{fin}(A)$ such that $f(n)Rf(n+1)$, i.e. a finitely supported function $f : \mathbb{N} \to \wp_{fin}(A)$ with the property that $f(n) \subsetneq f(n+1)$ for all $n \in \mathbb{N}$. Then f is injective. Thus, $f(\mathbb{N})$ is an infinite family of finite subsets of A (an infinite family of elements in $\wp_{fin}(A)$) with the property that all the elements of $f(\mathbb{N})$ are supported by the same finite set $supp(f)$. This is because, according to Proposition 2.9, $\pi \star f(n) = f(\pi \diamond n) = f(n)$ for all $n \in \mathbb{N}$ and all $\pi \in Fix(supp(f))$, where $\diamond$ represents the trivial S_A-action on $\mathbb{N}$. However, there are only finitely many elements of $\wp_{fin}(A)$ supported by $supp(f)$, namely those subsets of $supp(f)$. Therefore, **DC** fails in FSM.
- We prove that **SIP** fails in FSM. Let us consider the invariant set $(\wp_{fs}(A), \star)$ and the trivial invariant set $(\mathbb{N}, \diamond)$. Define $f : \wp_{fs}(A) \to \mathbb{N}$ by $f(X) = |supp(X)|$ for any $X \in \wp_{fs}(A)$, where $|supp(X)|$ represents the number of elements of

$supp(X)$. Let us consider the equivariant family $(X_i)_{i\geq 1}$ where each X_i is the set of all i-sized subsets of A. Since A is infinite, it follows that each $X_i, i \geq 1$ is non-empty. Furthermore, for any fixed $n \in \mathbb{N}$ we have $f(\{x_1,\dots,x_n\}) = |supp(\{x_1,\dots,x_n\})| = n$, for all $\{x_1,\dots,x_n\} \in X_n$ (see Proposition 2.5), and so the image of f under X_n is $f(X_n) = n$ for all $n \in \mathbb{N}$. Thus, $\mathbb{N} = f(\underset{i\in\mathbb{N}}{\cup} X_i) \subseteq f(\wp_{fs}(A)) = Im(f)$, and so f is surjective. No choice principle is required because, for proving the surjectivity of f, we do not need to identify a set of representatives for the family $(X_i)_{i\geq 1}$. Now we prove that f is equivariant. According to Proposition 2.1 and because any permutation of atoms is one-to-one, we have $f(\pi\star X) = |supp(\pi\star X)| = |\pi(supp(X))| = |supp(X)| = f(X) = \pi\diamond f(X)$ for all $\pi \in S_A$ and $X \in \wp_{fs}(A)$. From Proposition 2.9, f is equivariant. Suppose by contradiction that **SIP** is true in FSM. Then f has a finitely supported right inverse g, that is $f\circ g = 1_{\mathbb{N}}$. Therefore, $g:\mathbb{N}\to\wp_{fs}(A)$ is injective. Then $g(\mathbb{N})$ is an infinite family of subsets of A with the property that all the elements of $g(\mathbb{N})$ are supported by the same finite set $supp(g)$. This is because $\pi\star g(n) = g(\pi\diamond n) = g(n)$ for all $n\in\mathbb{N}$ and all $\pi\in Fix(supp(g))$. However, there are only finitely many subsets of A supported by $supp(g)$, namely the subsets of $supp(g)$ and the supersets of $A\setminus supp(g)$, and so we get a contradiction.

- We prove that **FPE** fails in FSM. For this we consider the invariant sets $(A,\cdot)$ and $(\wp_{fin}(A),\star)$ and we prove that there does not exist a finitely supported surjection from A onto $\wp_{fin}(A)$. Suppose by contradiction that there is a finitely supported surjection $f:A\to\wp_{fin}(A)$. Let us fix an element $a\in A$ with $a\notin supp(f)$. Let b be an arbitrary element from $A\setminus supp(f)$ with $b\neq a$. Since $b\notin supp(f)$, we have that the transposition $(a\,b)$ fixes every element from $supp(f)$. However, $supp(f)$ supports f, and so according to Proposition 2.8, we have $f((a\,b)(c)) = (a\,b)\star f(c)$ for all $c\in A$. In particular, $f(b) = f((a\,b)(a)) = (a\,b)\star f(a)$. If $f(a)$ is an n-sized subset of A, then, because transpositions are injective functions, $f(b)$ is another n-sized subset of A. Thus, because b has been chosen arbitrarily from $A\setminus supp(f)$, we have $|f(a)| = |f(b)|$ for all $a,b\in A\setminus supp(f)$. However $supp(f)$ is finite. If we assume $|supp(f)| = k$ (k is finite), then $Im(f)$ can contain subsets of A of at most (finite) $k+1$ different sizes. This means f cannot be surjective.
- We show that **GCH** fails in FSM by proving that there exists an invariant set X and an invariant set Y placed between X and $\wp_{fs}(X)$ such that there is no finitely supported bijection between Y and X and no finitely supported bijection between Y and $\wp_{fs}(X)$. Let us consider $X = A$ and $Y = \wp_{fin}(A)$. The mapping $i:A\to\wp_{fin}(A)$ defined by $i(x) = \{x\}$ is obviously an equivariant injective mapping from $(A,\cdot)$ to $(\wp_{fin}(A),\star)$. However, from the above item there does not exist a finitely supported surjection from A onto $\wp_{fin}(A)$.
  We obviously have the equivariant identity injection from $\wp_{fin}(A)$ to $\wp_{fs}(A)$. We prove below that there does not exist a finitely supported injective mapping from $\wp_{fs}(A)$ onto one of its finitely supported proper subsets, i.e. any finitely supported injection $f:\wp_{fs}(A)\to\wp_{fs}(A)$ is also surjective. Let us consider a finitely supported injection $f:\wp_{fs}(A)\to\wp_{fs}(A)$. Suppose by contradiction that $Im(f)\subsetneq\wp_{fs}(A)$. This means that there exists $X_0\in\wp_{fs}(A)$ such that $X_0\notin Im(f)$.

Since f is injective, we can define an infinite sequence $\mathscr{F} = (X_n)_n$ starting from X_0, with distinct terms of form $X_{n+1} = f(X_n)$ for all $n \in \mathbb{N}$. Furthermore, according to Proposition 2.9, for a fixed $k \in \mathbb{N}$ and $\pi \in Fix(supp(f) \cup supp(X_k))$, we have $\pi \star X_{k+1} = \pi \star f(X_k) = f(\pi \star X_k) = f(X_k) = X_{k+1}$. Then, $supp(X_{n+1}) \subseteq supp(f) \cup supp(X_n)$ for all $n \in \mathbb{N}$, and by induction on n we have that $supp(X_n) \subseteq supp(f) \cup supp(X_0)$ for all $n \in \mathbb{N}$. We obtained that each element $X_n \in \mathscr{F}$ is supported by the same finite set $S := supp(f) \cup supp(X_0)$. However, there could exist only finitely many subsets of A (i.e. only finitely many elements in $\wp_{fs}(A)$) supported by S, namely the subsets of S and the supersets of $A \setminus S$ (where a superset of $A \setminus S$ is of form $A \setminus X$ with $X \subseteq S$). We contradict the statement that the infinite sequence $(X_n)_n$ never repeats. Thus, f is surjective, and so there could not exist a bijection between $\wp_{fin}(A)$ and $\wp_{fs}(A)$. □

Remark 3.1 According to [25], the following implication is valid in the ZF framework: ***AC(fin)*** *implies that 'every infinite set X has an infinite subset Y such that $X \setminus Y$ is also infinite'*. A similar result holds for **Fin** according to [39]. However, we cannot directly conclude that such an implication holds in (can be reformulated in) FSM where only finitely supported objects are allowed. We cannot say (without a proof which is well-constructed with respect to the finite support requirement from FSM) that the following statement is valid: " 'Given any invariant set X, and any finitely supported family $\mathscr{F}$ of non-empty finite subsets of X, there exists a finitely supported choice function on $\mathscr{F}$ (i.e. **AC(Fin)** in FSM)' implies 'Every infinite invariant set X has an infinite, finitely supported subset Y such that $X \setminus Y$ is also finitely supported and infinite'." Therefore, we cannot directly conclude that **AC(fin)** is false in FSM just because the statement "Every infinite invariant set X has an infinite, finitely supported subset Y such that $X \setminus Y$ is also finitely supported and infinite" is false in FSM. Such a result requires a separate proof reformulated in terms of finitely supported objects.

More explicitly, an amorphous set in FSM is the set A of atoms where A is a fixed ZF set formed by elements whose internal structure is not taken into consideration (alternatively, if FSM is defined over ZFA, the set of atoms can be considered the ZFA set of elements with no internal structure obtained by modifying the ZF Axiom of Extensionality). However, A is not an ordinary ZF set (since the internal structure of its elements is irrelevant), but an atomic one endowed with the canonical action $(\pi, a) \mapsto \pi(a)$. If we use, for example, (the proof of) Brunner's result in ZF (without reformulating it in FSM) [25] we would obtain that there exists an atomic ZF set not satisfying **AC(fin)**, i.e. an atomic ZF set X having an infinite family $\mathscr{F}$ of finite subsets with no choice function. However, we do not know if these X and $\mathscr{F}$ are finitely supported subsets of invariant sets (under the canonical action of S_A), and so we actually do not contradict immediately **AC(fin)** in FSM which should be valid only for finitely supported families (under the canonical hierarchical S_A-actions) and not for those atomic ZF families which are not finitely supported.

In fact, Remark 3.1 states that we cannot prove an FSM result only by employing a ZF result (without an additional proof made according to the finite support requirement). There exist a lot of results which are valid in the ZF framework, but fail in FSM. Related details are in Section 1.4.

Note that, according to the definition of an FM set, the previous choice principles can as well be reformulated in terms of FM sets by informally replacing 'finitely supported subset of an invariant set' with 'FM set'. For example, **AC** can be reformulated in the form: "Given any finitely supported family $\mathscr{F}$ of non-empty FM sets, there exists a finitely supported choice function on $\mathscr{F}$", and so on. The non-validity of various choice principles in the Fraenkel-Mostowski cumulative universe (which is a model of axiomatic FM set theory) can be proved similarly as we proved the inconsistency of the related choice principles in FSM (FM sets are actually hereditary finitely supported FSM sets). Formally we can express this as:

Corollary 3.1 *None of the choice principles* ***AC, ZL, HP, DC, CC, PCC, AC(fin), Fin, PIT, UFT, OP, KW, RKW, OEP, SIP, FPE*** *and* ***GCH*** *is consistent with the FM axiomatic set theory.*

Since nominal sets [44] are defined in the same way as invariant sets with the requirement that the set of atoms is countable, by particularizing the results in this book (which were presented by assuming an arbitrary infinite set of atoms) for a countable infinite set of atoms, and by noting that in the proof of Theorem 3.1 the counterexamples constructed for proving the inconsistency of choice principles are actually represented by invariant sets rather than arbitrary finitely supported subsets of invariant sets, we obtain the following result:

Corollary 3.2 *All the choice principles* ***AC, ZL, HP, DC, CC, PCC, AC(fin), Fin, PIT, UFT, OP, KW, RKW, OEP, SIP, FPE*** *and* ***GCH*** *are consistent in the framework of nominal sets.*

Since the theory of invariant sets makes sense regardless the countability of A, the inconsistency results in FSM presented in this section (adequately reformulated when the fixed ZF set A is replaced by the set of atoms in ZFA), do not overlap on some related properties in the basic or in the second Fraenkel models of ZFA set theory which are defined using countable sets of atoms [37]. Furthermore, the results in this section do not follow immediately from [44] because in [44] the nominal sets are defined over countable sets of atoms, while we defined invariant sets over infinite (not necessarily countable) sets of atoms; in the viewpoint from [44] (i.e. if the set of atoms is countable) the inconsistency of the countable choice principles would be trivial. Since no information about the countability of the set of atoms is available in a general theory of invariant sets, the consistency of **CC(fin)** in FSM remains an open problem.

Chapter 4
Connections with Tarski's Concept of Logicality

Abstract Our goal is to emphasize a connection between the approach regarding finitely supported structures and Tarski's definition of logicality requiring invariance under the one-to-one transformations of a universe of discourse onto itself. We also provide a connection with the Erlangen Program of Felix Klein for the classification of various geometries according to invariants under suitable groups of transformations. As they are invariant under (some) atomic permutations, finitely supported sets satisfy a certain form of logicality. Particularly, invariant sets are logical notions in Tarski sense.

4.1 Tarski's Logical Notions

The plurality of geometrical systems raises the question which of these systems is the 'true' geometry. Felix Klein developed a mathematical framework in which, by studying the properties remaining invariant under different transformations, it is possible to classify systematically various geometries [38]. In his Erlangen Program, Klein classified the notions to be studied in various geometries (such as Euclidean, affine and projective geometry) according to the groups of (one-to-one and onto) transformations under which they are invariant. With logic thought of as the most general theory, Alfred Tarski defined the logical notions as those invariant under all possible one-to-one transformations of the universe of discourse onto itself which means that if they are described set-theoretically, they should have the same meaning (independent of the set-theoretical universe) [49]. The question is what are the set theoretical notions appropriate to the Tarski logical notions. The answer is important because the language of set theory is able to express the whole mathematics. In some sense, the question is whether certain foundation mathematical notions are logical.

According to [47], invariance under permutations reflects the formality of notions from logic. Thus, invariant notions are formal in the sense that they do not depend on the identity of objects. For example, in a formal language the extension of the existential quantifier $\exists$ consists of all non-empty subsets of a specific domain.

A. Alexandru, G. Ciobanu, *Foundations of Finitely Supported Structures*,
https://doi.org/10.1007/978-3-030-52962-8_4

Obviously, all the one-to-one mappings of this domain onto itself transform any non-empty subset of the domain in another non-empty subset of the domain. Therefore, the interpretation of the existential quantifier is invariant under any permutation.

Tarski's thesis regarding logicality makes sense for objects in the finite relational type structure over a domain of basic objects D, where the objects at each level are relations of one or more arguments between objects of lower levels. Rather than considering the entire set of permutations of a universe built as a cumulative hierarchy over D, we can consider only those permutations of the objects in D. Thus, the logical notions (and the logical operations) are those invariant under arbitrary permutations of objects from D.

Based on the previous remark and the approach in [47], we can provide a practical method of verifying logicality in cumulative hierarchies. The general idea is to study the effect of permutations over sets. More precisely, we consider an initial domain D of basic objects, and construct a hierarchy of sets starting with the objects in D. After that we consider any permutation of objects from D, and see what effect have these permutations on the sets of various levels. The sets which are fixed under all permutations are exactly the sets that can be denoted by a logical symbol. It is easy to see that both the identity relation between basic objects and its negation are fixed under every permutation. At a higher level, the sets of sets that are fixed include the set of all non-empty sets (which is related to the existential quantifier) and the set consisting of just the empty set (which is related to the negated universal quantifier). This means that all of them can be denoted by logical symbols, and so they represent formal notions.

Formally, Tarski's logicality criterion can be expressed as

"Given a domain D of basic objects, an operation f in the type hierarchy over D is logical if and only if it is invariant under all permutations on D."

4.2 Logical Notions in FSM

The Fraenkel-Mostowski approach corresponds to Tarski's view. In order to define the cumulative Fraenkel-Mostowski universe $FM(A)$, we started with a collection of basic objects (the set A of atoms) and constructed a cumulative hierarchy of sets above them. According to Proposition 2.11, a bijection of A onto itself is finitely supported if and only if it leaves unchanged all but finitely many atoms of A. In the Fraenkel-Mostowski universe $FM(A)$, the set S_A of all finite permutations of A is exactly the set of all bijections of A which belong to $FM(A)$. Thus, in $FM(A)$, S_A coincides with the set of all bijections of A in Tarski's view (i.e. with the set of all one-to-one transformations of A onto itself). According to the recursive definition of the S_A action $\cdot$ on $FM(A)$, we can say that an element having the form $\pi \cdot x$ (where $\pi \in S_A$ and $x \in FM(A)$) is a new element $y \in FM(A)$ obtained by replacing each atom a from the structure of x by $\pi(a)$. Thus, an element of the form $\pi \cdot x$ can be associated with 'the effect of the transformation π on the element x' in Tarski's view. We conclude that the empty supported elements in $FM(A)$ (i.e. those invariant sets defined in the FM cumulative universe) are invariant under all permutations of A, and so the related elements are *logical notions*.

Theorem 4.1 *Invariant sets defined in the FM cumulative hierarchy are logical (in Tarski sense). In particular, the nominal set S_A of all finite permutations of A is logical (in Tarski sense).*

The FM sets, i.e. the arbitrary elements from the FM cumulative universe, satisfy only a 'weaker' form of logicality, meaning that they are fixed only by those permutations of atoms satisfying an additional requirement. More precisely, an FM set x is invariant under all permutations of atoms fixing its support pointwise. Furthermore, this is the 'strongest' possible form of invariance because the support of an element is the least set supporting it. However, given an invariant set X from the FM cumulative universe, the set of all finitely supported subsets of X (generally denoted by $\wp_{fs}(X)$) is logical, i.e. invariant under all permutations of atoms. This follows from Proposition 2.1 which states than for any finitely supported subset Y of X we have that $\pi \star Y$ is supported by $\pi \star supp(Y)$ for any $\pi \in S_A$. We describe this formally in the following result.

Proposition 4.1 *Let X be an invariant set from the FM cumulative hierarchy. Then the set $\wp_{fs}(X)$ of all finitely supported subsets of X is logical (in Tarski sense).*

Invariant sets defined as ZF sets equipped with canonical actions of the group of all permutations of the fixed ZF set A are the natural ZF correspondent of the invariant sets defined in the FM framework (when the fixed ZF set A is replaced by the set of atoms in ZFA) as those empty supported elements from $FM(A)$. Actually a ZF invariant set is necessarily equipped with a permutation action hierarchically constructed under the rules in Proposition 2.2 (for atoms, powersets, Cartesian products, disjoint unions etc). Whenever an atom a appears in (the construction of) an invariant set $(X, \cdot)$, the effect of a permutation of atoms π on a under $\cdot$ is $\pi(a)$. Thus, we can say that all the invariant sets in FSM are logical in Tarski sense.

A new FSM quantifier, denoted by И, is introduced in Definition 11.3. Formally, if P is a predicate over A, we say that $\text{И}a.P(a)$ is true if $P(a)$ is true for all but finitely many elements of A. In a formal language, the extension of the quantifier И consists of all cofinite subsets of the domain A. Obviously, all the one-to-one mappings of A onto itself transform any cofinite subset of the domain in another cofinite subset of the domain. Therefore, the interpretation of the quantifier И is invariant under every permutation of atoms. We describe this formally in the following result.

Theorem 4.2 *The quantifier И is logical (in Tarski sense).*

As a direct application, the quantifier И can be used in order to describe new semantics for various process calculi, where transition rules are expressed by employing a mixture of logical operators instead of side conditions. More precisely, by using invariant sets, we were able to present a more compact semantics (formed by transition rules presented without assuming additional freshness conditions) for the fusion calculus [42]. The central idea was to use atoms to represent variable symbols, and the FSM abstraction defined in Chapter 12 to represent the binding operators. The terms (processes) in each of these process calculi form an invariant set, and the set of terms modulo α-conversion in each of these process calculi can also be represented as an invariant set [4].

We present an example of how a certain transition rule in the (monadic version of the) fusion calculus is rephrased in FSM by using Tarski logical symbols. The rule U-PASS in the original semantics *uc* of the fusion calculus

$$\frac{P \xrightarrow[uc]{[x/y]} P'}{(z)P \xrightarrow[uc]{[x/y]} (z)P'}, z \neq x, y; x \neq y$$

becomes the following rule in the new FM semantics *nuc* of the fusion calculus:

$$\forall x.\text{И}y.\text{И}z.\forall P,P'.\frac{P \xrightarrow[nuc]{[x/y]} P'}{[z]P \xrightarrow[nuc]{[x/y]} [z]P'}.$$

The update action in the monadic fusion calculus, generally denoted by $[z/t]$ indicates the replacement of all t by z in both *uc* and *nuc*. In *uc*, the scope operator $(x)Q$ limits the scope of x to Q; scopes can be used to delimit the extent of updates (specifically, the update effects with respect to x are limited to Q). In *nuc*, the bindings represented by the scope operator are associated to FSM abstractions. We were able to prove that the new semantics are equivalent with (i.e. they have the same expressive power as) the original semantics of the related process calculi. However, the newly defined semantics of these process calculi were presented by involving only logical notions and symbols, whereas the original semantics of these process calculi were presented by assuming additional freshness conditions for each transition rule.

Chapter 5
Partially Ordered Sets in Finitely Supported Mathematics

Abstract We introduce and study finitely supported partially ordered sets. We study the notion of 'cardinality' for a finitely supported set, proving several properties related to this concept. Some properties are naturally extended from the non-atomic Zermelo-Fraenkel framework into the world of atomic structures with finite supports. In this sense, we prove that the Cantor theorem and the Cantor-Schröder-Bernstein theorem for cardinalities are still valid in the world of atomic finitely supported sets. Several other cardinality arithmetic properties are preserved from the classical Zermelo-Fraenkel set theory. However, the dual of the Cantor-Schröder-Bernstein theorem (where cardinalities are ordered via surjective mappings) is no longer valid in this framework. Other specific order properties of cardinalities that do not have related Zermelo-Fraenkel correspondents are also proved. Finally, we present a collection of fixed point theorems in the framework of finitely supported ordered structures, preserving the validity of several classical Zermelo-Fraenkel fixed point theorems such as the Bourbaki-Witt theorem, the Scott theorem and the Tarski-Kantorovitch theorem. We also prove several specific fixed point properties in the framework of finitely supported algebraic structures (especially fixed point properties of mappings defined on finitely supported sets that do not contain infinite uniformly supported subsets), results that are not reformulations of some corresponding Zermelo-Fraenkel results.

5.1 Finitely Supported Partially Ordered Sets

Order theory is involved in several aspects of mathematics and computer science. In particular, partially ordered sets are employed in logic, domain theory, formal methods and static analysis. An FSM theory for partially ordered sets was developed first by Shinwell [48] in order to describe a denotational semantics for a functional programming language incorporating facilities for manipulating syntax involving names and binding operations. For this, Shinwell presents the solution of the Scott recursive domain equation $D \cong (D \to D)$ within the world of finitely supported

A. Alexandru, G. Ciobanu, *Foundations of Finitely Supported Structures*,
https://doi.org/10.1007/978-3-030-52962-8_5

structures. This work was continued by Alexandru and Ciobanu in order to present a theory of abstract interpretations in FSM. Some calculability properties (obtained from the narrowing and widening techniques of approximation) of the fixed points of a class of finitely supported functions were proved in [8]. Since there exist FSM complete lattices failing to be ZF complete, in [8] it is proved that there may also exist abstract interpretations of some programming languages that can be easier described by using invariant sets. Upper and lower approximations of finitely supported subsets of invariant sets, derived by involving rough sets theory and several results regarding invariant Boolean lattices and invariant Galois connections, were described in [7]. Using invariant complete lattices, the properties of the family of all finitely supported fuzzy sets over an invariant set can be described [10]. In this book, our goal is to enrich the existing development regarding order relations in the world of finitely supported structures by presenting new (especially cardinality and fixed point) results regarding invariant partially ordered sets and invariant lattices. The results presented in this chapter were presented first by the authors in the journal articles [11, 12, 14].

Definition 5.1
- An *invariant partially ordered set (invariant poset)* is an invariant set $(P,\cdot)$ together with an equivariant partial order relation $\sqsubseteq$ on P. An invariant poset is denoted by $(P,\sqsubseteq,\cdot)$ or simply P.
- A *finitely supported partially ordered set (finitely supported poset)* is a finitely supported subset X of an invariant set $(P,\cdot)$ together with a partial order relation $\sqsubseteq$ on X that is finitely supported as a subset of $P\times P$.

A partial order relation $\sqsubseteq$ on P is a subset of the Cartesian product $P\times P$; this relation is reflexive, anti-symmetric and transitive. According to Definition 2.2, $\sqsubseteq$ is equivariant (or finitely supported) if it is equivariant (finitely supported) as a subset of the Cartesian product $P\times P$ in the sense of Definition 2.4. This means that $\sqsubseteq$ is equivariant (supported by a set S of atoms) if and only if for each pair $(x,x')\in\sqsubseteq$ and each $\pi\in S_A$ (respectively for each $\pi\in Fix(S)$) we have that $\pi\otimes(x,x')\in\sqsubseteq$ (where $\otimes$ represents the canonical S_A-action on the Cartesian product $P\times P$). If we write '$(x,x')\in\sqsubseteq$' as '$x\sqsubseteq x'$', the equivariance property of $\sqsubseteq$ can be expressed by: "$x\sqsubseteq x'$ implies $\pi\cdot x\sqsubseteq\pi\cdot x'$, whenever $\pi\in S_A$ (respectively whenever $\pi\in Fix(S)$)".

5.2 Order of Cardinalities in FSM

The notion of cardinality cannot be developed in the world of atomic structures by involving the classical notion of *ordinal*. This is because ordinals are defined within non-atomic ZF, that is they do not contain atoms. Despite this, we are able to present a specific notion of cardinality in FSM.

Definition 5.2 Two FSM sets X and Y are called *FSM equipollent* if there exists a finitely supported bijection $f:X\to Y$.

Theorem 5.1 *The equipollence relation is an equivariant equivalence relation on the family of all FSM sets.*

Proof 1. The equipollence relation is equivariant.
For any FSM sets X and Y, whenever there is a finitely supported bijection $f : X \to Y$, for any $\pi \in S_A$ we have that $\pi \widetilde{\star} f : \pi \star X \to \pi \star Y$ (where by $\star$ we generically denoted the possibly different S_A-actions on the powersets of the invariant sets containing X and Y, respectively) defined by $(\pi \widetilde{\star} f)(\pi \cdot x) = \pi \cdot f(x)$ for all $x \in X$ (where by $\cdot$ we generically denoted the possibly different S_A-actions on the invariant sets containing X and Y, respectively) is bijective and finitely supported by $\pi(supp(f)) \cup \pi(supp(X)) \cup \pi(supp(Y))$. Indeed, according to Proposition 2.1 we have that $\pi(supp(X))$ supports $\pi \star X$ and $\pi(supp(Y))$ supports $\pi \star Y$. Let $\sigma \in Fix(\pi(supp(f)) \cup \pi(supp(X)) \cup \pi(supp(Y)))$. Thus, $\sigma(\pi(a)) = \pi(a)$ for all $a \in supp(f)$. Therefore, $\pi^{-1}(\sigma(\pi(a))) = \pi^{-1}(\pi(a)) = a$ for all $a \in supp(f)$. So we get $\pi^{-1} \circ \sigma \circ \pi \in Fix(supp(f))$; from Proposition 2.9 this means $(\pi^{-1} \circ \sigma \circ \pi) \cdot x \in X$ and $f((\pi^{-1} \circ \sigma \circ \pi) \cdot x) = (\pi^{-1} \circ \sigma \circ \pi) \cdot f(x)$ for all $x \in X$. Fix an arbitrary $x \in X$. We have that $\sigma \cdot (\pi \cdot x) \in \pi \star X$, i.e. there exists $x' \in X$ such that $(\sigma \circ \pi) \cdot x = \pi \cdot x'$, and so $x' = (\pi^{-1} \circ \sigma \circ \pi) \cdot x$. According to Proposition 2.9, we have $(\pi \widetilde{\star} f)(\sigma \cdot (\pi \cdot x)) = (\pi \widetilde{\star} f)(\pi \cdot x') = \pi \cdot f(x') = \pi \cdot f((\pi^{-1} \circ \sigma \circ \pi) \cdot x) = \pi \cdot ((\pi^{-1} \circ \sigma \circ \pi) \cdot f(x)) = (\sigma \circ \pi) \cdot f(x) = \sigma \cdot (\pi \cdot f(x)) = \sigma \cdot (\pi \widetilde{\star} f)(\pi \cdot x)$. Now, from Proposition 2.9 we conclude that $\pi \widetilde{\star} f$ is finitely supported. The bijectivity of $\pi \widetilde{\star} f$ is obvious. Thus, $\pi \star X$ is equipollent with $\pi \star Y$ whenever X is equipollent with Y.

2. The equipollence relation is reflexive because for each FSM set X, the identity of X is a finitely supported (by $supp(X)$) bijection from X to X.
3. The equipollence relation is symmetric because for any FSM sets X and Y, whenever there exists a finitely supported bijection $f : X \to Y$, we have that $f^{-1} : Y \to X$ is bijective and supported by $supp(f) \cup supp(X) \cup supp(Y)$. Indeed, let $\pi \in Fix(supp(f) \cup supp(X) \cup supp(Y))$, and consider an arbitrary $y \in Y$. Since $\pi^{-1} \in Fix(supp(f) \cup supp(X) \cup supp(Y))$, we have $f^{-1}(\pi \cdot y) = z \Leftrightarrow f(z) = \pi \cdot y \Leftrightarrow \pi^{-1} \cdot f(z) = y \Leftrightarrow f(\pi^{-1} \cdot z) = y \Leftrightarrow \pi^{-1} \cdot z = f^{-1}(y) \Leftrightarrow z = \pi \cdot f^{-1}(y)$. Therefore, $f^{-1}(\pi \cdot y) = \pi \cdot f^{-1}(y)$ for all $y \in Y$, which in the view of Proposition 2.9 means that f^{-1} is finitely supported.
4. The equipollence relation is transitive because for any FSM sets X, Y and Z, whenever there are two finitely supported bijections $f : X \to Y$ and $g : Y \to Z$, there exists a bijection $g \circ f : X \to Z$ which is finitely supported by $supp(f) \cup supp(g)$. Indeed, let $\pi \in Fix(supp(f) \cup supp(g))$. According to Proposition 2.9, we get $\pi \cdot x \in X$, $\pi \cdot f(x) \in Y$, $\pi \cdot g(f(x)) \in Z$ and $(g \circ f)(\pi \cdot x) = g(f(\pi \cdot x)) = g(\pi \cdot f(x)) = \pi \cdot g(f(x)) = \pi \cdot (g \circ f)(x)$ for all $x \in X$, and so the conclusion follows by involving again Proposition 2.9. □

It is worth noting that the equipollence relation is actually a relation of the family of all FSM sets which is actually a class. However, invariant classes can be defined as well in FSM, by updating Definition 2.2 accordingly.

Definition 5.3 The *FSM cardinality* of X is defined as the equivalence class of all FSM sets equipollent to X, and is denoted by $|X|$.

Theorem 5.2 *The family of FSM cardinalities can be organized as an invariant family under the S_A-action $\diamond$ defined by $\pi \diamond |X| = |\pi \star X|$ for any cardinality $|X|$.*

Proof Firstly, we prove that the above definition is correct (i.e. it does not depend on the chosen representatives for cardinalities). Assume that for two FSM sets we have $|X| = |Y|$, i.e. there is a finitely supported bijection $f : X \to Y$. According to item 1 of the proof of Theorem 5.1, for any $\pi \in S_A$ we have that $\pi \widetilde{\star} f : \pi \star X \to \pi \star Y$, defined by $(\pi \widetilde{\star} f)(\pi \cdot x) = \pi \cdot f(x)$ for all $x \in X$, is bijective and finitely supported. Thus, $|\pi \star X| = |\pi \star Y|$, where by $\star$ we generically denoted the (possibly different) S_A-actions on the powersets of the invariant sets containing X and Y, respectively. Therefore, $\pi \diamond |X| = \pi \diamond |Y|$. By checking the properties in Definition 2.2(1), the mapping $\diamond$ is clearly a group action of S_A on the family of cardinalities (see Remark 5.1), and for each FSM set X, we have that $supp(X)$ supports $|X|$. □

Remark 5.1 According to the above theorem the family of all FSM cardinalities has the same properties as an invariant set described in Definition 2.2. Formally, this family is a not a set, but a class. However, what it matters are the properties described in Definition 2.2 that are also properties of *invariant classes.*

Lemma 5.1 (Cantor-Schröder-Bernstein Theorem)

Let $(B, \cdot)$ and $(C, \diamond)$ be two invariant sets. If there exist a finitely supported injective mapping $f : B \to C$ and a finitely supported injective mapping $g : C \to B$, then there exists a finitely supported bijective mapping $h : B \to C$. Furthermore, $supp(h) \subseteq supp(f) \cup supp(g)$.

Proof Let us define $F : \wp_{fs}(B) \to \wp_{fs}(B)$ by

$F(X) = B - g(C - f(X))$ for all finitely supported subsets X of B.

Claim 1: F is correctly defined, i.e. $Im(F) \subseteq \wp_{fs}(B)$.

For every finitely supported subset X of B, we have that $f(X)$ is supported by $supp(f) \cup supp(X)$. Indeed, let $\pi \in Fix(supp(f) \cup supp(X))$. Let y be an arbitrary element from $f(X)$; then $y = f(x)$ for some $x \in X$. However, because $\pi \in Fix(supp(X))$, it follows that $\pi \cdot x \in X$ and so, because $supp(f)$ supports f and π fixes $supp(f)$ pointwise, from Proposition 2.9 we get $\pi \diamond y = \pi \diamond f(x) = f(\pi \cdot x) \in f(X)$. Thus, $\pi \widetilde{\star} f(X) = f(X)$, where $\widetilde{\star}$ is the S_A-action on $\wp_{fs}(C)$ defined as in Proposition 2.2. Similarly, $g(Y)$ is finitely supported by $supp(g) \cup supp(Y)$ for all $Y \in \wp_{fs}(C)$. It is easy to remark that for every finitely supported subset X of B we have that $C - f(X)$ is also supported by $supp(f) \cup supp(X)$, $g(C - f(X))$ is supported by $supp(g) \cup supp(f) \cup supp(X)$, and $B - g(C - f(X))$ is supported by $supp(g) \cup supp(f) \cup supp(X)$. Thus, F is well-defined.

Claim 2: F is a finitely supported function.

We prove that F is finitely supported by $supp(f) \cup supp(g)$. Let us consider $\pi \in Fix(supp(f) \cup supp(g))$. Since $\pi \in Fix(supp(f))$ and $supp(f)$ supports f, according to Proposition 2.9 we have that $f(\pi \cdot x) = \pi \diamond f(x)$ for all $x \in B$. Thus, for every finitely supported subset X of B we have $f(\pi \star X)=\{f(\pi \cdot x) \mid x \in X\} = \{\pi \diamond f(x) \mid x \in X\} = \pi \tilde{\star} f(X)$, where $\star$ is the S_A-action on $\wp_{fs}(B)$ and $\tilde{\star}$ is the S_A-action on $\wp_{fs}(C)$. Similarly, $g(\pi \tilde{\star} Y) = \pi \star g(Y)$ for any finitely supported subset Y of C. Therefore, $F(\pi \star X) = B - g(C - f(\pi \star X)) = B - g(C - \pi \tilde{\star} f(X)) \overset{\pi \tilde{\star} C = C}{=} B - g(\pi \tilde{\star}(C - f(X))) = B - (\pi \star g(C - f(X))) \overset{\pi \star B = B}{=} \pi \star (B - g(C - f(X))) = \pi \star F(X)$. From Proposition 2.9 it follows that F is finitely supported. Moreover, because $supp(F)$ is the least set of atoms supporting F, we have $supp(F) \subseteq supp(f) \cup supp(g)$.

Claim 3: For any $X, Y \in \wp_{fs}(B)$ with $X \subseteq Y$, we have $F(X) \subseteq F(Y)$. This remark follows by direct calculation.

Claim 4: The set $S := \{X \mid X \in \wp_{fs}(B), X \subseteq F(X)\}$ is a non-empty, finitely supported subset of $\wp_{fs}(B)$. Obviously, $\emptyset \in S$. We claim that S is supported by $supp(F)$. Let $\pi \in Fix(supp(F))$, and $X \in S$. Then $X \subseteq F(X)$. From the definition of $\star$ (see Proposition 2.2) we have $\pi \star X \subseteq \pi \star F(X)$. According to Proposition 2.9, because $supp(F)$ supports F, we have $\pi \star X \subseteq \pi \star F(X) = F(\pi \star X)$, and so $\pi \star X \in S$. It follows that S is finitely supported, and $supp(S) \subseteq supp(F)$.

Claim 5: $T := \underset{X \in S}{\cup} X$ is finitely supported by $supp(S)$.
Let $\pi \in Fix(supp(S))$, and $t \in T$. Since $T = \underset{X \in S}{\cup} X$, we have that there exists $Z \in S$ such that $t \in Z$. Therefore, $\pi \cdot t \in \pi \star Z$. However, since π fixes $supp(S)$ pointwise and $supp(S)$ supports S, we have that $\pi \star Z \in S$. Thus, there exists $Y \in S$ such that $\pi \star Z = Y$. Therefore $\pi \cdot t \in Y$, and so $\pi \cdot t \in \underset{X \in S}{\cup} X$. It follows that $\underset{X \in S}{\cup} X$ is finitely supported, and so $T = \underset{X \in S}{\cup} X \in \wp_{fs}(B)$. Furthermore, $supp(T) \subseteq supp(S)$.

Claim 6: We prove that $F(T) = T$.
Let $X \in S$ arbitrary. We have $X \subseteq F(X) \subseteq F(T)$. By taking the supremum on S, this leads to $T \subseteq F(T)$. However, because $T \subseteq F(T)$, from Claim 3 we also have $F(T) \subseteq F(F(T))$. Furthermore, $F(T)$ is supported by $supp(F) \cup supp(T)$ (i.e. by $supp(f) \cup supp(g)$), and so $F(T) \in S$. According to the definition of T, we get $F(T) \subseteq T$.

We get $T = B - g(C - f(T))$, or equivalently $B - T = g(C - f(T))$. Since g is injective, we obtain that for each $x \in B - T$, $g^{-1}(x)$ is a set containing exactly one element. Let us define $h : B \to C$ by

$$h(x) = \begin{cases} f(x), & \text{for } x \in T; \\ g^{-1}(x), & \text{for } x \in B - T. \end{cases}$$

Claim 7: We claim that h is supported by the set $supp(f) \cup supp(g) \cup supp(T)$ (more exactly, by $supp(f) \cup supp(g)$, according to the previous claims). Let $\pi \in Fix(supp(f) \cup supp(g) \cup supp(T))$, and x an arbitrary element of B.

If $x \in T$, because $\pi \in Fix(supp(T))$ and $supp(T)$ supports T, we have $\pi \cdot x \in T$. Thus, from Proposition 2.9 we get $h(\pi \cdot x) = f(\pi \cdot x) = \pi \diamond f(x) = \pi \diamond h(x)$.

If $x \in B - T$, we have $\pi \cdot x \in B - T$. Otherwise, we would obtain the contradiction $x = \pi^{-1} \cdot (\pi \cdot x) \in T$ because π^{-1} also fixes $supp(T)$ pointwise. Thus, because g is finitely supported, according to Proposition 2.9 we have $h(\pi \cdot x) = g^{-1}(\pi \cdot x) = \{y \in C \mid g(y) = \pi \cdot x\} = \{y \in C \mid \pi^{-1} \cdot g(y) = x\} = \{y \in C \mid g(\pi^{-1} \diamond y) = x\} \overset{\pi^{-1} \diamond y := z}{=} \{\pi \diamond z \in$

$C \mid g(z)=x\} = \pi \diamond \{z \in C \mid g(z)=x\} = \pi \diamond g^{-1}(x) = \pi \diamond h(x)$. We obtained $h(\pi \cdot x) = \pi \diamond h(x)$ for all $\pi \in Fix(supp(f) \cup supp(g) \cup supp(T))$ and all $x \in B$. According to Proposition 2.9, we get that h is finitely supported. Furthermore, we also have that $supp(h) \subseteq supp(f) \cup supp(g) \cup supp(T) \overset{Claim\,5}{\subseteq} supp(f) \cup supp(g) \cup supp(S) \overset{Claim\,4}{\subseteq} supp(f) \cup supp(g) \cup supp(F) \overset{Claim\,2}{\subseteq} supp(f) \cup supp(g)$.

Claim 8: h is a bijective function.

Firstly, we prove that h is injective. Let us suppose that $h(x) = h(y)$. We claim that either $x, y \in T$ or $x, y \in B - T$. Indeed, let us suppose that $x \in T$ and $y \notin T$ (the case $x \notin T$, $y \in T$ is similar). We have $h(x) = f(x)$ and $h(y) = g^{-1}(y)$. If we denote $g^{-1}(y) = z$, we have $g(z) = y$. However, we supposed that $y \in B - T$, and so there exists $u \in C - f(T)$ such that $y = g(u)$. Since $y = g(z)$, from the injectivity of g we get $u = z$. This is a contradiction because $u \notin f(T)$, while $z = f(x) \in f(T)$. Since we proved that both x, y are contained either in T or in $B - T$, the injectivity of h follows from the injectivity of f or g, respectively.

Now we prove that h is surjective. Let $y \in C$ be arbitrarily chosen. If $y \in f(T)$, then there exists $z \in T$ such that $y = f(z)$, and so $y = h(z)$. If $y \in C - f(T)$, and because $g(C - f(T)) = B - T$, there exists $x \in B - T$ such that $g(y) = x$. Thus, $y \in g^{-1}(x)$. Since g is injective, and so $g^{-1}(x)$ is a one-element set, we can say that $g^{-1}(x) = y$ with $x \in B - T$. Thus, we have $y = h(x)$. □

Lemma 5.2 *Let $(B, \cdot)$ and $(C, \diamond)$ be two invariant sets (in particular, B and C could coincide), B_1 a finitely supported subset of B and C_1 a finitely supported subset of C. If there exist a finitely supported injective mapping $f : B_1 \to C_1$ and a finitely supported injective mapping $g : C_1 \to B_1$, then there exists a finitely supported bijective mapping $h : B_1 \to C_1$. Furthermore, $supp(h) \subseteq supp(f) \cup supp(g) \cup supp(B_1) \cup supp(C_1)$.*

Proof We follow the proof of Lemma 5.1. We define $F : \wp_{fs'}(B_1) \to \wp_{fs'}(B_1)$ by $F(X) = B_1 - g(C_1 - f(X))$ for all $X \in \wp_{fs'}(B_1)$, where $\wp_{fs'}(B_1)$ is a finitely supported subset of the invariant set $\wp_{fs}(B)$ (supported by $supp(B_1)$) defined by $\wp_{fs'}(B_1) = \{X \in \wp_{fs}(B) \mid X \subseteq B_1\}$. As in the previous lemma, but using Proposition 2.9, we get that F is well-defined, i.e. for every $X \in \wp_{fs'}(B_1)$ we have that $F(X)$ is supported by $supp(f) \cup supp(g) \cup supp(B_1) \cup supp(C_1) \cup supp(X)$ which means $F(X) \in \wp_{fs'}(B_1)$. Moreover, F is itself finitely supported (in the sense of Definition 2.6) by $supp(f) \cup supp(g) \cup supp(B_1) \cup supp(C_1)$. The set $S := \{X \mid X \in \wp_{fs'}(B_1),\ X \subseteq F(X)\}$ is contained in $\wp_{fs'}(B_1)$ and it is supported by $supp(F)$ as a subset of $\wp_{fs}(B)$. The set $T := \underset{X \in S}{\cup} X \in \wp_{fs'}(B_1)$ is finitely supported by $supp(S)$, and it is a fixed point of F.

As in the proof of Lemma 5.1, we define the bijection $h : B_1 \to C_1$ by

$$h(x) = \begin{cases} f(x), & \text{for } x \in T; \\ g^{-1}(x), & \text{for } x \in B_1 - T. \end{cases}$$

According to Proposition 2.9, we obtain that h is finitely supported by $supp(f) \cup supp(g) \cup supp(B_1) \cup supp(C_1) \cup supp(T)$, and $supp(h) \subseteq supp(f) \cup supp(g) \cup$

$supp(B_1) \cup supp(C_1)$. Therefore, h is the required finitely supported bijection between B_1 and C_1. □

Lemma 5.3 (failure of the dual Cantor-Schröder-Bernstein Theorem)

There are two invariant sets B and C such that there exist both a finitely supported surjective mapping $f : C \to B$ and a finitely supported surjective mapping $g : B \to C$, but it does not exist a finitely supported bijective mapping $h : B \to C$.

Proof Let us consider the invariant set $(A, \cdot)$ of atoms. The family $T_{fin}(A) = \{(x_1, \dots, x_m) \subseteq (A \times \dots \times A) \,|\, m \geq 0\}$ of all finite injective tuples from A (including the empty tuple denoted by $\bar{\emptyset}$) is an S_A-set with the S_A-action $\star : S_A \times T_{fin}(A) \to T_{fin}(A)$ defined by $\pi \star \bar{\emptyset} = \bar{\emptyset}$ for all $\pi \in S_A$ and $\pi \star (x_1, \dots, x_m) = (\pi \cdot x_1, \dots, \pi \cdot x_m)$ for all non-empty tuples $(x_1, \dots, x_m) \in T_{fin}(A)$ and all $\pi \in S_A$. Since A is an invariant set, we have that $T_{fin}(A)$ is an invariant set. This statement follows from Corollary 2.5. Whenever X is an invariant set, we have that each injective tuple $(x_1, \dots, x_m)$ of elements belonging to X is finitely supported and furthermore, $supp(x_1, \dots, x_m) = supp(x_1) \cup \dots \cup supp(x_m)$. Particularly, we obtain that $supp(a_1, \dots, a_m) = \{a_1, \dots, a_m\}$, for any injective tuple of atoms $(a_1, \dots, a_m)$ (similarly as in Proposition 2.5).

Since $supp(\bar{\emptyset}) = \emptyset$, it follows that $T^*_{fin}(A) = T_{fin}(A) \setminus \bar{\emptyset}$ is an equivariant subset of $T_{fin}(A)$, and it is itself an invariant set. Let us fix an atom $a \in A$.

We define $f : T_{fin}(A) \to T_{fin}(A) \setminus \bar{\emptyset}$ by

$$f(y) = \begin{cases} y, & \text{if } y \text{ is an injective non-empty tuple;} \\ (a), & \text{if } y = \bar{\emptyset}. \end{cases}$$

Clearly, f is surjective. We claim that f is supported by $supp(a)$. Let us consider $\pi \in Fix(supp(a))$, i.e. $a = \pi(a) = \pi \star (a)$. If y is a non-empty tuple of atoms, we obviously have $f(\pi \star y) = \pi \star y = \pi \star f(y)$. If $y = \bar{\emptyset}$, we have $\pi \star y = \bar{\emptyset}$, and so $f(\pi \star y) = (a) = \pi(a) = \pi \star f(y)$. Thus, $f(\pi \star y) = \pi \star f(y)$ for all $y \in T_{fin}(A)$. According to Proposition 2.9, we have that f is finitely supported.

We define an equivariant surjective function $g : T_{fin}(A) \setminus \bar{\emptyset} \to T_{fin}(A)$ by

$$g(y) = \begin{cases} \bar{\emptyset}, & \text{if } y \text{ is a tuple with exactly one element;} \\ y', & \text{otherwise;} \end{cases}$$

where y' is a new tuple formed by deleting the first element in tuple y (the first position in a finite injective tuple exists without requiring any form of choice).

Clearly, g is surjective. Indeed, $\bar{\emptyset} = g((a))$ for some one-element tuple (a) (A is non-empty, and so it has at least one atom). For a fixed finite injective non-empty m-tuple y, we have that y can be seen as being 'contained' in an injective $(m+1)$-tuple z of form (b, y) (whose first element is a certain atom b, and the following elements are precisely the elements of y). The related atom b exists because y is finite, while A is infinite (generally, always we can find an atom $b \notin supp(y) = \{y\}$ according to the finite support requirement in FSM; more details in Section 2.9 of [7]). We get $y = g(z)$. For proving the surjectivity of g we do not need to 'choose' a precise element b (we do not need to define an inverse function for g); it is sufficient to show that $g(b,y) = y$ for every $b \in A \setminus \{y\}$ and $A \setminus \{y\}$ is non-empty (the axiom of choice is not required because for proving only the surjectivity of g we do not involve the construction of a system of representatives for the family $(g^{-1}(y))_{y \in T_{fin}(A)}$).

We claim now that g is equivariant. Let (x) be a one-element tuple from A and π an arbitrary permutation from S_A. We have that $\pi \star (x) = (\pi(x))$ is a one-element tuple from A, and so $g(\pi \star (x)) = \bar{\emptyset} = \pi \star \bar{\emptyset} = \pi \star g((x))$. Now, let us consider $(x_1,\dots,x_m) \in T_{fin}(A), m \geq 2$ and $\pi \in S_A$. We have $g(\pi \star (x_1,\dots,x_m)) = g((\pi \cdot x_1,\dots,\pi \cdot x_m)) = g((\pi(x_1),\dots,\pi(x_m))) = (\pi(x_2),\dots,\pi(x_m)) = \pi \star (x_2,\dots,x_m) = \pi \star g(x_1,\dots,x_m)$. According to Proposition 2.9, we have that g is empty supported (equivariant).

We prove by contradiction that there could not exist a finitely supported injective $h : T_{fin}(A) \to T_{fin}(A) \setminus \bar{\emptyset}$. Let us suppose there is a finitely supported injection $h : T_{fin}(A) \to T_{fin}(A) \setminus \bar{\emptyset}$. We have $\bar{\emptyset} \notin Im(h)$ because $Im(h) \subseteq T_{fin}(A) \setminus \bar{\emptyset}$. We can form an infinite sequence $\mathscr{F}$ which has the first term $y_0 = \bar{\emptyset}$, and the general term $y_{n+1} = h(y_n)$ for all $n \in \mathbb{N}$. Since $\bar{\emptyset} \notin Im(h)$, it follows that $\bar{\emptyset} \neq h(\bar{\emptyset})$. Since h is injective and $\bar{\emptyset} \notin Im(h)$, we obtain by induction that $h^n(\bar{\emptyset}) \neq h^m(\bar{\emptyset})$ for all $n,m \in \mathbb{N}$ with $n \neq m$.

We prove now that for each $n \in \mathbb{N}$ we have that y_{n+1} is supported by $supp(h) \cup supp(y_n)$. Let $\pi \in Fix(supp(h) \cup supp(y_n))$. According to Proposition 2.9, because $\pi \in Fix(supp(h))$ we have $h(\pi \star y_n) = \pi \star h(y_n)$. Since $\pi \in Fix(supp(y_n))$ we have $\pi \star y_n = y_n$, and so $h(y_n) = \pi \star h(y_n)$. Thus, $\pi \star y_{n+1} = \pi \star h(y_n) = h(y_n) = y_{n+1}$. Furthermore, because $supp(y_{n+1})$ is the least set supporting y_{n+1}, we have $supp(y_{n+1}) \subseteq supp(h) \cup supp(y_n)$ for all $n \in \mathbb{N}$. Since each y_n is a finite injective tuple of atoms, it follows that $supp(y_n) = \{y_n\}$ for all $n \in \mathbb{N}$ (where by $\{y_n\}$ we denoted the set of atoms forming y_n). We get $\{y_{n+1}\} = supp(y_{n+1}) \subseteq supp(h) \cup supp(y_n) = supp(h) \cup \{y_n\}$. By repeatedly applying this result, we get $\{y_n\} \subseteq supp(h) \cup \{y_0\} = supp(h) \cup \emptyset = supp(h)$ for all $n \in \mathbb{N}$. Since $supp(h)$ has only a finite number of subsets, we contradict the statement that the infinite sequence $(y_n)_n$ never repeats. Thus, there does not exist a finitely supported bijection between $T_{fin}(A) \setminus \bar{\emptyset}$ and $T_{fin}(A)$. □

Lemma 5.4 *If X is an infinite ordinary ZF set, then for any finitely supported function $f : A \to X$ and any finitely supported function $g : X \to A$, $Im(f)$ and $Im(g)$ are finite. As a direct consequence there are no finitely supported injective mappings and no finitely supported surjective mappings between A and X.*

Proof Let us consider a finitely supported mapping $f : A \to X$. If $supp(f) = \{a_1,\dots,a_n\}$, then, from Proposition 2.10, $Im(f) = \{f(a_1)\} \cup \ldots \cup \{f(a_n)\} \cup \{f(b)\}$ where b is an arbitrary element from $A \setminus supp(f)$. Thus, $Im(f)$ is finite (because it is a finite union of singletons).

Let $g : X \to A$ be a finitely supported function. Assume by contradiction that $Im(g)$ is infinite. Pick any atom $a \in Im(g) \setminus supp(g)$ (such an atom exists because $supp(g)$ is finite). There exists $x \in X$ such that $g(x) = a$. Now pick any atom $b \in$

$Im(g) \setminus (supp(g) \cup \{a\})$, The transposition (ab) fixes $supp(g)$ pointwise, and so $g(x) = g((ab) \diamond x) = (ab) \cdot g(x) = (ab)(a) = b$, contradicting the fact that g is a function. Thus, $Im(g)$ is finite. □

According to Definition 5.3 for two FSM sets X and Y we have $|X| = |Y|$ if and only if there exists a finitely supported bijection $f : X \to Y$. On the family of cardinalities we can define the relations:

- $\leq$ by: $|X| \leq |Y|$ if and only if there is a finitely supported injective (one-to-one) mapping $f : X \to Y$.
- $\leq^*$ by: $|X| \leq^* |Y|$ if and only if there is a finitely supported surjective (onto) mapping $f : Y \to X$.

Theorem 5.3

1. *The relation $\leq$ is equivariant, reflexive, anti-symmetric and transitive, but it is not total.*
2. *The relation $\leq^*$ is equivariant, reflexive and transitive, but it is not anti-symmetric, nor total.*

Proof • Firstly, we prove that $\leq$ and $\leq^*$ over cardinalities are well-defined, i.e. their definitions do not depend on the chosen representatives for cardinalities. Assume that there exist FSM sets X, Y and X', Y', such that $|X| = |X'|$ and $|Y| = |Y'|$ (i.e. there exist two finitely supported bijections $f : X \to X'$ and $g : Y \to Y'$). If $|X| \leq |Y|$, i.e. if there is a finitely supported injection $h : X \to Y$, then the function $g \circ h \circ f^{-1} : X' \to Y'$ is injective and finitely supported by $supp(f^{-1}) \cup supp(h) \cup supp(g) = supp(f) \cup supp(h) \cup supp(g)$. Thus, $|X'| \leq |Y'|$.
If $|X| \leq^* |Y|$, i.e. if there is a finitely supported surjection $h' : Y \to X$, then the function $f \circ h' \circ g^{-1} : Y' \to X'$ is injective and finitely supported by $supp(f) \cup supp(h') \cup supp(g^{-1}) = supp(f) \cup supp(h') \cup supp(g)$. Thus, $|X'| \leq^* |Y'|$.

- $\leq$ and $\leq^*$ are equivariant because for any FSM sets X and Y, whenever there is a finitely supported injection/surjection $f : X \to Y$, according to Proposition 2.1, we have that the function $g_f : \pi \star X \to \pi \star Y$, defined by $g_f(\pi \cdot x) = \pi \cdot f(x)$ for all $x \in X$, is a finitely supported injective/surjective mapping, and so $\pi \star X$ is comparable with $\pi \star Y$ (under $\leq$ or $\leq^*$, after case). Technical details of how to prove the equivariance of the related relations are similar to those mentioned at item 1 in the proof of Theorem 5.1.
- $\leq$ and $\leq^*$ are clearly reflexive because for each FSM set X, the identity of X is a finitely supported (by $supp(X)$) bijection from X to X.
- $\leq$ and $\leq^*$ are transitive because for any FSM sets X, Y and Z, whenever there are two finitely supported injections/surjections $f : X \to Y$ and $g : Y \to Z$, there exists an injection/surjection $g \circ f : X \to Z$ which is finitely supported by $supp(f) \cup supp(g)$.
- The anti-symmetry of $\leq$ follows from Lemma 5.1 and Lemma 5.2, because FSM sets are actually finitely supported subsets of invariant sets.
- $\leq^*$ is not anti-symmetric. This follows from Lemma 5.3.
- $\leq$ and $\leq^*$ are not total. This follows from Lemma 5.4. □

Lemma 5.5 *Let X and Y be two FSM sets and $f : X \to Y$ a finitely supported surjective function. Then the mapping $g : \wp_{fs}(Y) \to \wp_{fs}(X)$ defined by $g(V) = f^{-1}(V)$ for all $V \in \wp_{fs}(Y)$ is well-defined, injective and finitely supported by $supp(f) \cup supp(X) \cup supp(Y)$.*

Proof Let V be an arbitrary element from $\wp_{fs}(Y)$. We claim that $f^{-1}(V) \in \wp_{fs}(X)$. Indeed we prove that the set $f^{-1}(V)$ is supported by $supp(f) \cup supp(V) \cup supp(X) \cup supp(Y)$. Let us consider $\pi \in Fix(supp(f) \cup supp(V) \cup supp(X) \cup supp(Y))$, and $x \in f^{-1}(V)$. This means $f(x) \in V$. According to Proposition 2.9, and because π fixes $supp(f)$ pointwise and $supp(f)$ supports f, we have $f(\pi \cdot x) = \pi \cdot f(x) \in \pi \star V = V$, and so $\pi \cdot x \in f^{-1}(V)$ (we denoted the actions on X and Y generically by $\cdot$, and the actions on their powersets by $\star$). Therefore, $f^{-1}(V)$ is finitely supported, and so the function g is well-defined. We claim that g is supported by $supp(f) \cup supp(X) \cup supp(Y)$. Let $\pi \in Fix(supp(f) \cup supp(X) \cup supp(Y))$. For any arbitrary $V \in \wp_{fs}(Y)$ we get $\pi \star V \in \wp_{fs}(Y)$ and $\pi \star g(V) \in \wp_{fs}(X)$, and by Proposition 2.9 we have that $\pi^{-1} \in Fix(supp(f))$, and so $f(\pi^{-1} \cdot x) = \pi^{-1} \cdot f(x)$ for all $x \in X$. For any arbitrary $V \in \wp_{fs}(Y)$, we have that $z \in g(\pi \star V) = f^{-1}(\pi \star V) \Leftrightarrow f(z) \in \pi \star V \Leftrightarrow \pi^{-1} \cdot f(z) \in V \Leftrightarrow f(\pi^{-1} \cdot z) \in V \Leftrightarrow \pi^{-1} \cdot z \in f^{-1}(V) \Leftrightarrow z \in \pi \star f^{-1}(V) = \pi \star g(V)$. If follows that $g(\pi \star V) = \pi \star g(V)$ for all $V \in \wp_{fs}(Y)$, and so g is finitely supported. Moreover, because f is surjective, a simple calculation shows us that g is injective. Indeed, let us suppose that $g(U) = g(V)$ for some $U, V \in \wp_{fs}(Y)$. We have $f^{-1}(U) = f^{-1}(V)$, and so $f(f^{-1}(U)) = f(f^{-1}(V))$. Since f is surjective, we get $U = f(f^{-1}(U)) = f(f^{-1}(V)) = V$. □

Corollary 5.1 *There exist two invariant sets B and C such that there is a finitely supported bijection between $\wp_{fs}(B)$ and $\wp_{fs}(C)$, but there is no finitely supported bijection between B and C.*

Proof As in Lemma 5.3, we consider the sets $B = T_{fin}(A) \setminus \bar{\emptyset}$ and $C = T_{fin}(A)$. According to Lemma 5.3, there exists a finitely supported surjective function $f : C \to B$ and a finitely supported (equivariant) surjection $g : B \to C$. Thus, according to Lemma 5.5, there exist a finitely supported injective function $f' : \wp_{fs}(B) \to \wp_{fs}(C)$ and a finitely supported injective function $g' : \wp_{fs}(C) \to \wp_{fs}(B)$. According to Lemma 5.1, there is a finitely supported bijection between $\wp_{fs}(B)$ and $\wp_{fs}(C)$. However, we proved in Lemma 5.3 that there is no finitely supported bijection between $B = T_{fin}(A) \setminus \bar{\emptyset}$ and $C = T_{fin}(A)$. □

The following result communicated in [40] by Levy for non-atomic ZF sets can be reformulated in the world of finitely supported atomic structures.

Corollary 5.2 *Let X and Y be two invariant sets with the property that whenever $|2^X_{fs}| = |2^Y_{fs}|$ we have $|X| = |Y|$. If $|X| \leq^\star |Y|$ and $|Y| \leq^\star |X|$, then $|X| = |Y|$.*

Proof According to the hypothesis and to Lemma 5.5 there exist two finitely supported injective functions $f : \wp_{fs}(Y) \to \wp_{fs}(X)$ and $g : \wp_{fs}(X) \to \wp_{fs}(Y)$. According to Lemma 5.1, there is a bijective mapping $h : \wp_{fs}(X) \to \wp_{fs}(Y)$. According to Theorem 2.3, we get $|2^X_{fs}| = |2^Y_{fs}|$, and so $|X| = |Y|$. □

Theorem 5.4 (Cantor Theorem)

Let X be a finitely supported subset of an invariant set $(Y,\cdot)$. Then $|X| \lneq |\wp_{fs}(X)|$ and $|X| \lneq^{*} |\wp_{fs}(X)|$

Proof Firstly, we prove that there is no finitely supported bijection between X and $\wp_{fs}(X)$, and so their cardinalities cannot be equal. Assume by contradiction that there is a finitely supported surjective mapping $f : X \to \wp_{fs}(X)$. Let us consider $Z = \{x \in X \,|\, x \notin f(x)\}$. We claim that $supp(X) \cup supp(f)$ supports Z. Let $\pi \in Fix(supp(X) \cup supp(f))$. Let $x \in Z$. Then $\pi \cdot x \in X$ and $\pi \cdot x \notin \pi \star f(x) = f(\pi \cdot x)$. Thus, $\pi \cdot x \in Z$, and so $Z \in \wp_{fs}(X)$. Therefore, since f is surjective there is $x_0 \in X$ such that $f(x_0) = Z$. However, from the definition of Z we have $x_0 \in Z$ if and only if $x_0 \notin f(x_0) = Z$, which is a contradiction.

Now, it is clear that the mapping $i : X \to \wp_{fs}(X)$ defined by $i(x) = \{x\}$ is injective and supported by $supp(X)$. Thus, $|X| \lneq |\wp_{fs}(X)|$. Let us fix an atom $y \in X$. We define $s : \wp_{fs}(X) \to X$ by

$$s(U) = \begin{cases} u, & \text{if}\, U \text{ is a one-element set } \{\text{u}\}\ ; \\ y, & \text{if}\, U \text{ has more than one element}. \end{cases}$$

Clearly, s is surjective. We claim that s is supported by $supp(y) \cup supp(X)$. Let $\pi \in Fix(supp(y) \cup supp(X))$. Thus, $y = \pi \cdot y$. If U is of form $U = \{u\}$, we obviously have $s(\pi \star U) = s(\{\pi \cdot u\}) = \pi \cdot u = \pi \cdot s(U)$. If U has more than one element, then $\pi \star U$ has more than one element, and we have $s(\pi \star U) = y = \pi \cdot y = \pi \cdot s(U)$. Thus, $\pi \star U \in \wp_{fs}(X)$, $\pi \cdot s(U) \in X$, and $s(\pi \star U) = \pi \cdot s(U)$ for all $U \in \wp_{fs}(X)$. According to Proposition 2.9, we have that s is finitely supported. Therefore, $|X| \lneq^{*} |\wp_{fs}(X)|$.□

In Theorem 5.4 we used a technique for constructing a surjection starting from an injection defined in the opposite way, that can be generalized as follows.

Proposition 5.1 *Let X and Y be finitely supported subsets of invariant sets.*

If $|X| \leq |Y|$, then $|X| \leq^{\star} |Y|$.

The converse is not valid. However, if $|X| \leq^{\star} |Y|$, then $|X| \leq |\wp_{fs}(Y)|$.

Proof We generically denote the (possibly different) actions of the invariant sets containing X,Y by $\cdot$, and the actions of powersets by $\star$. Suppose there exists a finitely supported injective mapping $f : X \to Y$. We consider the case $Y \neq \emptyset$ (otherwise, the result follows trivially). Fix $x_0 \in X$. Define the mapping $f' : Y \to X$ by

$$f'(y) = \begin{cases} f^{-1}(y), & \text{if}\, y \in Im(f)\ ; \\ x_0, & \text{if}\, y \notin Im(f)\,. \end{cases}$$

Since f is injective, it follows that $f^{-1}(y)$ is a one-element set for each $y \in Im(f)$, and so f' is a function. Clearly, f' is surjective. We claim that f' is supported by the set $supp(f) \cup supp(x_0) \cup supp(X) \cup supp(Y)$. Indeed, let us consider $\pi \in Fix(supp(f) \cup supp(x_0) \cup supp(X) \cup supp(Y))$. Whenever $y \in Im(f)$ we have $y = f(z)$ for some $z \in X$ and $\pi \cdot y = \pi \cdot f(z) = f(\pi \cdot z) \in Im(f)$, which means $Im(f)$

is finitely supported by $supp(f)$. Consider an arbitrary $y_0 \in Im(f)$, and so $\pi \cdot y_0 \in Im(f)$. Then $f'(y_0) = f^{-1}(y_0) = z_0$ with $f(z_0)=y_0$, and so $f(\pi \cdot z_0) = \pi \cdot f(z_0) = \pi \cdot y_0$, which means $f'(\pi \cdot y_0) = f^{-1}(\pi \cdot y_0) = \pi \cdot z_0 = \pi \cdot f^{-1}(y_0) = \pi \cdot f'(y_0)$. Now, for $y \notin Im(f)$ we have $\pi \cdot y \notin Im(f)$, which means $f'(\pi \cdot y) = x_0 = \pi \cdot x_0 = \pi \cdot f(y)$ since π fixes x_0 pointwise. Thus, $|X| \leq^{\star} |Y|$. Conversely, from the proof of Lemma 5.3 we know that there is a finitely supported surjection $g : T_{fin}(A) \setminus \bar{\emptyset} \to T_{fin}(A)$, but there does not exist a finitely supported injection $h : T_{fin}(A) \to T_{fin}(A) \setminus \bar{\emptyset}$.

Assume now there is a finitely supported surjective mapping $f : Y \to X$. We proceed similarly as in the proof of Lemma 5.5. Fix $x \in X$. Then $f^{-1}(\{x\})$ is supported by $supp(f) \cup supp(x) \cup supp(X)$. Indeed, let $\pi \in Fix(supp(f) \cup supp(x) \cup supp(X))$, and $y \in f^{-1}(\{x\})$. This means $f(y) = x$. According to Proposition 2.9, we have $f(\pi \cdot y) = \pi \cdot f(y) = \pi \cdot x = x$, and so $\pi \cdot y \in f^{-1}(\{x\})$. Define $g : X \to \wp_{fs}(Y)$ by $g(x) = f^{-1}(\{x\})$. We claim that g is supported by $supp(f) \cup supp(X)$. Let $\pi \in Fix(supp(f) \cup supp(X))$. For any arbitrary $x \in X$, we have that $z \in g(\pi \cdot x) = f^{-1}(\{\pi \cdot x\}) \Leftrightarrow f(z) = \pi \cdot x \Leftrightarrow \pi^{-1} \cdot f(z) = x \Leftrightarrow f(\pi^{-1} \cdot z) = x \Leftrightarrow \pi^{-1} \cdot z \in f^{-1}(\{x\}) \Leftrightarrow z \in \pi \star f^{-1}(\{x\}) = \pi \star g(x)$. From Proposition 2.9 it follows that g is finitely supported. Since g is also injective, we get $|X| \leq |\wp_{fs}(Y)|$. □

As in the ZF case, we can define operations between FSM cardinalities. Let X and Y be finitely supported subsets of invariant sets. We define

- $|X| + |Y| = |X + Y|$; (cardinalities sum)
- $|X| \cdot |Y| = |X \times Y|$; (cardinalities product)
- $|Y|^{|X|} = |Y^X_{fs}| = |\{f : X \to Y \,|\, f \textit{ is finitely supported}\}|$. (cardinalities exponential)

We prove that the above definitions are correct (i.e. they do not depend on the chosen representatives for equivalence classes). Let us assume that there exist the finitely supported sets X', Y' with $|X| = |X'|$ and $|Y| = |Y'|$. We generically denote the (possibly different) actions on the invariant sets containing X, Y, X', Y' by $\cdot$, the actions on functions spaces by $\star$, the actions on Cartesian products by $\otimes$ and the actions on disjoint unions by $\diamond$.

1. There exist two finitely supported bijective mappings $f : X \to X'$ and $g : Y \to Y'$. Define $h : X + Y \to X' + Y'$ by $h((0,x)) = (0, f(x))$ for all $x \in X$ and $h((1,y)) = (1, g(y))$ for all $y \in Y$. Clearly, h is bijective. Let $\pi \in Fix(supp(f) \cup supp(g))$. According to Proposition 2.9 we have $h(\pi \diamond (0,x)) = h((0, \pi \cdot x)) = (0, f(\pi \cdot x)) = (0, \pi \cdot f(x)) = \pi \diamond (0, f(x)) = \pi \diamond h((0,x))$ for all $x \in X$, and similarly, $h(\pi \diamond (1,y)) = h((1, \pi \cdot y)) = (1, g(\pi \cdot y)) = (1, \pi \cdot g(y)) = \pi \diamond h((1,y))$ for all $y \in Y$. According to Proposition 2.9, we get that h is supported by $supp(f) \cup supp(g)$, and so $|X + Y| = |X' + Y'|$.
2. There exist two finitely supported bijective mappings $f : X \to X'$ and $g : Y \to Y'$. Define $h : X \times Y \to X' \times Y'$ by $h(x,y) = (f(x), g(y))$ for all $x \in X$ and all $y \in Y$. Clearly, h is bijective. Let $\pi \in Fix(supp(f) \cup supp(g))$. According to Proposition 2.9, we have $h(\pi \otimes (x,y)) = h(\pi \cdot x, \pi \cdot y) = (f(\pi \cdot x), g(\pi, \cdot y)) = (\pi \cdot f(x), \pi \cdot g(y)) = \pi \otimes (f(x), g(y)) = \pi \otimes h(x,y)$ for all $x \in X$ and all $y \in Y$.

According to Proposition 2.9, we get that h is supported by $supp(f) \cup supp(g)$, and so $|X \times Y| = |X' \times Y'|$.

3. There exist two finitely supported bijective mappings $f : X \to X'$ and $g : Y \to Y'$. Define $\varphi : Y^X_{fs} \to Y'^{X'}_{fs}$ by $\varphi(h) = g \circ h \circ f^{-1}$ for any finitely supported mapping $h : X \to Y$. Clearly, φ is bijective. Let $\pi \in Fix(supp(f) \cup supp(g))$ and h an arbitrary finitely supported mapping from X to Y. Fix an arbitrary $x' \in X'$. According to Proposition 2.9 and because f^{-1} is also supported by $supp(f)$, we have $\varphi(\pi \star h)(x') = (g \circ (\pi \star h) \circ f^{-1})(x') = g((\pi \star h)(f^{-1}(x'))) = g(\pi \cdot h(\pi^{-1} \cdot f^{-1}(x'))) = g(\pi \cdot h(f^{-1}(\pi^{-1} \cdot x'))) = \pi \cdot g(h(f^{-1}(\pi^{-1} \cdot x'))) = \pi \cdot \varphi(h)(\pi^{-1} \cdot x') = (\pi \star \varphi(h))(x')$. Therefore $\varphi(\pi \star h) = \pi \star \varphi(h)$ for all $h \in Y^X_{fs}$, and so φ is finitely supported according to Proposition 2.9, which means $|Y|^{|X|} = |Y'|^{|X'|}$.

Based on Theorem 2.4, Theorem 2.5 and Theorem 2.6, the following results hold.

Proposition 5.2 *Let X, Y, Z be finitely supported subsets of invariant sets. The following arithmetic properties hold in FSM.*

1. $|Z|^{|X| \cdot |Y|} = (|Z|^{|Y|})^{|X|}$;
2. $|Z|^{|X|+|Y|} = |Z|^{|X|} \cdot |Z|^{|Y|}$;
3. $(|X| \cdot |Y|)^{|Z|} = |X|^{|Z|} \cdot |Y|^{|Z|}$.

Proposition 5.3 *Let X, Y, Z be finitely supported subsets of invariant sets. The following order properties hold in FSM.*

1. If $|X| \leq |Y|$, then $|X| + |Z| \leq |Y| + |Z|$;
2. If $|X| \leq |Y|$, then $|X| \cdot |Z| \leq |Y| \cdot |Z|$;
3. If $|X| \leq |Y|$, then $|X^Z_{fs}| \leq |Y^Z_{fs}|$;
4. If $|X| \leq |Y|$ and $Z \neq \emptyset$, then $|Z^X_{fs}| \leq |Z^Y_{fs}|$;
5. $|X| + |Y| \leq |X| \cdot |Y|$ whenever both X and Y have more than two elements.

Proof We generically denote the (possibly different) actions of the invariant sets containing X, Y, Z by $\cdot$, the actions on function spaces by $\widetilde{\star}$, the actions on Cartesian products by $\otimes$ and the actions on disjoint unions by $\diamond$.

1. Suppose there exists a finitely supported injective mapping $f : X \to Y$. Define the injection $g : X + Z \to Y + Z$ by

$$g(u) = \begin{cases} (0, f(x)), & \text{if}\, u = (0,x)\, \text{with}\, x \in X; \\ (1,z), & \text{if}\, u = (1,z)\, \text{with}\, z \in Z. \end{cases}$$

Since f is finitely supported we have that $f(\pi \cdot x) = \pi \cdot f(x)$ for all $x \in X$ and $\pi \in Fix(supp(f))$. Using Proposition 2.9, i.e. verifying that $g(\pi \diamond u) = \pi \diamond g(u)$ for all $u \in X + Z$ and all $\pi \in Fix(supp(f) \cup supp(X) \cup supp(Y) \cup supp(Z))$, we have that g is also finitely supported.

2. Suppose there exists a finitely supported injective mapping $f : X \to Y$. Define the injection $g : X \times Z \to Y \times Z$ by $g((x,z)) = (f(x), z)$ for all $(x,z) \in X \times Z$. Clearly, g is injective. Since f is finitely supported we have that $f(\pi \cdot x) = \pi \cdot f(x)$

for all $x \in X$ and $\pi \in Fix(supp(f))$, and so $g(\pi \otimes (x,z)) = g((\pi \cdot x, \pi \cdot z)) = (f(\pi \cdot x), \pi \cdot z) = (\pi \cdot f(x), \pi \cdot z) = \pi \otimes g((x,z))$ for all $(x,z) \in X \times Z$ and $\pi \in Fix(supp(f) \cup supp(X) \cup supp(Y) \cup supp(Z))$, which means g is supported by $supp(f) \cup supp(X) \cup supp(Y) \cup supp(Z)$.

3. Suppose there exists a finitely supported injective mapping $f : X \to Y$. Define $g : X^Z_{fs} \to Y^Z_{fs}$ by $g(h) = f \circ h$. We get that g is injective, and for any $\pi \in Fix(supp(f))$ we have $\pi \widetilde{\star} f = f$, where the notation $\widetilde{\star}$ is used to denote the canonical actions on function spaces constructed as in Proposition 2.7. Thus, $g(\pi \widetilde{\star} h) = f \circ (\pi \widetilde{\star} h) = (\pi \widetilde{\star} f) \circ (\pi \widetilde{\star} h) = \pi \widetilde{\star} (f \circ h) = \pi \widetilde{\star} g(h)$ for all $h \in X^Z_{fs}$. We used the relation $(\pi \widetilde{\star} f) \circ (\pi \widetilde{\star} h) = \pi \widetilde{\star} (f \circ h)$ for all $\pi \in S_A$ that can be proved as follows. Fixing $x \in Z$, we have $(\pi \widetilde{\star} (f \circ h))(x) = \pi \cdot (f(h(\pi^{-1} \cdot x)))$. Also, if we denote $(\pi \widetilde{\star} h)(x) = y$, then we get $y = \pi \cdot (h(\pi^{-1} \cdot x))$ and $((\pi \widetilde{\star} f) \circ (\pi \widetilde{\star} h))(x) = (\pi \widetilde{\star} f)(y) = \pi \cdot (f(\pi^{-1} \cdot y)) = \pi \cdot (f((\pi^{-1} \circ \pi) \cdot h(\pi^{-1} \cdot x))) = \pi \cdot (f(h(\pi^{-1} \cdot x)))$. Finally, it follows that g is supported by $supp(f) \cup supp(X) \cup supp(Y) \cup supp(Z)$.

4. Suppose there exists a finitely supported injective mapping $f : X \to Y$. According to Proposition 5.1, there is a finitely supported surjective mapping $f' : Y \to X$. Define the injective mapping $g : Z^X_{fs} \to Z^Y_{fs}$ by $g(h) = h \circ f'$. As in item 3, we can prove that g is finitely supported by $supp(f') \cup supp(X) \cup supp(Y) \cup supp(Z)$.

5. Fix $x_0, x_1 \in X$, $x_0 \neq x_1$ and $y_0, y_1 \in Y$, $y_0 \neq y_1$. We define $g : X + Y \to X \times Y$ by

$$g(u) = \begin{cases} (x, y_0), & \text{if } u = (0,x) \text{ with } x \in X, x \neq x_0; \\ (x_0, y), & \text{if } u = (1,y) \text{ with } y \in Y; \\ (x_1, y_1), & \text{if } u = (0, x_0) \end{cases}$$

It follows that g is supported by $supp(x_0) \cup supp(y_0) \cup supp(x_1) \cup supp(y_1) \cup supp(X) \cup supp(Y)$, and g is injective. □

It is easy to verify that the properties of $\leq$ presented in Proposition 5.3 (1), (2) and (4) also hold for $\leq^\star$; we left the details to the reader.

Theorem 5.5 *There exists an invariant set X (particularly the set A of atoms) having the following properties:*

1. $|X \times X| \nleq^* |\wp_{fs}(X)|$;
2. $|X \times X| \nleq |\wp_{fs}(X)|$;
3. $|X \times X| \nleq^* |X|$;
4. $|X \times X| \nleq |X|$;
5. *For each $n \in \mathbb{N}, n \geq 2$ we have $|X| \lneq |\wp_n(X)| \lneq |\wp_{fs}(X)|$, where $\wp_n(X)$ is the family of all n-sized subsets of X;*
6. *For each $n \in \mathbb{N}$ we have $|X| \lneq^* |\wp_n(X)| \lneq^* |\wp_{fs}(X)|$;*
7. $|X| \lneq |\wp_{fin}(X)| \lneq |\wp_{fs}(X)|$;
8. $|X| \lneq^* |\wp_{fin}(X)| \lneq^* |\wp_{fs}(X)|$;
9. $|\wp_{fs}(X) \times \wp_{fs}(X)| \nleq^* |\wp_{fs}(X)|$;
10. $|\wp_{fs}(X) \times \wp_{fs}(X)| \nleq |\wp_{fs}(X)|$;
11. $|X + X| \lneq^* |X \times X|$;
12. $|X + X| \lneq |X \times X|$.

Proof 1. We prove that there does not exist a finitely supported surjective mapping $f : \wp_{fs}(A) \to A \times A$. Suppose by contradiction that there is a finitely supported surjective mapping $f : \wp_{fs}(A) \to A \times A$. Let us consider two atoms $a, b \notin supp(f)$ with $a \neq b$. These atoms exist because A is infinite, while $supp(f) \subseteq A$ is finite. It follows that the transposition (ab) fixes each element from $supp(f)$, i.e. $(ab) \in Fix(supp(f))$. Since f is surjective, it follows that there exists an element $X \in \wp_{fs}(A)$ such that $f(X) = (a,b)$. Since $supp(f)$ supports f and $(ab) \in Fix(supp(f))$, from Proposition 2.9 we have $f((ab) \star X) = (ab) \otimes f(X) = (ab) \otimes (a,b) = ((ab)(a),(ab)(b)) = (b,a)$. Due to the functionality of f we should have $(ab) \star X \neq X$. Otherwise, we would obtain $(a,b) = (b,a)$.
We claim that if both $a, b \in supp(X)$, then $(ab) \star X = X$. Indeed, suppose $a, b \in supp(X)$. Since X is a finitely supported subset of A, then X is either finite or cofinite. If X is finite, then $supp(X) = X$, and so $a, b \in X$. Moreover, $(ab)(a) = b$, $(ab)(b) = a$, and $(ab)(c) = c$ for all $c \in X$ with $c \neq a, b$. Therefore, $(ab) \star X = \{(ab)(x) \,|\, x \in X\} = \{(ab)(a)\} \cup \{(ab)(b)\} \cup \{(ab)(c) \,|\, c \in X \setminus \{a,b\}\} = \{b\} \cup \{a\} \cup (X \setminus \{a,b\}) = X$. Informally, we remarked that, when $a, b \in X$, we have that the transposition (ab) interchanges a with b in X and leaves the other atoms in X (different from a and b) unchanged. Essentially, X is left unchanged under (ab) because the order of a and b in the set X is, obviously, non-important. Now, if X is cofinite, then $supp(X) = A \setminus X$, and so $a, b \in A \setminus X$. Since $a, b \notin X$, we have $a, b \neq x$ for all $x \in X$, and so $(ab)(x) = x$ for all $x \in X$. Thus, in this case we also have $(ab) \star X = X$.
Since when both $a, b \in supp(X)$ we have $(ab) \star X = X$, it follows that one of a or b does not belong to $supp(X)$. Suppose $b \notin supp(X)$ (the other case is similar). Let us consider $c \neq a, b$, $c \notin supp(f)$, $c \notin supp(X)$. Then $(bc) \in Fix(supp(X))$, and because $supp(X)$ supports X, we have $(bc) \star X = X$. Furthermore, $(bc) \in Fix(supp(f))$, and by Proposition 2.9 we have $(a,b) = f(X) = f((bc) \star X) = (bc) \otimes f(X) = (bc) \otimes (a,b) = ((bc)(a),(bc)(b)) = (a,c)$ which is a contradiction because $b \neq c$. Thus, $|A \times A| \nleq^* |\wp_{fs}(A)|$.

2. We prove that there does not exist a finitely supported injective mapping $f : A \times A \to \wp_{fs}(A)$. Suppose by contradiction that there is a finitely supported injective mapping $f : A \times A \to \wp_{fs}(A)$. According to Proposition 5.1 one can define a finitely supported surjection $g : \wp_{fs}(A) \to A \times A$. This contradicts the above item. Thus, $|A \times A| \nleq |\wp_{fs}(A)|$.
3. We prove that there does not exist a finitely supported surjection $f : A \to A \times A$. Since there exists a surjection s from $\wp_{fs}(A)$ onto A defined by

$$s(X) = \begin{cases} x, \text{ if} X \text{ is a one-element set } \{x\}\ ; \\ a, \text{ if} X \text{ is not a one-element set} \end{cases}$$

where a is a fixed atom, and s is finitely supported (by$\{a\}$), the result follows from item 1. Thus, $|A \times A| \nleq^* |A|$.

4. We prove that there does not exist a finitely supported injection $f : A \times A \to A$. Since there exists an equivariant injection from A into $\wp_{fs}(A)$ defined as $x \mapsto \{x\}$, the result follows from item 2. Thus, $|A \times A| \nleq |A|$;
Alternatively, one can prove that there does not exist a one-to-one mapping from $A \times A$ to A (and so neither a finitely supported one). Suppose by contradiction that there is a an injective mapping $i : A \times A \to A$. Let us fix two atoms x and y with $x \neq y$. The sets $\{i(a,x) \,|\, a \in A\}$ and $\{i(a,y) \,|\, a \in A\}$ are disjoint and infinite. Thus, $\{i(a,x) \,|\, a \in A\}$ is an infinite and coinfinite subset of A, which contradicts the fact that any subset of A is either finite or cofinite.
5. We prove that $|A| \lneq |\wp_n(A)| \lneq |\wp_{fs}(A)|$ for all $n \in \mathbb{N}, n \geq 2$.
Consider $a_1, a_2, \ldots, a_{n-1}, a_1^1, \ldots, a_1^n, \ldots, a_{n-1}^1, \ldots, a_{n-1}^n \in A$ a family of pairwise different elements. Then $i : A \to \wp_n(A)$ defined by

$$i(x) = \begin{cases} \{x, a_1, a_2, \ldots, a_{n-1}\}, & \text{if} x \neq a_1, \ldots, a_{n-1}\,; \\ \{a_1^1, \ldots, a_1^n\}, & \text{if} x = a_1 \\ \vdots & \\ \{a_{n-1}^1, \ldots, a_{n-1}^n\}, & \text{if} x = a_{n-1} \end{cases}$$

is obviously an injective mapping from $(A, \cdot)$ to $(\wp_n(A), \star)$. Furthermore, i is supported by the finite set $\{a_1, a_2, \ldots, a_{n-1}, a_1^1, \ldots, a_1^n, \ldots, a_{n-1}^1, \ldots, a_{n-1}^n\}$, and so $|A| \leq |\wp_n(A)|$ in FSM. We claim that there does not exist a finitely supported injection from $\wp_n(A)$ into A. Assume on the contrary that there exists a finitely supported injection $f : \wp_n(A) \to A$.
Firstly, we claim that for any $Y \in \wp_n(A)$ which is disjoint from $supp(f)$, we have $f(Y) \notin Y$. Assume by contradiction that $f(Y) \in Y$ for a fixed Y with $Y \cap supp(f) = \emptyset$. Let π be a permutation of atoms which fixes $supp(f)$ pointwise, and interchanges all the elements of Y (e.g. π is a cyclic permutation of Y). Since π permutes all the elements of Y, we have $\pi \cdot f(Y) = \pi(f(Y)) \neq f(Y)$. However, $\pi \star Y = \{\pi(a_1), \ldots, \pi(a_n)\} = \{a_1, \ldots, a_n\} = Y$. Since π fixes $supp(f)$ pointwise and $supp(f)$ supports f, we have $\pi(f(Y)) = \pi \cdot f(Y) = f(\pi \star Y) = f(Y)$, a contradiction.
Since $supp(f)$ is finite, there are infinitely many such Y with the property that $Y \cap supp(f) = \emptyset$. Thus, because it is injective, f takes infinitely many values on those Y. Since $supp(f)$ is finite, there should exist at least one element in $\wp_n(A)$, denoted by Z such that $Z \cap supp(f) = \emptyset$ and $f(Z) \notin supp(f)$. Thus, $f(Z) = a$ for some $a \in A \setminus (Z \cup supp(f))$. Let $b \in A \setminus (supp(f) \cup Z \cup \{a\})$ and also let $\pi = (a\,b)$. Then $\pi \in Fix(supp(f) \cup Z)$, and hence $f(Z) = f((ab) \star Z) = (ab)(f(Z)) = b$, a contradiction. We obtained that $|A| \neq |\wp_n(A)|$ in FSM, and so $|A| < |\wp_n(A)|$.
We obviously have $|\wp_n(A)| \leq |\wp_{fs}(A)|$. According to the proof of Theorem 3.1 (the part regarding the inconsistency of **GCH** with FSM), there does not exist a finitely supported injective mapping from $\wp_{fs}(A)$ onto one of its finitely supported proper subsets. Thus, because $\wp_n(A)$ is a particular subset of $\wp_{fs}(A)$, there could not exist a bijection $f : \wp_{fs}(A) \to \wp_n(A)$, and so $|\wp_n(A)| \neq |\wp_{fs}(A)|$.
6. Fix $n \in \mathbb{N}$. As in the above item there do not exist neither a finitely supported bijection between $\wp_n(A)$ and $\wp_{fs}(A)$, nor a finitely supported bijection between A

and $\wp_n(A)$. However, there exists a finitely supported injection $i: A \to \wp_n(A)$. Fix an atom $a \in A$. The mapping $s: \wp_n(A) \to A$ defined by

$$s(X) = \begin{cases} i^{-1}(X), & \text{if} X \in Im(i) \,; \\ a, & \text{if} X \notin Im(i) \end{cases}$$

is supported by $supp(i) \cup \{a\}$ and is surjective. Now, fix n atoms $x_1, \dots, x_n$. The mapping $g: \wp_{fs}(A) \to \wp_n(A)$ defined by

$$g(X) = \begin{cases} X, & \text{if} X \in \wp_n(A) \,; \\ \{x_1, \dots x_n\}, & \text{if} X \notin \wp_n(A) \end{cases}$$

is supported by $\{x_1, \dots x_n\}$, and surjective.

7. We prove that $|A| \lneq |\wp_{fin}(A)| \lneq |\wp_{fs}(A)|$. We obviously have that $|A| \leq |\wp_{fin}(A)|$ by taking the equivariant injective mapping $f: A \to \wp_{fin}(A)$ defined by $f(a) = \{a\}$ for all $a \in A$. We prove by contradiction that there is no finitely supported surjection from A onto $\wp_{fin}(A)$. Assume that $g: A \to \wp_{fin}(A)$ is a finitely supported surjection. Let us fix two atoms x and y in order to define the function $h: \wp_{fin}(A) \to \wp_2(A)$ by $h(X) = \begin{cases} X, & \text{if} |X| = 2 \,; \\ \{x, y\}, & \text{if} |X| \neq 2 \,. \end{cases}$. Since for every $\pi \in S_A$ and $X \in \wp_{fin}(A)$ we have $|\pi \star X| = |X|$, we conclude that h is finitely supported by $\{x, y\}$. Thus, $h \circ g$ is a surjection from A onto $\wp_2(A)$ supported by $supp(g) \cup \{x, y\}$, which contradicts the previous item. Therefore, $|A| < |\wp_{fin}(A)|$. Since every element in $\wp_{fin}(A)$ belongs to $\wp_{fs}(A)$, but there does not exist a finitely supported injective mapping from $\wp_{fs}(A)$ onto one of its finitely supported proper subsets, we also have $|\wp_{fin}(A)| < |\wp_{fs}(A)|$.
8. As in the above item there do not exist neither a finitely supported bijection between $\wp_{fin}(A)$ and $\wp_{fs}(A)$, nor a finitely supported bijection between A and $\wp_{fin}(A)$. Fix an atom $a \in A$. The mapping $s: \wp_{fin}(A) \to A$ defined by

$$s(X) = \begin{cases} x, & \text{if} X \text{ is a one-element set } \{x\} \,; \\ a, & \text{if} X \text{ is not a one-element set} \end{cases}$$

is supported by $\{a\}$ and is surjective. Now, fix an atom b. The mapping $g: \wp_{fs}(A) \to \wp_{fin}(A)$ defined by

$$g(X) = \begin{cases} X, & \text{if} X \in \wp_{fin}(A) \,; \\ \{b\}, & \text{if} X \notin \wp_{fin}(A) \end{cases}$$

is supported by $\{b\}$, and surjective.

9. According to Theorem 5.5(1) there is no finitely supported surjection from $\wp_{fs}(A)$ onto $A \times A$. Suppose there is a finitely supported surjective mapping $f: \wp_{fs}(A) \to \wp_{fs}(A) \times \wp_{fs}(A)$. Obviously, there exists a supported surjection $s: \wp_{fs}(A) \to A$ defined by

$$s(X) = \begin{cases} a, & \text{if} X \text{ is a one-element set } \{\text{a}\} \,; \\ x, & \text{if} X \text{ has more than one element} \,. \end{cases},$$

where x is a fixed atoms of A. The surjection s is supported by $supp(x) = x$. Thus, we can define a surjection $g : \wp_{fs}(A) \times \wp_{fs}(A) \to A \times A$ by $g(X,Y) = (s(X), s(Y))$ for all $X, Y \in \wp_{fs}(A)$. Let $\pi \in Fix(supp(s))$. Since $supp(s)$ supports s, by Proposition 2.9 we have $g(\pi \otimes_\star (X,Y)) = g(\pi \star X, \pi \star Y) = (s(\pi \star X), s(\pi \star Y)) = (\pi \cdot s(X), \pi \cdot s(Y)) = \pi \otimes (s(X), s(Y))$ for all $X, Y \in \wp_{fs}(A)$, where $\otimes_\star$ and $\otimes$ represent the S_A-actions on $\wp_{fs}(A) \times \wp_{fs}(A)$ and $A \times A$, respectively. Thus, $supp(s)$ supports g, and so $supp(g) \subseteq supp(s)$. Furthermore, the function $h = g \circ f : \wp_{fs}(A) \to A \times A$ is surjective and finitely supported by $supp(s) \cup supp(f)$. This is a contradiction, and so $|\wp_{fs}(A) \times \wp_{fs}(A)| \nleq^* |\wp_{fs}(A)|$.

10. Suppose by contradiction that there is a finitely supported injective mapping $f : \wp_{fs}(A) \times \wp_{fs}(A) \to \wp_{fs}(A)$. In the view of Proposition 5.1, let us fix two finitely supported subsets of A, namely U and V. We define the function $g : \wp_{fs}(A) \to \wp_{fs}(A) \times \wp_{fs}(A)$ by

$$g(X) = \begin{cases} f^{-1}(X), & \text{if}\, X \in Im(f)\,; \\ (U,V), & \text{if}\, X \notin Im(f)\,. \end{cases}$$

Clearly, g is surjective. Furthermore, g is supported by $supp(f) \cup supp(U) \cup supp(V)$ (the proof uses the fact that $Im(f)$ is a subset of $\wp_{fs}(A)$ supported by $supp(f)$). This contradicts the above item, and so $|\wp_{fs}(A) \times \wp_{fs}(A)| \nleq |\wp_{fs}(A)|$.

11. In the view of Proposition 5.3(5) there is a finitely supported injection from $A + A$ into $A \times A$, and a finitely supported surjection from $A \times A$ onto $A + A$ according to Proposition 5.1. Thus, $|A + A| \leq |A \times A|$ and $|A + A| \leq^* |A \times A|$. Fix three different atoms $a, b, c \in A$. Define the mapping $f : A + A \to \wp_{fs}(A)$ by

$$f(u) = \begin{cases} \{x\}, & \text{if}\, u = (0,x)\, \text{with}\, x \in A; \\ \{a,y\}, & \text{if}\, u = (1,y)\, \text{with}\, y \in A, y \neq a; \\ \{b,c\}, & \text{if}\, u = (1,a) \end{cases}$$

One can directly prove that f is injective and supported by $\{a,b,c\}$. According to Proposition 5.1, we have $|A + A| \leq^* |\wp_{fs}(A)|$. If $|A \times A| = |A + A|$, we would obtain $|A \times A| \leq^* |\wp_{fs}(A)|$, which contradicts item 1.

12. According to the above item $|A + A| \leq |\wp_{fs}(A)|$. If we had $|A \times A| = |A + A|$, we would obtain $|A \times A| \leq |\wp_{fs}(A)|$, which contradicts item 2. □

Proposition 5.4 *There exists an invariant set X having the following properties.*

1. $|X| \lneq |X| + |X|$;
2. $|X| \lneq^* |X| + |X|$.

Proof 1. Firstly, we prove that in FSM we have $|\wp_{fs}(A)| = 2|\wp_{fin}(A)|$. Let us consider the function $f : \wp_{fin}(A) \to \wp_{cofin}(A)$ defined by $f(U) = A \setminus U$ for all $U \in \wp_{fin}(A)$. Clearly, f is bijective. We claim that f is equivariant. Indeed, let $\pi \in S_A$. To prove that $f(\pi \star U) = \pi \star f(U)$ for all $U \in \wp_{fin}(A)$, we have to prove that $A \setminus (\pi \star U) = \pi \star (A \setminus U)$ for all $U \in \wp_{fin}(A)$. Let $y \in A \setminus (\pi \star U)$.

We can express y as $y = \pi \cdot (\pi^{-1} \cdot y)$. If $\pi^{-1} \cdot y \in U$, then $y \in \pi \star U$, which is a contradiction. Thus, $\pi^{-1} \cdot y \in (A \setminus U)$, and so $y \in \pi \star (A \setminus U)$. Conversely, if $y \in \pi \star (A \setminus U)$, then $y = \pi \cdot x$ with $x \in A \setminus U$. Suppose $y \in \pi \star U$. Then $y = \pi \cdot z$ with $z \in U$. Thus, $x = z$ which is a contradiction, and so $y \in A \setminus (\pi \star U)$. Since f is equivariant and bijective, it follows that $|\wp_{fin}(A)| = |\wp_{cofin}(A)|$. However, every finitely supported subset of A is either finite or cofinite, and so $\wp_{fs}(A)$ is the union of the disjoint subsets $\wp_{fin}(A)$ and $\wp_{cofin}(A)$. Thus, $|\wp_{fs}(A)| = 2|\wp_{fin}(A)|$. Moreover, there exists an equivariant injection $i : \wp_{fin}(A) \to \wp_{fs}(A)$ defined by $i(U) = U$ for all $U \in \wp_{fin}(A)$. However, there does not exist a finitely supported one-to-one mapping from $\wp_{fs}(A)$ onto one of its finitely supported proper subsets. Thus, there could not exist a bijection $f : \wp_{fs}(A) \to \wp_{fin}(A)$. Therefore, $|\wp_{fin}(A)| \neq |\wp_{fs}(A)| = 2|\wp_{fin}(A)|$. We can consider $X = \wp_{fin}(A)$ or $X = \wp_{cofin}(A)$.
2. It remains to prove that there is a finitely supported surjection from $\wp_{fs}(A)$ onto $\wp_{fin}(A)$. We either use Proposition 5.1 or we effectively construct the surjection as below. Fix $a \in A$. We define $g : \wp_{fs}(A) \to \wp_{fin}(A)$ by

$$g(U) = \begin{cases} U, & \text{if} U \in \wp_{fin}(A)\ ; \\ \{a\}, & \text{if} U \notin \wp_{fin}(A)\,. \end{cases}$$

Clearly, g is supported by $\{a\}$ and surjective. We can consider $X = \wp_{fin}(A)$ or $X = \wp_{cofin}(A)$. □

5.3 Fixed Points Results

Theorem 5.6 (Bourbaki-Witt Theorem)

Let $(X, \sqsubseteq, \cdot)$ be an invariant partially ordered set with the property that every finitely supported totally ordered subset Z (every finitely supported chain Z) of X has a least upper bound $\sqcup Z$. Let $f : X \to X$ be a finitely supported function with the property that $x \sqsubseteq f(x)$ for all $x \in X$. Then there exists $x \in X$ such that $f(x) = x$.

Proof Fix an element $x_0 \in X$. Let $X' = \{x \in X \,|\, x_0 \sqsubseteq x\}$. We claim that X' is supported by $supp(x_0)$. Let us consider $\pi \in Fix(supp(x_0))$. Since $supp(x_0)$ supports x_0, we have $\pi \cdot x_0 = x_0$. Since $\sqsubseteq$ is an equivariant order relation on X, for each $x \in X'$ we have $x_0 \sqsubseteq x$, and so $x_0 = \pi \cdot x_0 \sqsubseteq \pi \cdot x$, which means $\pi \cdot x \in X'$. Thus, $\pi \star X' = X'$ for all $\pi \in Fix(supp(x_0))$ (where $\star$ represents the canonical permutation action on $\wp(X)$), and so $supp(X') \subseteq supp(x_0)$. The least upper bound of a finitely supported chain belonging to X' also belongs to X'.

A subset $Y \subseteq X$ is called FSM admissible if

- Y is finitely supported as a subset of X and $Y \subseteq X'$.
- $x_0 \in Y$;
- If $y \in Y$, then $f(y) \in Y$;

- If $C \subseteq Y$ is a finitely supported chain, then $\sqcup C \in Y$;

We prove that the family $\mathscr{F}$ of all FSM admissible subsets of X' is supported by $supp(f) \cup supp(x_0)$. For this, we have to prove that any permutation $\pi \in Fix(supp(f) \cup supp(x_0))$ leaves $\mathscr{F}$ invariant, that is $\pi \star Y \in \mathscr{F}$ for all $\pi \in Fix(supp(f) \cup supp(x_0))$ and all $Y \in \mathscr{F}$. Fix $\pi \in Fix(supp(f) \cup supp(x_0))$ and an FSM admissible set $Y \subseteq X'$.

- According to Proposition 2.1 we have that $\pi \star Y$ is a finitely supported subset of X because Y is finitely supported. Furthermore, because $Y \subseteq X'$ and $\star$ is equivariant, we get $\pi \star Y \subseteq \pi \star X'$. However, because $supp(X') \subseteq supp(x_0)$ and π fixes $supp(x_0)$ pointwise, we have $\pi \star X' = X'$, which means $\pi \star Y \subseteq X'$;
- Since $\pi \in Fix(supp(x_0))$ and $supp(x_0)$ supports x_0, we have $x_0 = \pi \cdot x_0 \in \pi \star Y$;
- Let $y \in \pi \star Y$. Then $\pi^{-1} \cdot y \in \pi^{-1} \star (\pi \star Y) = (\pi^{-1} \circ \pi) \star Y = Y$. Since Y is FSM admissible, we have $f(\pi^{-1} \cdot y) \in Y$. Thus, because π fixes $supp(f)$ pointwise and $supp(f)$ supports f, according to Proposition 2.9 we get $\pi^{-1} \cdot f(y) = f(\pi^{-1} \cdot y) \in Y$, and so $f(y) = (\pi \circ \pi^{-1}) \cdot f(y) = \pi \cdot (\pi^{-1} \cdot f(y)) \in \pi \star Y$;
- Let $C \subseteq \pi \star Y$ be a finitely supported chain. According to Proposition 2.1, we have that $\pi^{-1} \star C$ is finitely supported, and because $\subseteq$ is an equivariant subset of $\wp_{fs}(X) \times \wp_{fs}(X)$, we obtain $\pi^{-1} \star C \subseteq \pi^{-1} \star (\pi \star Y) = (\pi^{-1} \circ \pi) \star Y = Y$. Furthermore, $\pi^{-1} \star C$ is a chain because $\sqsubseteq$ is equivariant. Since Y is FSM admissible, we have $\sqcup(\pi^{-1} \star C) \in Y$. Moreover, we claim that $\pi^{-1} \cdot \sqcup C = \sqcup(\pi^{-1} \star C)$. Let $x \in C$. We have $x \sqsubseteq \sqcup C$. Since $\sqsubseteq$ is equivariant, we have $\pi^{-1} \cdot x \sqsubseteq \pi^{-1} \cdot \sqcup C$. Therefore, $\sqcup(\pi^{-1} \star C) \sqsubseteq \pi^{-1} \cdot \sqcup C$ according to the definition of a least upper bound. However, the relation $\sqcup(\sigma \star Z) \sqsubseteq \sigma \cdot \sqcup Z$ can be proved for any $\sigma \in S_A$ and Z a finitely supported chain of X. We can also apply this relation for the permutation $\sigma = \pi$ and for the finitely supported chain $Z = \pi^{-1} \star C$ (which is a finitely supported chain of X due to the equivariance of $\sqsubseteq$), and we obtain $\sqcup C = \sqcup(\pi \star (\pi^{-1} \star C)) \sqsubseteq \pi \cdot \sqcup(\pi^{-1} \star C)$. Since $\sqsubseteq$ is equivariant, we obtain $\pi^{-1} \cdot \sqcup C \sqsubseteq \sqcup(\pi^{-1} \star C)$. Finally, we get $\sqcup C = \pi \cdot \sqcup(\pi^{-1} \star C) \in \pi \star Y$.

We proved above that $\mathscr{F}$ is finitely supported. We claim that $\bigcap_{Z \in \mathscr{F}} Z$ is supported by $supp(\mathscr{F}) \subseteq supp(f) \cup supp(x_0)$. Let $\pi \in Fix(supp(\mathscr{F}))$. Let $x \in \bigcap_{Z \in \mathscr{F}} Z$. Then $x \in Z$ for all $Z \in \mathscr{F}$. We have to prove that $\pi \cdot x \in Z$ for all $Z \in \mathscr{F}$. Let us consider an arbitrary $T \in \mathscr{F}$. Since $\pi^{-1} \in Fix(supp(\mathscr{F}))$ (and so π^{-1} leaves $\mathscr{F}$ invariant), we have $\pi^{-1} \star T \in \mathscr{F}$, and hence there exists $V \in \mathscr{F}$ such that $V = \pi^{-1} \star T$. Since x belongs to any element of $\mathscr{F}$, we have $x \in V$, and so $\pi \cdot x \in \pi \star V = T$. Thus, because T has been arbitrarily chosen from $\mathscr{F}$, we have $\pi \cdot x \in Z$ for all $Z \in \mathscr{F}$. Thus, $\pi \cdot x \in \bigcap_{Z \in \mathscr{F}} Z$, and so $\pi \star \bigcap_{Z \in \mathscr{F}} Z = \bigcap_{Z \in \mathscr{F}} Z$.

Let $M = \bigcap_{Z \in \mathscr{F}} Z$. Then M is finitely supported. As in the ZF framework, we show that M satisfies the conditions to be admissible, and so it is FSM admissible.

- M is finitely supported, and $M \subseteq X'$ because $Z \subseteq X'$ for all $Z \in \mathscr{F}$;
- $x_0 \in M$ because $x_0 \in Z$ for all $Z \in \mathscr{F}$;
- If $y \in M$, then $y \in Z$ for all $Z \in \mathscr{F}$. Since every $Z \in \mathscr{F}$ is FSM admissible, we get $f(y) \in Z$ for all $Z \in \mathscr{F}$, and so $f(y) \in M$;

- If C is a finitely supported chain in M, then C is a finitely supported chain contained in any element of $\mathscr{F}$. Thus, $\sqcup C \in Z$ for all $Z \in \mathscr{F}$, and so $\sqcup C \in M$;

Furthermore, according to the construction of M, x_0 is the least element of M.

A point $c \in M$ is called extremal if whenever $x \in M$ with $x \sqsubset c$ we have $f(x) \sqsubseteq c$. For any extremal point c we define $M_c = \{x \in M \mid x \sqsubseteq c\} \cup \{x \in M \mid f(c) \sqsubseteq x\}$. We claim that M_c is FSM admissible for any extremal point $c \in M$. Fix an extremal point $c \in M$.

- M_c is finitely supported by $supp(c) \cup supp(f) \cup supp(M)$. Let us take $\pi \in Fix(supp(c) \cup supp(f) \cup supp(M))$. Let $x \in M_c$. Then either $x \sqsubseteq c$ or $f(c) \sqsubseteq x$. If $x \sqsubseteq c$, because $\sqsubseteq$ is equivariant and π fixes $supp(c)$ pointwise (and so π also fixes c), we have $\pi \cdot x \sqsubseteq \pi \cdot c = c$. If $f(c) \sqsubseteq x$, then we get $f(c) = f(\pi \cdot c) = \pi \cdot f(c) \sqsubseteq \pi \cdot x$ by Proposition 2.9, because π fixes $supp(f)$ pointwise and $supp(f)$ supports f. Thus, since we also have $\pi \cdot x \in M$, it follows that $\pi \cdot x \in M_c$, and so $\pi \star M_c = M_c$. Obviously, $M_c \subseteq X'$ because $M_c \subseteq M$ and $M \subseteq X'$.
- $x_o \in M_c$ (because x_0 is the least element of M, and so $x_0 \sqsubseteq c$ for the fixed $c \in M$).
- Since for $x \in M$, because c is extremal, we have that $x \sqsubset c$ implies $f(x) \sqsubseteq c$ and $x = c$ implies $f(c) \sqsubseteq f(x)$, we obtain $f(\{x \in M \mid x \sqsubseteq c\}) \subseteq M_c$. Moreover, because for all $x \in M$ with the property that $f(c) \sqsubseteq x$ we have $f(c) \sqsubseteq x \sqsubseteq f(x)$, we obtain that $f(\{x \in M \mid f(c) \sqsubseteq x\}) \subseteq \{x \in M \mid f(c) \sqsubseteq x\}$. Therefore, $f(M_c) \subseteq M_c$.
- If C is a finitely supported chain in M_c, then either $x \sqsubseteq c$ for all $x \in C$, and so $\sqcup C \sqsubseteq c$ (according to the definition of supremum), or $f(c) \sqsubseteq x$ for some $x \in C$ which implies $f(c) \sqsubseteq \sqcup C$. Thus, $\sqcup C \in M_c$.

Let $S \subseteq M$ be the set of all extremal points. We prove now that the mapping $c \mapsto M_c$ defined on S is finitely supported, and so the association of an M_c to each extremal point $c \in M$ is consistent with FSM, which means the family $(M_c)_{c \in S}$ exists in FSM. According to Proposition 2.9, we should prove that there is a finite $B \subseteq A$ such that, for all $\pi \in Fix(B)$, we have $\pi \cdot c \in S$ for all $c \in S$, and $M_{\pi \cdot c} = \pi \star M_c$ for all $c \in S$. We consider $B = supp(f) \cup supp(M)$. Let $\pi \in Fix(supp(f) \cup supp(M))$ and $c \in S$.

- Since $c \in S$, it satisfies the condition of being an extremal point which means whenever $y \in M$ with $y \sqsubset c$ we have $f(y) \sqsubseteq c$. Let $x \in M$ be an arbitrary element such that $x \sqsubset \pi \cdot c$. Since the order relation on X is equivariant, $\cdot$ is a group action, $c \in S$, π fixes $supp(f)$ pointwise and f is supported by $supp(f)$, then according to Proposition 2.9 we have $x \sqsubset \pi \cdot c \Rightarrow \pi^{-1} \cdot x \sqsubset c \Rightarrow f(\pi^{-1} \cdot x) \sqsubseteq c \Rightarrow \pi^{-1} \cdot f(x) \sqsubseteq c \Rightarrow f(x) \sqsubseteq \pi \cdot c$. Thus, $\pi \cdot c \in S$. Therefore, because π also fixes $supp(M)$ pointwise, we get $\pi \star S = S$, which means that $supp(f)$ supports S.
- According to Proposition 2.9 and because $\sqsubseteq$ is equivariant, we have $M_{\pi \cdot c} = \{x \in M \mid x \sqsubseteq \pi \cdot c\} \cup \{x \in M \mid f(\pi \cdot c) \sqsubseteq x\} = \{x \in M \mid \pi^{-1} \cdot x \sqsubseteq c\} \cup \{x \in M \mid \pi \cdot f(c) \sqsubseteq x\} = \{x \in M \mid \pi^{-1} \cdot x \sqsubseteq c\} \cup \{x \in M \mid f(c) \sqsubseteq \pi^{-1} \cdot x\} \overset{\pi^{-1} \cdot x := z}{=} \{\pi \cdot z \in M \mid z \sqsubseteq c\} \cup \{\pi \cdot z \in M \mid f(c) \sqsubseteq z\} = (\pi \star \{z \in M \mid z \sqsubseteq c\}) \cup (\pi \star \{z \in M \mid f(c) \sqsubseteq z\}) = \pi \star M_c$.

We conclude that the family $(M_c)_c$ is well-defined in FSM. Fix a point $c \in S$. Since M_c is FSM admissible, we have $M \subseteq M_c$ according to the definition of M (M

is the least FSM admissible subset of X). Furthermore, according to the definition of M_c we also have $M_c \subseteq M$. We obtained $M = M_c$ for any extremal point c.

We claim that the set S of all extremal points is FSM admissible.

- S is finitely supported by $supp(f) \cup supp(M)$. Let $\pi \in Fix(supp(f) \cup supp(M))$. Let us consider an arbitrary $c \in S$. As above, we obtain $\pi \cdot c \in S$, and so $\pi \star S = S$. Moreover, obviously, $S \subseteq X'$ because $S \subseteq M$ and $M \subseteq X'$.
- Obviously, $x_0 \in S$ (because x_0 is the least element of M, there does not exist $x \in M$ with $x \sqsubset x_0$).
- As in the ZF framework we verify that $f(S) \subseteq S$. Let $c \in S$. Suppose $x \in M$ with $x \sqsubset f(c)$. Since $M = M_c$, it follows that $x \in M_c$, and so $x \sqsubseteq c$ or $f(c) \sqsubseteq x$. Thus, because $x \sqsubset f(c)$, we get $x \sqsubseteq c$. If $x \sqsubset c$, then because $c \in S$, we have $f(x) \sqsubseteq c \sqsubseteq f(c)$. If $x = c$, then $f(x) = f(c)$. In either case we get $f(x) \sqsubseteq f(c)$. Thus, for any $x \in M$ we have that $x \sqsubset f(c)$ implies $f(x) \sqsubseteq f(c)$, which means $f(c) \in S$.
- Let C be a finitely supported chain in S. Suppose $x \in M$ with $x \sqsubset \sqcup C$. Since for all $c \in S$ we have $M = M_c$, either $f(c) \sqsubseteq x$ for all $c \in C$, or $x \sqsubseteq c$ for some $c \in C$. In the first case, due to the hypothesis, x is an upper bound of C, and so $\sqcup C \sqsubseteq x$. This contradicts the assumption $x \sqsubset \sqcup C$. In the remaining case we have $x \sqsubseteq c$ for some $c \in C$. If $x \sqsubset c$, then because c is extremal we have $f(x) \sqsubseteq c \sqsubseteq \sqcup C$. Suppose now $x = c$ with $c \in C$. Since C is a finitely supported chain in S and $S \subseteq M$, we have that C is a finitely supported chain in M, and so $\sqcup C \in M$ because M is FSM admissible. Since $M = M_c$ we have $\sqcup C \in M_c$, that is $\sqcup C \sqsubseteq c$ or $f(c) \sqsubseteq \sqcup C$. However, since $x \sqsubset \sqcup C$ and $x = c$, we have $c \sqsubset \sqcup C$, and so it remains that $f(x) = f(c) \sqsubseteq \sqcup C$. Thus, for all $x \in M$, we have that $x \sqsubset \sqcup C$ implies $f(x) \sqsubseteq \sqcup C$. Therefore, $\sqcup C \in S$.

Thus, $M = S$. Let $x, y \in M$. Then $x \in S$. Since $M \subseteq M_x$, we get $y \in M_x$. Thus, $y \sqsubseteq x$ or $f(x) \sqsubseteq y$. Since $x \sqsubseteq f(x)$ by hypothesis, we obtain $y \sqsubseteq x$ or $x \sqsubseteq y$. Thus, M is totally ordered. Since M is finitely supported, we have that there exists $\sqcup M$. Since M is FSM admissible and M is totally ordered we have $\sqcup M \in M$. Since M is FSM admissible and $\sqcup M \in M$, we also have $f(\sqcup M) \in M$, and so $f(\sqcup M) \sqsubseteq \sqcup M$. Since, by hypothesis, we also have $\sqcup M \sqsubseteq f(\sqcup M)$, we obtained that $\sqcup M$ is the required fixed point of f. □

Corollary 5.3 *Let $(X, \sqsubseteq, \cdot)$ be an invariant partially ordered set with the property that every finitely supported totally ordered subset (every finitely supported chain) of X has a least upper bound. Let $f : X \to X$ be a finitely supported function with the property that $x \sqsubseteq f(x)$ for all $x \in X$. Then for any $y \in Y$ there exists $u \in X$ such that $f(u) = u$ and $y \sqsubseteq u$.*

Proof In the proof of Theorem 5.6 we considered an arbitrary $x_0 \in X$ and we obtained a fixed point of f defined as the least upper bound of a finitely supported chain in X whose least element is x_0. □

Theorem 5.7 *Let $(X, \sqsubseteq, \cdot)$ be an invariant partially ordered set containing no infinite uniformly supported subset. Let $f : X \to X$ be a finitely supported function with* *the property that $x \sqsubseteq f(x)$ for all $x \in X$. Then for each $x \in X$, there exists some $m \in \mathbb{N}$ such that $f^m(x)$ is a fixed point of f.*

Proof Let us fix an arbitrary element $x \in X$. We consider the ascending sequence $(x_n)_{n\in\mathbb{N}}$ which has the first term $x_0 = x$ and the general term $x_{n+1} = f(x_n)$ for all $n \in \mathbb{N}$. We prove by induction that $supp(x_n) \subseteq supp(f) \cup supp(x)$ for all $n \in \mathbb{N}$. Clearly, $supp(x_0) = supp(x) \subseteq supp(f) \cup supp(x)$. Assume that $supp(x_k) \subseteq supp(f) \cup supp(x)$. Let $\pi \in Fix(supp(f) \cup supp(x))$. Thus, $\pi \cdot x_k = x_k$ according to the inductive hypothesis. According to Proposition 2.9, because π fixes $supp(f)$ pointwise and $supp(f)$ supports f, we get $\pi \cdot x_{k+1} = \pi \cdot f(x_k) = f(\pi \cdot x_k) = f(x_k) = x_{k+1}$. Since $supp(x_{k+1})$ is the least set supporting x_{k+1}, we obtain $supp(x_{k+1}) \subseteq supp(f) \cup supp(x)$. Thus, $(x_n)_{n\in\mathbb{N}} \subseteq X$ is uniformly supported by $supp(f) \cup supp(x)$, and so $(x_n)_{n\in\mathbb{N}}$ must be finite. Since, by hypothesis we have $x_0 \sqsubseteq x_1 \sqsubseteq \ldots \sqsubseteq x_n \sqsubseteq \ldots$, there should exist $m \in \mathbb{N}$ such that $x_m = x_{m+1}$, i.e. $f^m(x) = f^{m+1}(x) = f(f^m(x))$, and so the result follows. □

Corollary 5.4 *Let $f : \wp_{fin}(A) \to \wp_{fin}(A)$ be a finitely supported function with the property that $X \subseteq f(X)$ for all $X \in \wp_{fin}(A)$. Then there exists $X \in \wp_{fin}(A)$ such that $f(X) = X$.*

Proof The result follows from Theorem 5.7 since $\wp_{fin}(A)$ does not contain an infinite uniformly supported subset (the only elements in $\wp_{fin}(A)$ supported by a certain finite set S are the subsets of S). □

We can even prove a more general result than Corollary 5.4.

Proposition 5.5 *Let $f : \wp_{fin}(A) \to \wp_{fin}(A)$ be a finitely supported function with the property that $X \subseteq f(X)$ for all $X \in \wp_{fin}(A)$. There are infinitely many fixed points of f, namely the finite subsets of A containing all the elements of $supp(f)$.*

Proof Let $X \in \wp_{fin}(A)$. Since the support of a finite subset of atoms coincides with the related subset, we have $supp(X) = X$ and $supp(f(X)) = f(X)$. According to Proposition 2.9, for any permutation $\pi \in Fix(supp(f) \cup supp(X)) = Fix(supp(f) \cup X)$ we have $\pi \star f(X) = f(\pi \star X) = f(X)$ which means $supp(f) \cup X$ supports $f(X)$, that is $f(X) = supp(f(X)) \subseteq supp(f) \cup X$ (claim 1). Since we also have $X \subseteq f(X)$, we get $X \setminus supp(f) \subseteq f(X) \setminus supp(f) \subseteq X \setminus supp(f)$, that is $X \setminus supp(f) = f(X) \setminus supp(f)$ (claim 2). If $supp(f) = \emptyset$, the result follows obviously. Let $supp(f) = \{a_1, \ldots, a_n\}$. According to (claim 1) we have $supp(f) \subseteq f(supp(f)) \subseteq supp(f)$, and so $f(supp(f)) = supp(f)$. If X has the form $X = \{a_1, \ldots, a_n, b_1, \ldots, b_m\}$ with $b_1, \ldots, b_m \in A \setminus supp(f)$, $m \geq 1$, we should have by hypothesis that $a_1, \ldots, a_n \in f(X)$, and by (claim 2) $f(X) \setminus supp(f) = \{b_1, \ldots, b_m\}$. Since no other elements different from $a_1, \ldots, a_n$ are in $supp(f)$, from (claim 1) we get $f(X) = \{a_1, \ldots, a_n, b_1, \ldots, b_m\}$. □

Definition 5.4 1. Let $(X, \sqsubseteq, \cdot)$ be an invariant poset. A *finitely supported (countable) sequence* in X is a (countable) sequence $(x_n)_{n\in\mathbb{N}} \subseteq X$ such that the mapping $n \mapsto x_n$ is a finitely supported function from $\mathbb{N}$ to X.

2. Let $(X, \sqsubseteq, \cdot)$ be an invariant poset. A *finitely supported countable chain* in X is a totally ordered sequence $(x_n)_{n\in\mathbb{N}} \in X$ such that the mapping $n \mapsto x_n$ is a finitely supported function from $\mathbb{N}$ to X.
3. Let $(X, \sqsubseteq_X, \cdot_X)$ and $(Y, \sqsubseteq_Y, \cdot_Y)$ be two invariant posets. A finitely supported function $f : X \to Y$ is *s-continuous* if and only if for each countable finitely supported chain $(x_n)_{n\in\mathbb{N}}$ in X which has a least upper bound $\underset{n\in\mathbb{N}}{\sqcup} x_n$, we have that $f((x_n)_{n\in\mathbb{N}})$ has a least upper bound in Y and $f(\underset{n\in\mathbb{N}}{\sqcup} x_n) = \underset{n\in\mathbb{N}}{\sqcup}(f(x_n))$.

Proposition 5.6 *Let $(X, \sqsubseteq, \cdot)$ be an invariant poset. A sequence $(x_n)_{n\in\mathbb{N}} \in X$ is finitely supported if and only if there exists a finite set $S \subset A$ such that $supp(x_n) \subseteq S$ for all $n \in \mathbb{N}$.*

Proof A sequence $(x_n)_{n\in\mathbb{N}} \in X$ is finitely supported if and only if the mapping $f : \mathbb{N} \to X$ with $f(n) = x_n$ for all $n \in \mathbb{N}$ is finitely supported. If there exists such a finitely supported mapping, then according to Proposition 2.9, it follows that for any $\pi \in Fix(supp(f))$ we have $\pi \cdot x_n = \pi \cdot f(n) = f(\pi \diamond n) = f(n) = x_n$ for all $n \in \mathbb{N}$, where $\diamond$ represents the trivial S_A-action on $\mathbb{N}$. Thus, x_n is supported by $supp(f)$ for all $n \in \mathbb{N}$. Conversely, if there exists a finite set $S \subset A$ such that $supp(x_n) \subseteq S$, by Proposition 2.9 we have that the function $f : \mathbb{N} \to X$ defined by $f(n) = x_n$ for all $n \in \mathbb{N}$ is supported by S. $\square$

Theorem 5.8 (Tarski-Kantorovitch Theorem)

Let $(X, \sqsubseteq, \cdot)$ be an invariant poset and $f : X \to X$ a finitely supported s-continuous function. Assume that there exists $x_0 \in X$ having the following properties:

- *$x_0 \sqsubseteq f(x_0)$;*
- *every finitely supported countable chain in the set $X' = \{x \in X \mid x_0 \sqsubseteq x\}$ has a least upper bound in X.*

Then f possesses a fixed point $fp(f)$, defined in a constructive manner, with the property that $supp(fp(f)) \subseteq supp(f) \cup supp(x_0)$.

Proof Firstly, we prove that $X' = \{x \in X \mid x_0 \sqsubseteq x\}$ is a finitely supported subset of X, and so its construction is consistent with FSM. Let $\pi \in Fix(supp(x_0))$. Then $\pi \cdot x_0 = x_0$ because $supp(x_0)$ supports x_0. Since $\sqsubseteq$ is an equivariant order relation on X, for each $x \in X'$ we have $x_0 \sqsubseteq x$, and so $x_0 = \pi \cdot x_0 \sqsubseteq \pi \cdot x$, which means $\pi \cdot x \in X'$. Thus, $\pi \star X' = X'$ for all $\pi \in Fix(supp(x_0))$ (where by $\star$ we denoted the S_A-action on $\wp(X)$), and so $supp(X') \subseteq supp(x_0)$.

In follows directly that any s-continuous function f is monotone. Since $x_0 \sqsubseteq f(x_0)$ and f is monotone, we have that $(f^n(x_0))_{n\in\mathbb{N}}$ is an ascending chain, where $f^n(x_0) = f(f^{n-1}(x_0))$ and $f^0(x_0) = x_0$. We prove that $(f^n(x_0))_{n\in\mathbb{N}}$ is finitely supported, that there exists a finite $S \subset A$ such that $supp(f^n(x_0)) \subseteq S$ for each $n \in \mathbb{N}$. We claim that $S = supp(f) \cup supp(x_0)$, and we prove by induction on n that $supp(f^n(x_0)) \subseteq supp(f) \cup supp(x_0)$ for each $n \in \mathbb{N}$.

Obviously, $supp(f^0(x_0)) = supp(x_0) \subseteq supp(f) \cup supp(x_0)$. Let us assume that $supp(f^n(x_0)) \subseteq supp(f) \cup supp(x_0)$ for some $n \in \mathbb{N}$. We have to prove the relation $supp(f^{n+1}(x_0)) \subseteq supp(f) \cup supp(x_0)$. So, we have to prove that each permutation of atoms π which fixes $supp(f) \cup supp(x_0)$ pointwise also fixes $f^{n+1}(x_0)$. Let π be such a permutation of atoms, i.e. $\pi \in Fix(supp(f) \cup supp(x_0))$. According to the inductive hypothesis, we have that $\pi \in Fix(supp(f^n(x_0)))$, and so $\pi \cdot f^n(x_0) = f^n(x_0)$ because $supp(f^n(x_0))$ supports $f^n(x_0)$. According to Proposition 2.9, because π fixes $supp(f)$ pointwise and $supp(f)$ supports f, we have $\pi \cdot f^{n+1}(x_0) = \pi \cdot f(f^n(x_0)) = f(\pi \cdot f^n(x_0)) = f(f^n(x_0)) = f^{n+1}(x_0)$. Thus, $(f^n(x_0))_{n\in\mathbb{N}}$ is uniformly supported by $supp(f) \cup supp(x_0)$. Furthermore, $(f^n(x_0))_{n\in\mathbb{N}}$ is contained in $\{x \in X \mid x_0 \sqsubseteq x\}$ because $x_0 \sqsubseteq f(x_0)$ and f is monotone. Thus, there exists $\bigsqcup_{n\in\mathbb{N}} f^n(x_0)$.

We prove that $supp(\bigsqcup_{n\in\mathbb{N}} f^n(x_0)) \subseteq supp(f) \cup supp(x_0)$. We have to prove that any permutation of atoms π fixing $supp(f) \cup supp(x_0)$ pointwise, also fixes $\bigsqcup_{n\in\mathbb{N}} f^n(x_0)$. Let us fix a permutation of atoms $\pi \in Fix(supp(f) \cup supp(x_0))$. We have that $\pi \cdot f^n(x_0) = f^n(x_0)$ for all $n \in \mathbb{N}$. Hence $\bigsqcup_{n\in\mathbb{N}} \pi \cdot f^n(x_0) = \bigsqcup_{n\in\mathbb{N}} f^n(x_0)$ (1). Due to the equivariance of $\sqsubseteq$ we have that $\bigsqcup_{n\in\mathbb{N}} \sigma \cdot d_n \sqsubseteq \sigma \cdot \bigsqcup_{n\in\mathbb{N}} d_n$ (2) whenever $\sigma \in Fix(supp(f) \cup supp(x_0))$ and $(d_n)_n$ is a finitely supported countable chain in X' for which $(\sigma \cdot d_n)_n$ is also a finitely supported countable chain in X'. Indeed, for such a σ and such a $(d_n)_n$ we have $d_n \sqsubseteq \bigsqcup_{n\in\mathbb{N}} d_n$ for all $n \in \mathbb{N}$, and so $\sigma \cdot d_n \sqsubseteq \sigma \cdot \bigsqcup_{n\in\mathbb{N}} d_n$ (3) for all $n \in \mathbb{N}$. Relation (2) follows by taking the supremum of $(\sigma \cdot d_n)_n$ in relation (3). By applying (2), firstly for $\sigma = \pi$ and $(d_n)_n = (f^n(x_0))_n$, and secondly for $\sigma = \pi^{-1} \in Fix(supp(f) \cup supp(x_0))$ and $(d_n)_n = (f^n(x_0))_n = (\pi \cdot f^n(x_0))_n$, we obtain $\bigsqcup_{n\in\mathbb{N}} \pi \cdot f^n(x_0) = \pi \cdot \bigsqcup_{n\in\mathbb{N}} f^n(x_0)$, which means π fixes $\bigsqcup_{n\in\mathbb{N}} f^n(x_0)$ according to relation (1).

It remains to prove that $\bigsqcup_{n\in\mathbb{N}} f^n(x_0)$ is a fixed point of f. This is easy, because from the s-continuity of f we get $f(\bigsqcup_{n\in\mathbb{N}} f^n(x_0)) = \bigsqcup_{n\in\mathbb{N}} f^{n+1}(x_0) = \bigsqcup_{n\in\mathbb{N}} f^n(x_0)$. □

Definition 5.5 • An *invariant cpo* X is an invariant poset with the additional condition that any finitely supported countable chain $(x_n)_{n\in\mathbb{N}}$ in D has a least upper bound denoted by $\bigsqcup_{n\in\mathbb{N}} x_n$.

- An *invariant cppo* X is an invariant cpo with a least element $\perp$.

From (the proof of) Theorem 5.8 we obtain the following result.

Corollary 5.5 (Scott Theorem) *Let $(X, \sqsubseteq, \cdot, \perp)$ be an invariant cppo. Then each finitely supported s-continuous function $f : X \to X$ possesses a least fixed point $lfp(f)$ which has the property that $supp(lfp(f)) \subseteq supp(f)$.*

Theorem 5.9 *Let $(X, \sqsubseteq, \cdot, \perp)$ be an invariant partially ordered set, with a least element $\perp$, containing no infinite uniformly supported subset. Each finitely supported monotone function $f : X \to X$ possesses a least fixed point $lfp(f)$ which has the property that $supp(lfp(f)) \subseteq supp(f)$.*

Proof Since $\bot \sqsubseteq f(\bot)$ and f is monotone, we have that $(f^n(\bot))_{n\in\mathbb{N}}$ is an ascending chain. By definition, we have $\bot \sqsubseteq \pi \cdot \bot$ and $\bot \sqsubseteq \pi^{-1} \cdot \bot$ for each $\pi \in S_A$, which means $\bot = \pi \cdot \bot$ and $supp(\bot) = \emptyset$. By induction, we prove that $supp(f^n(\bot)) \subseteq supp(f)$ for all $n \in \mathbb{N}$. Clearly, $supp(f^0(\bot)) = supp(\bot) \subseteq supp(f)$. Let us suppose that $supp(f^k(\bot)) \subseteq supp(f)$ for some $k \in \mathbb{N}$. Now we prove that $supp(f^{k+1}(\bot)) \subseteq supp(f)$; for this we have to prove that each permutation of atoms π which fixes $supp(f)$ pointwise also fixes $f^{k+1}(\bot)$. Let $\pi \in Fix(supp(f))$. From the inductive hypothesis, we have that $supp(f^k(\bot)) \subseteq supp(f)$, and so $\pi \cdot f^k(\bot) = f^k(\bot)$. According to Proposition 2.9, we have $\pi \cdot f^{k+1}(\bot) = \pi \cdot f(f^k(\bot)) = f(\pi \cdot f^k(\bot)) = f(f^k(\bot)) = f^{k+1}(\bot)$. Therefore, $(f^n(\bot))_{n\in\mathbb{N}}$ is uniformly supported by $supp(f)$. Thus, $(f^n(\bot))_{n\in\mathbb{N}}$ has to be finite according to theorem's hypothesis. Since it is an ascending chain it follows that there exists $n_0 \in \mathbb{N}$ such that $f^n(\bot) = f^{n_0}(\bot)$ for all $n \geq n_0$. Thus, $f(f^{n_0}(\bot)) = f^{n_0+1}(\bot) = f^{n_0}(\bot)$, and so $f^{n_0}(\bot)$ is a fixed point of f, and furthermore, it is supported by $supp(f)$. If x is another fixed point of f, it follows from the monotony of f and from the relation $\bot \sqsubseteq x$ that $f^{n_0}(\bot) = \bigsqcup_{n\in\mathbb{N}} f^n(\bot) \sqsubseteq x$, and so $f^{n_0}(\bot) = lfp(f)$. □

For any fixed finite set $S \subseteq A$, there are only finitely many subsets of A supported by S, namely the subsets of S and the supersets of $A \setminus S$; therefore $\wp_{fs}(A)$ does not have an infinite uniformly supported subset. Similarly, $\wp_{fin}(A)$ does not have an infinite uniformly supported subset. According to Proposition 2.5 which states that $supp(Y) = \bigcup_{y\in Y} supp(y)$ whenever Y is a finite subset of an invariant set X, and because $\wp_{fs}(A)$ does not have an infinite uniformly supported subset, we obtain that $\wp_{fin}(\wp_{fs}(A))$ does not have an infinite uniformly supported subset. The following result (which is specific to FSM and does not have a ZF correspondent) follows from Theorem 5.9.

Corollary 5.6 *1. Let $f : \wp_{fin}(A) \to \wp_{fin}(A)$ be a finitely supported monotone function. Then there exists a least $X_0 \in \wp_{fin}(A)$ supported by $supp(f)$ such that $f(X_0) = X_0$.*

2. Let $f : \wp_{fin}(\wp_{fs}(A)) \to \wp_{fin}(\wp_{fs}(A))$ be a finitely supported monotone function. Then there exists a least $X_0 \in \wp_{fin}(\wp_{fs}(A))$ supported by $supp(f)$ such that $f(X_0) = X_0$.

Chapter 6
Lattices in Finitely Supported Mathematics

Abstract We introduce and study lattices in the framework of finitely supported structures. Various properties of lattices are obtained by extending the classical Zermelo-Fraenkel results from the world of non-atomic structures to the world of atomic finitely supported structures. We particularly prove that Tarski fixed point theorem for Zermelo-Fraenkel complete lattices remains valid in the world of finitely supported structures, and we present some calculability properties for specific fixed points of finitely supported monotone self-mappings defined on finitely supported complete lattices. Such results can be applied for the particular finitely supported complete lattices constructed in the next chapter. FSM Tarski fixed point theorem can also be applied in an FSM theory of abstract interpretation to prove the existence of least fixed points for specific finitely supported mappings (defined on invariant complete lattices of properties) modelling the transitions between properties of programming languages. This chapter is based on [14].

6.1 Basic Properties

Definition 6.1 An *invariant lattice* is an invariant set $(L, \cdot)$ together with an equivariant lattice order $\sqsubseteq$ on L.

From the general theory of lattices, we know that if $\sqsubseteq$ is a lattice order on L we can define two binary operations called *meet* and *join*:

- $\wedge : L \times L \to L$, $(x,y) \overset{\wedge}{\mapsto} z$, where $z = inf\{x,y\}$
- $\vee : L \times L \to L$, $(x,y) \overset{\vee}{\mapsto} z$, where $z = sup\{x,y\}$

such that $(L, \wedge, \vee)$ is a lattice (which means $\wedge$ and $\vee$ satisfy the properties of commutativity, associativity and absorption). According to the S-finite support principle, these binary operations make sense in FSM.

Proposition 6.1 *Let* $(L, \sqsubseteq, \cdot)$ *be an invariant lattice.*
Then the operators meet and join are equivariant functions on $L \times L$.

A. Alexandru, G. Ciobanu, *Foundations of Finitely Supported Structures*,
https://doi.org/10.1007/978-3-030-52962-8_6

Proof This result is a direct consequence of the equivariance principle. However, for readers not familiarized with higher-order logic we provide below the full calculation.

We know that, $L \times L$ is an invariant set with the S_A-action $\otimes : S_A \times (L \times L) \to (L \times L)$ defined by $\pi \otimes (x,y) = (\pi \cdot x, \pi \cdot y)$ for all $\pi \in S_A$ and all $x,y \in L$. According to Proposition 2.9, we have to prove that $\pi \cdot (x \wedge y) = (\pi \cdot x) \wedge (\pi \cdot y)$ and $\pi \cdot (x \vee y) = (\pi \cdot x) \vee (\pi \cdot y)$ for all $\pi \in S_A$ and all $x,y \in L$. We prove one of these assertions; the other one is similar. Let $x,y \in L$ and $\pi \in S_A$. Let $c = x \wedge y$. Then $c \sqsubseteq x$ and $c \sqsubseteq y$. Since $\sqsubseteq$ is equivariant, we have $\pi \cdot c \sqsubseteq \pi \cdot x$ and $\pi \cdot c \sqsubseteq \pi \cdot y$. Let $d \in L$ such that $d \sqsubseteq \pi \cdot x$ and $d \sqsubseteq \pi \cdot y$. Then $\pi^{-1} \cdot d \sqsubseteq x$ and $\pi^{-1} \cdot d \sqsubseteq y$. Since $c = inf\{x,y\}$, we have $\pi^{-1} \cdot d \sqsubseteq c$. Since $\sqsubseteq$ is equivariant, we obtain $d \sqsubseteq \pi \cdot c$. Therefore, $\pi \cdot c = inf\{\pi \cdot x, \pi \cdot y\} = (\pi \cdot x) \wedge (\pi \cdot y)$. □

Definition 6.2 An *invariant complete lattice* is an invariant partially ordered set $(L, \sqsubseteq, \cdot)$ such that every finitely supported subset $X \subseteq L$ has a least upper bound with respect to the order relation $\sqsubseteq$. The least upper bound of X is denoted by $\sqcup X$.

Theorem 6.1 *Let $(L, \sqsubseteq, \cdot)$ be an invariant complete lattice. Then every finitely-supported subset $X \subseteq L$ has a greatest lower bound with respect to the order relation $\sqsubseteq$. The greatest lower bound of X is denoted by $\sqcap X$.*

Proof Let X be a finitely supported subset of L. Let $D = \cap\{\downarrow x \,|\, x \in X\}$, where by $\downarrow x$ we denote the set $\{y \in L \,|\, y \sqsubseteq x\}$. Informally, D is the set of lower bounds of X with respect to the order relation $\sqsubseteq$. If X is empty, we take $D = L$. Firstly, we show that D is finitely supported. If we prove this, D has a least upper bound (denoted by $\sqcup D$). We know that X has a finite support $supp(X)$, and we show that $supp(X)$ supports D. Let π be a permutation of atoms that fixes $supp(X)$ pointwise, i.e. $\pi \in Fix(supp(X))$. Let $d \in D$ be arbitrarily chosen; then $d \sqsubseteq x$ for all $x \in X$. We claim that $\pi \cdot d \in D$, which is the same as saying that $\pi \cdot d$ is a lower bound of X. Indeed, let $y \in X$ be an arbitrary element of X. Since $\pi \in Fix(supp(X))$ and $supp(X)$ supports X, we get $\pi \star X = X$ (where the S_A-action $\star$ on $\wp(L)$ is defined as in Proposition 2.2). This means that for our $y \in X$ there exists $x \in X$ such that $\pi \cdot x = y$. However, $d \sqsubseteq x$, and because $\sqsubseteq$ is equivariant we also have $\pi \cdot d \sqsubseteq \pi \cdot x = y$. Hence $\pi \cdot d \in D$. Since d is chosen arbitrarily from D, we can say that $\pi \star D \subseteq D$ whenever $\pi \in Fix(supp(X))$. Let us suppose that there is $\pi \in Fix(supp(X))$ such that $\pi \star D \subsetneq D$. By induction, we get $\pi^n \star D \subsetneq D$ for all $n \geq 1$. However, π is a finitary bijection of atoms, and so there is $k \in \mathbb{N}$ such that $\pi^k = Id$. We obtain $D \subsetneq D$, a contradiction. It follows that $\pi \star D = D$ whenever $\pi \in Fix(supp(X))$, and hence $supp(X)$ supports D. Thus, there exists $\sqcup D$.

We prove now that $\sqcup D$ is the greatest lower bound of X. If $x \in X$, then x is an upper bound of D, and so $\sqcup D \sqsubseteq x$. Since x was chosen arbitrarily from X, we have $\sqcup D \in D$. Since $\sqcup D$ is maximal between the lower bounds of X and it is a lower bound of X, then $\sqcup D = \sqcap X$, where $\sqcap X$ represents the greatest lower bound of X. □

Proposition 6.2 *Let $(L, \sqsubseteq, \cdot)$ be an invariant complete lattice and X a finitely supported subset of L. Then $\pi \cdot \sqcup X = \sqcup(\pi \star X)$ for all $\pi \in S_A$, where $\star$ is the S_A-action on the powerset of L. Similarly, $\pi \cdot \sqcap X = \sqcap(\pi \star X)$ for all $\pi \in S_A$.*

Proof This result is a direct consequence of the equivariance principle. However, for readers not familiarized with higher-order logic we provide below the full calculation.

Let X be a finitely supported subset of L. Since L is an invariant complete lattice, $\sqcup X$ exists. According to Proposition 2.1, we have that $\pi \star X$ is finitely supported, and there exists $\sqcup(\pi \star X)$. Let $x \in X$. We have $x \sqsubseteq \sqcup X$. Since $\sqsubseteq$ is equivariant, we have $\pi \cdot x \sqsubseteq \pi \cdot \sqcup X$ for all $\pi \in S_A$. Therefore, $\sqcup(\sigma \star Y) \sqsubseteq \sigma \cdot \sqcup Y$ for all $\sigma \in S_A$ and all finitely supported subsets Y of L (1). Now, fix some $\pi \in S_A$ and $X \in \wp_{fs}(L)$. According to (1), we have $\sqcup(\pi \star X) \sqsubseteq \pi \cdot \sqcup X$. We can also apply (1) for π^{-1} and $\pi \star X$, and we obtain $\sqcup X \sqsubseteq \pi^{-1} \cdot \sqcup(\pi \star X)$. Since $\sqsubseteq$ is equivariant, we obtain $\pi \cdot \sqcup X \sqsubseteq \sqcup(\pi \star X)$. Whenever X is a finitely supported subset of L, there exists $\sqcap X$. We can provide a similar proof for the relation $\pi \cdot \sqcap X = \sqcap(\pi \star X)$ for all $\pi \in S_A$. □

Corollary 6.1 *Let $(L, \sqsubseteq, \cdot)$ be an invariant complete lattice and X a finitely supported subset of L. Then $\sqcup X$ and $\sqcap X$ are both supported by $supp(X)$.*

Proof Let $\pi \in Fix(supp(X))$. Then $\pi \star X = X$. According to Proposition 6.2 we have $\pi \cdot \sqcup X = \sqcup(\pi \star X) = \sqcup X$ and $\pi \cdot \sqcap X = \sqcap(\pi \star X) = \sqcap X$. Therefore, the result follows from Definition 2.2. □

Definition 6.3 An *invariant modular lattice* is an invariant lattice $(L, \sqsubseteq, \cdot)$ that satisfies the classical ZF modularity law: $x \sqsubseteq z$ implies $x \vee (y \wedge z) = (x \vee y) \wedge z$ for all $x, y, z \in L$.

Definition 6.4 An invariant lattice $(L, \sqsubseteq, \cdot)$ is called an *invariant Boolean lattice* if the following conditions are satisfied:

- L is a distributive lattice;
- L is bounded by a unique least element denoted by 0 and a unique greatest element denoted by 1;
- L is uniquely complemented, that is for each element $x \in L$ there exists a unique element $x' \in L$ such that $x \wedge x' = 0$ and $x \vee x' = 1$.

Proposition 6.3 *Let $(L, \sqsubseteq, \cdot)$ be an invariant Boolean lattice. Then $(\pi \cdot x)' = \pi \cdot x'$ for each $\pi \in S_A$ and each $x \in L$. Moreover, $supp(x') = supp(x)$.*

Proof Since $\sqsubseteq$ is equivariant and $0, 1$ are unique, we have $\pi \cdot 0 = 0$ and $\pi \cdot 1$=1. Let $x \in L$. According to Proposition 6.1, we have $(\pi \cdot x) \wedge (\pi \cdot x') = \pi \cdot (x \wedge x') = \pi \cdot 0 = 0$ and $(\pi \cdot x) \vee (\pi \cdot x') = \pi \cdot (x \vee x') = \pi \cdot 1 = 1$. Therefore, since the complement of each element in L is unique, we have $(\pi \cdot x)' = \pi \cdot x'$, and $supp(x') = supp(x)$. □

Proposition 6.4 *Let $(L, \sqsubseteq, \cdot)$ be an invariant complete Boolean lattice and S a finitely supported subset of L. Then $(\sqcap S)' = (\sqcup S')$. Similarly, $(\sqcup S)' = (\sqcap S')$.*

Proof According to Proposition 6.3, whenever S is a finitely supported subset of L we have that $S' = \{x' \mid x \in S\}$ is also a finitely supported subset of L with the same support as S. Therefore, the statement of the property makes sense. We prove that $(\sqcap S) \wedge (\sqcup S') = 0$. Let $d \in L$ such that $d \sqsubseteq \sqcap S$ and $d \sqsubseteq \sqcup S'$. Then $d \sqsubseteq s$ for all $s \in S$, and so $s' \sqsubseteq d'$ for all $s \in S$. Therefore, $\sqcup S' \sqsubseteq d'$. Since $d \sqsubseteq \sqcup S'$, we obtain $d \sqsubseteq d'$, that is $d = 0$. Similarly, we prove that $(\sqcup S') \vee (\sqcap S) = 1$. Let $e \in L$ such that $\sqcup S' \sqsubseteq e$ and $\sqcap S \sqsubseteq e$. Then $s' \sqsubseteq e$ for all $s \in S$, and so $e' \sqsubseteq s$ for all $s \in S$. Therefore, $e' \sqsubseteq \sqcap S$. Since $\sqcap S \sqsubseteq e$ we get $e' \sqsubseteq e$, and it follows that $e = 1$. □

Definition 6.5 Let $(P, \sqsubseteq_P, \cdot_P)$ and $(Q, \sqsubseteq_Q, \cdot_Q)$ be two invariant posets. Let $f : P \to Q$.

- f is a *finitely supported join-morphism* if f is finitely supported, and whenever $a, b \in P$ and $a \vee b$ exists in P, then $f(a) \vee f(b)$ exists in Q and $f(a \vee b) = f(a) \vee f(b)$.
- f is an *finitely supported complete join-morphism* if f is finitely supported, and whenever $X \subseteq P$ is finitely supported and $\sqcup X$ exists in P, then $f(X)$ is finitely supported, $\sqcup f(X)$ exists in Q and $f(\sqcup X) = \sqcup f(X)$.

Invariant (complete) meet-morphisms are defined dually.

6.2 Fixed Points Results

Theorem 6.2 (Strong Tarski Theorem)

Let $(L, \sqsubseteq, \cdot)$ be an invariant complete lattice and $f : L \to L$ a finitely supported monotone function over L. Let P be the set of fixed points of f. Then $(P, \sqsubseteq, \cdot)$ is a non-empty, finitely supported (by $supp(f)$) complete sublattice of L.

Proof We prove that P is finitely supported by $supp(f)$. Since f is finitely supported by $supp(f)$, by Proposition 2.9 it follows that for all $\pi \in Fix(supp(f))$ and all $x \in L$ we have $f(\pi \cdot x) = \pi \cdot f(x)$. Thus, for $\pi \in Fix(supp(f))$, whenever x is a fixed point of f, we have $f(\pi \cdot x) = \pi \cdot f(x) = \pi \cdot x$, and so $\pi \cdot x$ is also a fixed point of f. Thus, $\pi \star P = P$ for all $\pi \in Fix(supp(f))$ (where $\star$ is the S_A-action on $\wp(L)$ defined as in Proposition 2.2), which means that $(P, \cdot|_P)$ is a finitely supported subset (supported by $supp(f)$) of the set $(L, \cdot)$.

We prove that P is non-empty. Let $D = \{d \in L \mid d \sqsubseteq f(d)\}$. Firstly, we remark that D is non-empty. This is obvious because at least $\perp$ (the least element of L) belongs to D. We prove that D is finitely supported. We claim that $supp(f)$ supports D. Let $\pi \in Fix(supp(f))$ and $d \in D$ be arbitrarily chosen. Then $d \sqsubseteq f(d)$ (or equivalently $(d, f(d)) \in \sqsubseteq$), and because $\sqsubseteq$ is equivariant we also have $\pi \otimes (d, f(d)) \in \sqsubseteq$, that is $\pi \cdot d \sqsubseteq \pi \cdot f(d)$, where $\otimes$ represents the canonical action on $D \times D$ defined as in Proposition 2.2. Since $\pi \in Fix(supp(f))$ and $supp(f)$ supports f, according to Proposition 2.9, we have $\pi \cdot d \sqsubseteq \pi \cdot f(d) = f(\pi \cdot d)$, and so $\pi \cdot d \in D$. Since d was chosen arbitrarily from D, we have $\pi \star D \subseteq D$. We prove by contradiction that $\pi \star D = D$. Let us suppose that $\pi \star D \subsetneq D$. By induction, we get $\pi^n \star D \subsetneq D$ for all

$n \geq 1$. However, π is a finitary bijection of atoms, and so there exists $k \in \mathbb{N}$ such that $\pi^k = Id$. We obtain $D \subsetneq D$, a contradiction. It follows that $\pi \star D = D$ whenever $\pi \in Fix(supp(f))$, and so $supp(f)$ supports D. Thus, D is finitely supported, and there exists the least upper bound of D, $d_0 = \sqcup D$. For each $d \in D$ we have $d \sqsubseteq d_0$. Since f preserves the order relation, we have $f(d) \sqsubseteq f(d_0)$. Since $d \in D$, it follows that $d \sqsubseteq f(d) \sqsubseteq f(d_0)$. Therefore, $d \sqsubseteq f(d_0)$ for each $d \in D$. According to the definition of a least upper bound, we have that $d_0 \sqsubseteq f(d_0)$, which means that $d_0 \in D$. However, because f preserves order, we have $f(x) \in D$ for each $x \in D$. Since $d_0 \in D$, it follows that $f(d_0) \in D$. Thus, $f(d_0) \sqsubseteq d_0$ because $d_0 = \sqcup D$. Therefore, we get $f(d_0) = d_0$, and so P is non-empty.

We prove now that P is a complete sublattice of L. We have to prove that any subset of P which is finitely supported as a subset of the invariant set L has a least upper bound in P. Let X be an arbitrary finitely supported subset of L that is contained in the finitely supported set P. We have to prove that X has a least upper bound in P. We already know that X has a least upper bound (denoted by $\sqcup X$) in L because $(L, \sqsubseteq, \cdot)$ is an invariant complete lattice.

Let $x \in X$ be an arbitrary element. We have that $x \sqsubseteq \sqcup X$, and so $f(x) \sqsubseteq f(\sqcup X)$. However, X contains only fixed points of f, and so $f(x) = x$ and $x \sqsubseteq f(\sqcup X)$. According to the definition of a least upper bound, it follows that $\sqcup X \sqsubseteq f(\sqcup X)$. Now, let $y \sqsupseteq \sqcup X$. Since f is a monotone function, we also have $f(y) \sqsupseteq f(\sqcup X)$. We have already proved that $\sqcup X \sqsubseteq f(\sqcup X)$, and hence $f(y) \sqsupseteq \sqcup X$. We get that $f(y) \sqsupseteq \sqcup X$ whenever $y \sqsupseteq \sqcup X$.

Let $D' = \{d \in L \,|\, f(d) \sqsubseteq d \text{ and } \sqcup X \sqsubseteq d\}$. We prove that $supp(f) \cup supp(\sqcup X)$ supports D', and so D' is a finitely supported set. We know that $supp(\sqcup X)$ exists because L is an invariant set and $\sqcup X \in L$. Let us consider $\pi \in Fix(supp(f) \cup supp(\sqcup X))$, and $d \in D'$ be arbitrarily chosen. Then $f(d) \sqsubseteq d$. Since $\sqsubseteq$ is equivariant, we also have $\pi \cdot f(d) \sqsubseteq \pi \cdot d$. Moreover, by Proposition 2.9, because f is supported by $supp(f)$ and π fixes $supp(f)$ pointwise, we have $\pi \cdot f(d) = f(\pi \cdot d)$. Thus, $f(\pi \cdot d) \sqsubseteq \pi \cdot d$. Since $d \in D'$, we also have $\sqcup X \sqsubseteq d$. Therefore, $\pi \cdot \sqcup X \sqsubseteq \pi \cdot d$. However, $\pi \cdot \sqcup X = \sqcup X$ because $\pi \in Fix(supp(\sqcup X))$. Finally, we obtain $\sqcup X \sqsubseteq \pi \cdot d$, and so $\pi \cdot d \in D'$, which means $\pi \star D' \subseteq D'$. Since each permutation of A is finitary, as above we obtain $\pi \star D' = D'$. Thus, $supp(f) \cup supp(\sqcup X)$ supports D', and so there exists the greatest lower bound of D' denoted by $e = \sqcap D'$. Then for each $d \in D'$, we have $e \sqsubseteq d$. Since f preserves the order relation, we also have $f(e) \sqsubseteq f(d)$. Since $d \in D'$, it follows that $f(e) \sqsubseteq f(d) \sqsubseteq d$. Therefore, $f(e) \sqsubseteq d$ for each $d \in D'$. According to the definition of a greatest lower bound, we have that $f(e) \sqsubseteq e$. Furthermore, $d \sqsupseteq \sqcup X$ for each $d \in D'$ implies $\sqcap D' \sqsupseteq \sqcup X$, which means $e \in D'$. However, because f is monotone and because $f(y) \sqsupseteq \sqcup X$ whenever $y \sqsupseteq \sqcup X$, we have that $f(x) \in D'$ for each $x \in D'$. Since $e \in D'$, it follows that $f(e) \in D'$, and so $e \sqsubseteq f(e)$ because $e = \sqcap D'$.

We proved that e is a fixed point of f such that $\sqcup X \sqsubseteq e$. Therefore, $e \in P$ is an upper bound for X. What remains to be proved is that e is the least upper bound for X in $(P, \sqsubseteq)$. Let $e' \in P$ be another upper bound for X. Then $\sqcup X \sqsubseteq e'$ (since $\sqcup X$ is the least upper bound for X in L, and clearly e' is an upper bound for X in L); it follows that $e' \in D'$. Since $e = \sqcap D'$, we get $e \sqsubseteq e'$. This means $e = \sqcup X$ in $(P, \sqsubseteq)$. □

According to the proof of Theorem 6.2 and to Corollary 6.1 we have:

Corollary 6.2 *Let $(L, \sqsubseteq, \cdot)$ be an invariant complete lattice and $f : L \to L$ a finitely supported monotone function over L. Then f has a least fixed point defined as $lfp(f) = \sqcap\{d \in L \mid f(d) \sqsubseteq d\}$ and a greatest fixed point defined as $gfp(f) = \sqcup\{d \in L \mid d \sqsubseteq f(d)\}$, which are both supported by $supp(f)$.*

Corollary 6.3 *Let $(L, \sqsubseteq, \cdot)$ be an invariant complete lattice and $f : L \to L$ an equivariant monotone function over L. Let P be the set of fixed points of f. Then $(P, \sqsubseteq, \cdot)$ is an invariant complete lattice.*

Proof This follows trivially from Theorem 6.2 since equivariant functions are empty supported. Thus, whenever $supp(f) = \emptyset$, we have $supp(P) = \emptyset$, which means P is an equivariant subset of L, i.e. P is itself an invariant set. □

Proposition 6.5 *Let $(P, \sqsubseteq, \cdot)$ be an invariant partially ordered set containing no infinite uniformly supported subset and $f : P \to P$ a finitely supported monotone function over P.*

- *If the set $D = \{d \in L \mid d \sqsubseteq f(d)\}$ is non-empty and totally ordered, then f has greatest fixed point defined as $gfp(f) = \sqcup D$.*
- *If the set $D' = \{d \in L \mid f(d) \sqsubseteq d\}$ is non-empty and totally ordered, then f has least fixed point defined as $lfp(f) = \sqcap D'$.*

In either of the above cases, f has only finitely many fixed points that form either a one-element set or a finitely supported complete lattice.

Proof As in the proof of Theorem 6.2 we have that D is finitely supported by $supp(f)$. If we prove that $\sqcup D$ exists, then it is the greatest fixed point of f. We claim that D is uniformly supported by $supp(f)$. Let $\pi \in Fix(supp(f))$ and let $d \in D$ an arbitrary element. Since π fixes $supp(f)$ pointwise and $supp(f)$ supports D, we obtain that $\pi \cdot d \in D$, and so we should have either $d \sqsubset \pi \cdot d$, or $d = \pi \cdot d$, or $\pi \cdot d \sqsubset d$. If $d \sqsubset \pi \cdot d$, then, because P is an invariant partially ordered set and because the mapping $z \mapsto \pi \cdot z$ is bijective from D to $\pi \star D = D$, we get $d \sqsubset \pi \cdot d \sqsubset \pi^2 \cdot d \sqsubset \ldots \sqsubset \pi^n \cdot d$ for all $n \in \mathbb{N}$. However, since any permutation of atoms has a finite order in the group S_A (because it interchanges only finitely many atoms), there exists $m \in \mathbb{N}$ such that $\pi^m = Id$. This means $\pi^m \cdot d = d$, and so we get $d \sqsubset \pi^m \cdot d = d$ which is a contradiction. Similarly, the assumption $\pi \cdot d \sqsubset d$ leads to the relation $\pi^n \cdot d \sqsubset \ldots \sqsubset \pi \cdot d \sqsubset d$ for all $n \in \mathbb{N}$ which is also a contradiction because π has finite order. Therefore, $\pi \cdot d = d$, and because d was arbitrary chosen form D, we obtain that D should be a uniformly supported subset of P. Thus, D is finite, and since it is totally ordered, we have that $\sqcup D$ exists. Similarly, D' is finite, and so $\sqcap D'$ exists and it is the least fixed point of f. Since any fixed point of f is a member of D and D is finite, we have that f has only finitely many fixed points. If there exist more than one fixed points, since these fixed points are totally ordered, they form a complete lattice which is finitely supported by $supp(f)$. □

Proposition 6.6 *Let $(L, \sqsubseteq, \cdot)$ be an invariant complete lattice containing no infinite uniformly supported subset. Let $f : L \to L$ be a finitely supported monotone function. Then there exist an element $n_0 \in \mathbb{N}$ such that $\sqcap\{x \in L \,|\, f(x) \sqsubseteq x\} = f^{n_0}(0)$ and an element $m_0 \in \mathbb{N}$ such that $\sqcup\{x \in L \,|\, x \sqsubseteq f(x)\} = f^{m_0}(1)$, where $0 = \sqcap L$ and $1 = \sqcup L$.*

Proof According to Theorem 6.2 we have that f has a least fixed point which is equal to $\sqcap\{x \in L \,|\, f(x) \sqsubseteq x\}$. As in the proof of Theorem 5.9, we have that the uniformly supported countable chain $(f^n(0))_{n\in\mathbb{N}}$ should be stationary, and so there exists $n_0 \in \mathbb{N}$ such that $f^n(0) = f^{n_0}(0)$ for all $n \geq n_0$. Therefore, $f^{n_0}(0)$ is the least fixed point of f. Since the least point of f is unique, it follows that $\sqcap\{x \in L \,|\, f(x) \sqsubseteq x\} = f^{n_0}(0)$.

Dually, from Theorem 6.2, f has a greatest fixed point equal to $\sqcup\{x \in L \,|\, x \sqsubseteq f(x)\}$. The descending countable chain $\ldots \sqsubseteq f^m(1) \sqsubseteq f^{m-1}(1) \sqsubseteq \ldots \sqsubseteq f(1) \sqsubseteq 1$ is uniformly supported by $supp(f) \cup supp(1) = supp(f) \cup \emptyset = supp(f)$. Therefore, because L cannot contain an infinite uniformly supported subset, the descending sequence $(f^m(1))_{m\in\mathbb{N}}$ is finite, i.e. there exists $m_0 \in \mathbb{N}$ such that $f^m(1) = f^{m_0}(1)$ for all $m \geq m_0$. Thus, $f^{m_0}(1)$ is a fixed point of f, and furthermore, it is the greatest one because whenever z is another fixed point of f, from $z \sqsubseteq 1$, we get $z = f^{m_0}(z) \sqsubseteq f^{m_0}(1)$. Since the greatest fixed point of f is unique we have $\sqcup\{x \in L \,|\, x \sqsubseteq f(x)\} = f^{m_0}(1)$. □

Widening and narrowing techniques for approximating fixed points of a finitely supported function defined on an invariant complete lattice are presented in Section 3.3.6 from [7].

6.3 A Slight Extension of the Previous Results

The results of this chapter (as well as the results of the previous chapter) can be easily transposed into the framework of (hereditary) finitely supported algebraic structures (instead of invariant algebraic structures). The proving methods are actually the same, but we need to consider the supports of the related finitely supported structures in order to find boundedness properties of the derived constructions. We present below some examples of such results. We think that this extending task is quite simple, and it does not deserve an in-depth study.

Theorem 6.3 *Let $(X, \sqsubseteq, \cdot, 0)$ be a finitely supported partially ordered set with a least element 0, containing no infinite uniformly supported subset. Each finitely supported order preserving function $f : X \to X$ possesses a least fixed point $lfp(f)$ which has the property that $supp(lfp(f)) \subseteq supp(f) \cup supp(X) \cup supp(\sqsubseteq)$.*

Proof Since $0 \sqsubseteq f(0)$ and f is order preserving, we have that $(f^n(0))_{n\in\mathbb{N}}$ is an ascending chain. By definition, we have $0 \sqsubseteq \pi \cdot 0$ and $0 \sqsubseteq \pi^{-1} \cdot 0$ for each $\pi \in Fix(supp(X))$, which means $0 = \pi \cdot 0$ when π additionally fixes $supp(\sqsubseteq)$ pointwise, and so $supp(0) \subseteq supp(X) \cup supp(\sqsubseteq)$. By induction, we prove $supp(f^n(0)) \subseteq$

$supp(f) \cup supp(X) \cup supp(\sqsubseteq)$ for all $n \in \mathbb{N}$. Clearly, $supp(f^0(0)) = supp(0) \subseteq supp(f) \cup supp(X) \cup supp(\sqsubseteq)$. Let us suppose that $supp(f^k(0)) \subseteq supp(f) \cup supp(X) \cup supp(\sqsubseteq)$ for some $k \in \mathbb{N}$. We have to prove that $supp(f^{k+1}(0)) \subseteq supp(f) \cup supp(X) \cup supp(\sqsubseteq)$. Thus, we have to prove that each permutation of atoms π which fixes $supp(f) \cup supp(X) \cup supp(\sqsubseteq)$ pointwise, also fixes $f^{k+1}(0)$. Let $\pi \in Fix(supp(f) \cup supp(X) \cup supp(\sqsubseteq))$. From the inductive hypothesis, we have that $supp(f^k(0)) \subseteq supp(f) \cup supp(X) \cup supp(\sqsubseteq)$, and so $\pi \cdot f^k(0) = f^k(0)$. According to Proposition 2.9, since π fixes $supp(f)$ pointwise we have $\pi \cdot f^{k+1}(0) = \pi \cdot f(f^k(0)) = f(\pi \cdot f^k(0)) = f(f^k(0)) = f^{k+1}(0)$. Therefore, $(f^n(0))_{n\in\mathbb{N}}$ is uniformly supported, and so it has to be finite according to theorem's hypothesis. Since it is an ascending chain it follows that there exists $n_0 \in \mathbb{N}$ such that $f^n(0) = f^{n_0}(0)$ for all $n \geq n_0$. Thus, $f(f^{n_0}(0)) = f^{n_0+1}(0) = f^{n_0}(0)$, and so $f^{n_0}(0)$ is a fixed point of f, and furthermore, it is supported by $supp(f)$. If x is another fixed point of f, it follows from the monotony of f and from the relation $0 \sqsubseteq x$ that $f^{n_0}(0) = \bigsqcup_{n\in\mathbb{N}} f^n(0) \sqsubseteq x$, and so $f^{n_0}(0) = lfp(f)$. □

Theorem 6.4 *Let $(L, \sqsubseteq, \cdot)$ be a finitely supported lattice having the property that every finitely supported subset $X \subseteq L$ has a least upper bound $\sqcup X$ with respect to the order relation $\sqsubseteq$. Then each finitely supported order preserving function $f : L \to L$ possesses a least fixed point $lfp(f)$ and a greatest fixed point $gfp(f)$ which are both supported by $supp(f) \cup supp(\sqsubseteq)$.*

Proof Firstly, we remark that every finitely supported subset of L also has a greatest lower bound w.r.t. $\sqsubseteq$. Let $U \in \wp_{fs}(L)$. Let $V = \cap\{\downarrow x \,|\, x \in U\}$, where $\downarrow x = \{y \in L \,|\, y \sqsubseteq x\}$. We claim that V is finitely supported by $supp(U) \cup supp(\sqsubseteq)$. Let $\pi \in Fix(supp(U) \cup supp(\sqsubseteq))$. Let $v \in V$, that is, $v \sqsubseteq x$ for all $x \in U$. We claim that $\pi \cdot v \in V$. Indeed, let $y \in U$ be an arbitrary element from U. Since $\pi \star U = U$, for our $y \in U$ there exists $x \in U$ such that $\pi \cdot x = y$. However, $v \sqsubseteq x$, and because π fixes $supp(\sqsubseteq)$ pointwise we also have $\pi \cdot v \sqsubseteq \pi \cdot x = y$. Hence $\pi \cdot v \in V$, and so $\pi \star V \subseteq V$. We prove by contradiction that $\pi \star V = V$. Let us suppose that $\pi \star V \subsetneq V$. By induction, we get $\pi^n \star V \subsetneq V$ for all $n \geq 1$. However, π is a (finite) permutation of atoms, and so it has a finite order. We obtain $V \subsetneq V$, a contradiction. Clearly, $\sqcup V$ is the greatest lower bound of U.

Let $Z = \{z \in L \,|\, z \sqsubseteq f(z)\}$. Firstly, we remark that Z is non-empty because the least element of L belongs to Z. We claim that $supp(f) \cup supp(\sqsubseteq)$ supports Z. Let $\pi \in Fix(supp(f) \cup supp(\sqsubseteq))$ and $z \in Z$ be arbitrarily chosen. Then $z \sqsubseteq f(z)$ (or, equivalently $(z, f(z)) \in \sqsubseteq$)), and because $\sqsubseteq$ is supported by $supp(\sqsubseteq)$, we also have $\pi \otimes (z, f(z)) \in \sqsubseteq$, that is $\pi \cdot z \sqsubseteq \pi \cdot f(z)$, where $\otimes$ represents the canonical action on Cartesian products defined as in Proposition 2.2. Since $\pi \in Fix(supp(f))$ and $supp(f)$ supports f, according to Proposition 2.9 we have $\pi \cdot z \sqsubseteq \pi \cdot f(z) = f(\pi \cdot z)$, and so $\pi \cdot z \in Z$. Thus, $\pi \star Z \subseteq Z$, and so $\pi \star Z = Z$. Therefore, $supp(f) \cup supp(\sqsubseteq)$ supports Z, and so there exists the least upper bound of Z, namely $z_0 = \sqcup Z$. As in ZF, we get $f(z_0) = z_0$ and z_0 is the greatest fixed point of f.

We prove that z_0 is supported by $supp(f) \cup supp(\sqsubseteq)$. Let $\pi \in Fix(supp(f) \cup supp(\sqsubseteq))$ and $X \in \wp_{fs}(L)$. According to Proposition 2.1, we have that $\pi \star X$ is finitely supported, and so there exists $\sqcup(\pi \star X)$. Let $x \in X$; we have $x \sqsubseteq \sqcup X$, and so $\pi \cdot x \sqsubseteq \pi \cdot \sqcup X$ because π fixes $supp(\sqsubseteq)$ pointwise. Thus, we have $\sqcup(\pi \star X) \sqsubseteq \pi \cdot \sqcup X$ (1). We apply (1), firstly for π and Z, and then for $\pi^{-1} \in Fix(supp(f) \cup supp(\sqsubseteq))$ and $\pi \star Z$ from which we obtain $\sqcup Z \sqsubseteq \pi^{-1} \cdot \sqcup(\pi \star Z)$. Since π fixes $supp(\sqsubseteq)$ pointwise, we finally get $\pi \cdot \sqcup Z = \sqcup(\pi \star Z)$. Since $supp(f) \cup supp(\sqsubseteq)$ supports Z, we have $\pi \star Z = Z$, and so $\pi \cdot z_0 = z_0$.

Similarly, f has a least fixed point defined as the greatest lower bound of the finitely supported set $\{z \in L \,|\, f(z) \sqsubseteq z\}$ (set supported by $supp(f) \cup supp(\sqsubseteq)$). □

Related to Theorem 6.4, the existence of fixed points can be proved even when relaxing the requirement "there exists a least upper bound for each finitely supported subset $X \subseteq L$" and imposing only the requirement "there exists a least upper bound for each uniformly supported subset $X \subseteq L$"; however, in this case we cannot prove the existence of least or greatest fixed points.

Theorem 6.5 *Let $(X, \sqsubseteq, \cdot)$ be a finitely supported partially ordered set having the property that every* uniformly supported *subset has a least upper bound. Then each finitely supported, order preserving function $f : X \to X$ for which there exists $x_0 \in X$ such that $x_0 \sqsubseteq f(x_0)$ has a fixed point.*

Proof Let $Z = \{z \in X \,|\, z \sqsubseteq f(z) \text{ and } supp(z) \subseteq supp(x_0) \cup supp(f) \cup supp(\sqsubseteq) \cup supp(X)\}$. We remark that Z is non-empty since $x_0 \in Z$. By its definition, Z is uniformly supported, and so there exists $z_0 = \sqcup Z$. We claim z_0 is supported by $supp(x_0) \cup supp(f) \cup supp(\sqsubseteq) \cup supp(X)$. Let $\pi \in Fix(supp(x_0) \cup supp(f) \cup supp(\sqsubseteq) \cup supp(X))$ and $Y \in \wp_{us}(X)$. According to Proposition 2.1 and because π fixes $supp(X)$ pointwise, we have that $\pi \star Y \in \wp_{us}(X)$, and so there exists $\sqcup(\pi \star Y)$. Let $y \in Y$. We have $y \sqsubseteq \sqcup Y$, and so $\pi \cdot y \sqsubseteq \pi \cdot \sqcup Y$ because π fixes $supp(\sqsubseteq)$ pointwise. Thus, we have $\sqcup(\pi \star Y) \sqsubseteq \pi \cdot \sqcup Y$ (claim 1). We apply (claim 1), firstly for π and Z, and then for $\pi^{-1} \in Fix(supp(f) \cup supp(x_0) \cup supp(\sqsubseteq) \cup supp(X))$ and $\pi \star Z \in \wp_{us}(X)$ from which we obtain $\sqcup Z \sqsubseteq \pi^{-1} \cdot \sqcup(\pi \star Z)$. Since π fixes $supp(\sqsubseteq)$ pointwise, we finally get $\pi \cdot \sqcup Z = \sqcup(\pi \star Z)$. Since $supp(x_0) \cup supp(f) \cup supp(\sqsubseteq) \cup supp(X)$ (uniformly) supports Z, we have $\pi \star Z = Z$, and so $\pi \cdot z_0 = z_0$. For each $z \in Z$ we have $z \sqsubseteq z_0$, and so $z \sqsubseteq f(z) \sqsubseteq f(z_0)$, from which $z_0 \sqsubseteq f(z_0)$, which means that $z_0 \in Z$. However, because f is order-preserving and $supp(f(z)) \subseteq supp(f) \cup supp(z) \subseteq supp(f) \cup supp(x_0) \cup supp(f) \cup supp(\sqsubseteq) \cup supp(X) = supp(x_0) \cup supp(f) \cup supp(\sqsubseteq) \cup supp(X)$ for all $z \in Z$, we have $f(z) \in Z$ for each $z \in Z$. Particularly, $f(z_0) \in Z$, and so $f(z_0) \sqsubseteq z_0$, which means that z_0 is a fixed point of f, but it is not necessarily the greatest one since there might exist another fixed point of f that does not belong to Z (i.e. a fixed point of f that is possibly not supported by $supp(x_0) \cup supp(f) \cup supp(\sqsubseteq) \cup supp(X)$). □

Dually, we have the following result.

Theorem 6.6 *Let $(X, \sqsubseteq, \cdot)$ be a finitely supported partially ordered set having the property that every* uniformly supported *subset has a greatest lower bound. Then each finitely supported, order preserving function $f : X \to X$ for which there exists $x_0 \in X$ such that $f(x_0) \sqsubseteq x_0$ has a fixed point.*

Proof A fixed point of f is $\sqcap Z$, where $Z = \{z \in X \,|\, f(z) \sqsubseteq z$ and $supp(z) \subseteq supp(x_0) \cup supp(f) \cup supp(\sqsubseteq) \cup supp(X)\}$. □

Theorem 6.7 *Let $(X, \sqsubseteq, \cdot)$ be a finitely supported partially ordered set containing no infinite uniformly supported subset. Let $f : X \to X$ be a finitely supported function with the property that $x \sqsubseteq f(x)$ for all $x \in X$. Then for each $x \in X$, there exists some $m \in \mathbb{N}$ such that $f^m(x)$ is a fixed point of f.*

Proof Let us fix an arbitrary element $x \in X$. We consider the ascending sequence $(x_n)_{n\in\mathbb{N}}$ which has the first term $x_0 = x$ and the general term $x_{n+1} = f(x_n)$ for all $n \in \mathbb{N}$. We prove by induction that $supp(x_n) \subseteq supp(f) \cup supp(x)$ for all $n \in \mathbb{N}$. Clearly, $supp(x_0) = supp(x) \subseteq supp(f) \cup supp(x)$. Assume that $supp(x_k) \subseteq supp(f) \cup supp(x)$. Let $\pi \in Fix(supp(f) \cup supp(x))$. Thus, $\pi \cdot x_k = x_k$ according to the inductive hypothesis. According to Proposition 2.9, because π fixes $supp(f)$ pointwise and $supp(f)$ supports f, we get $\pi \cdot x_{k+1} = \pi \cdot f(x_k) = f(\pi \cdot x_k) = f(x_k) = x_{k+1}$. Since $supp(x_{k+1})$ is the least set supporting x_{k+1}, we obtain $supp(x_{k+1}) \subseteq supp(f) \cup supp(x)$. Thus, $(x_n)_{n\in\mathbb{N}} \subseteq X$ is uniformly supported, and so $(x_n)_{n\in\mathbb{N}}$ must be finite. Since we have $x_0 \sqsubseteq x_1 \sqsubseteq \ldots \sqsubseteq x_n \sqsubseteq \ldots$ (by hypothesis), there should exist $m \in \mathbb{N}$ such that $x_m = x_{m+1}$, i.e. $f^m(x) = f^{m+1}(x) = f(f^m(x))$. □

We leave to the reader the full proofs of the following results (which are slight extensions of Theorem 5.6 and Theorem 5.8, respectively).

Theorem 6.8 *Let $(X, \sqsubseteq, \cdot)$ be a finitely supported partially ordered set with the property that every finitely supported totally ordered subset (every finitely supported chain) of X has a least upper bound. Let $f : X \to X$ be a finitely supported function with the property that $x \sqsubseteq f(x)$ for all $x \in X$. Then there exists $x \in X$ such that $f(x) = x$.*

Proof (Sketch) The reader should follow the proof of Theorem 5.6 by considering $supp(X) \cup supp(\sqsubseteq)$, i.e. by considering the unions between the original sets supporting the constructions in Theorem 5.6 and $supp(X) \cup supp(\sqsubseteq)$; such unions support the corresponding constructions related to the extended version of Theorem 5.6. □

Theorem 6.9 *Let $(X, \sqsubseteq, \cdot)$ be a finitely supported partially ordered set and $f : X \to X$ a finitely supported s-continuous function (i.e. a function having the property that $f(\bigsqcup_{n\in\mathbb{N}} x_n) = \bigsqcup_{n\in\mathbb{N}} (f(x_n))$ for any finitely supported countable chain $(x_n)_{n\in\mathbb{N}}$ contained in X). Assume that there exists $x_0 \in X$ having the following properties*

- *$x_0 \sqsubseteq f(x_0)$;*
- *every finitely supported countable chain in the set $X' = \{x \in X \,|\, x_0 \sqsubseteq x\}$ has a least upper bound in X.*

Then f possesses a fixed point $fp(f)$ (defined in a constructive manner) with the property that $supp(fp(f)) \subseteq supp(f) \cup supp(x_0) \cup supp(X) \cup supp(\sqsubseteq)$.

Proof (Sketch) The reader should follow the proof of Theorem 5.8 by considering $supp(X) \cup supp(\sqsubseteq)$, i.e. by considering the unions between the original sets supporting the constructions in Theorem 5.8 and $supp(X) \cup supp(\sqsubseteq)$.

For example $X' = \{x \in X \mid x_0 \sqsubseteq x\}$ is a finitely supported subset with $supp(X') \subseteq supp(x_0) \cup supp(X)$. The ascending chain $(f^n(x_0))_{n\in\mathbb{N}} \subseteq X'$ is uniformly supported by $supp(f) \cup supp(x_0)$. Thus, there exists $\bigsqcup_{n\in\mathbb{N}} f^n(x_0)$ which is proved to be supported by $supp(f) \cup supp(x_0) \cup supp(X) \cup supp(\sqsubseteq)$. □

Corollary 6.4 *Let $(X, \sqsubseteq, \cdot, \bot)$ be a finitely supported cppo. Then each finitely supported s-continuous function $f : X \to X$ possesses a least fixed point $lfp(f) = \bigsqcup_{n\in\mathbb{N}} f^n(\bot)$ having the property that $supp(lfp(f)) \subseteq supp(f) \cup supp(X) \cup supp(\sqsubseteq)$.*

Theorem 6.10 *Let $(X, \sqsubseteq, \cdot)$ be a finitely supported partially ordered set with the property that every uniformly supported subset has a least upper bound. If $f : X \to X$ is a finitely supported function having the properties that $f(\sqcup Y) = \sqcup f(Y)$ for every uniformly supported subset Y of X and there exist $x_0 \in X$ and $k \in \mathbb{N}^*$ such that $x_0 \sqsubseteq f^k(x_0)$, then f has a fixed point.*

Proof As in the proof of Theorem 5.8, the sequence $(f^n(x_0))_{n\in\mathbb{N}} \subseteq X$ is uniformly supported by $supp(f) \cup supp(x_0)$. Thus, there exists $\bigsqcup_{n\in\mathbb{N}} f^n(x_0)$ which is proved to be supported by $supp(f) \cup supp(x_0) \cup supp(X) \cup supp(\sqsubseteq)$. Since $f^0(x_0) = x_0 \sqsubseteq f^k(x_0)$, we have $\bigsqcup_{n\in\mathbb{N}} f^{n+1}(x_0) = \bigsqcup_{n\in\mathbb{N}^*} f^n(x_0) = \bigsqcup_{n\in\mathbb{N}} f^n(x_0)$, and so $f(\bigsqcup_{n\in\mathbb{N}} f^n(x_0)) = \bigsqcup_{n\in\mathbb{N}} f(f^n(x_0)) = \bigsqcup_{n\in\mathbb{N}} f^{n+1}(x_0) = \bigsqcup_{n\in\mathbb{N}} f^n(x_0)$, which means $\bigsqcup_{n\in\mathbb{N}} f^n(x_0)$ is a fixed point of f. □

Theorem 6.11 *Let $(L, \sqsubseteq, \cdot)$ be a finitely supported complete lattice and $f : L \to L$ a finitely supported monotone function over L. Let P be the set of fixed points of f. Then $(P, \sqsubseteq, \cdot)$ is a non-empty, finitely supported complete sublattice of L.*

Proof As in the proof of Theorem 6.2, P is supported by $supp(f) \cup supp(L)$. Moreover, P is non-empty according to Theorem 6.4. We prove now that P is a complete sublattice of L. Let X be an arbitrary finitely supported subset of P. We have to prove that X has a least upper bound in P. We already know that X has a least upper bound (denoted by $\sqcup X$) in L. Let $Z' = \{z \in L \mid f(z) \sqsubseteq z \text{ and } \sqcup X \sqsubseteq z\}$. We can easily prove that $supp(f) \cup supp(\sqcup X) \cup supp(L) \cup supp(\sqsubseteq)$ supports Z', and so there exists the greatest lower bound of Z' denoted by $z_0 = \sqcap Z'$ according to the first part of the proof of Theorem 6.4. As in the proof of Theorem 6.2, $z_0 = \sqcup X$ in $(P, \sqsubseteq)$. □

Chapter 7
Constructions of Lattices in Finitely Supported Mathematics

Abstract We present various fundamental examples of invariant lattices and study their properties. We particularly mention the finitely supported subsets of an invariant set, the finitely supported functions from an invariant set to an invariant complete lattice (i.e. the finitely supported L-fuzzy sets with L being an invariant complete lattice), the finitely supported subgroups of an invariant group, and the finitely supported fuzzy subgroups of an invariant group. For this specific invariant lattices the theorems presented in the previous chapter can provide new properties. The constructions in this chapter derive from our previous work [5, 6, 14].

7.1 Powersets

Theorem 7.1 *Let* $(X,\cdot)$ *be an invariant set.*
Then $(\wp_{fs}(X),\subseteq,\star)$ *is an invariant complete Boolean lattice.*

Proof According to the definition of the action $\star$ in Proposition 2.2, for any finitely supported subsets Y and Z of X with $Y\subseteq Z$, we have $\pi\star Y=\{\pi\cdot y\,|\,y\in Y\}\subseteq\{\pi\cdot z\,|\,z\in Z\}=\pi\star Z$. Thus, $\subseteq$ is an equivariant order relation on $\wp_{fs}(X)$.

Let $\mathscr{F}=(X_i)_{i\in I}$ be a finitely supported family of finitely supported subsets of X. We know that $\cup\mathscr{F}=\underset{i\in I}{\cup}X_i$ exists in X. We have to prove that $\underset{i\in I}{\cup}X_i\in\wp_{fs}(X)$. We claim that $supp(\mathscr{F})$ supports $\underset{i\in I}{\cup}X_i$. Let us consider $\pi\in Fix(supp(\mathscr{F}))$. Let $x\in\underset{i\in I}{\cup}X_i$. There exists $j\in I$ such that $x\in X_j$. Since $\pi\in Fix(supp(\mathscr{F}))$, we have $\pi\star X_j\in\mathscr{F}$, that there exists $k\in I$ such that $\pi\star X_j=X_k$. Therefore, $\pi\cdot x\in\pi\star X_j=X_k$, and so $\pi\cdot x\in\underset{i\in I}{\cup}X_i$. We obtain $\pi\star\underset{i\in I}{\cup}X_i=\underset{i\in I}{\cup}X_i$, and so $\underset{i\in I}{\cup}X_i$ is finitely supported. Therefore, $(\wp_{fs}(X),\subseteq,\star)$ is an invariant complete lattice.

The distributivity property over $(\wp_{fs}(X),\subseteq,\star)$ follows from the distributivity property on $\wp(X)$ because the intersection and the union of every two finitely supported subsets of X is finitely supported by the union of the related supports. The greatest lower bound and the least upper bound in $\wp_{fs}(X)$ are X and $\emptyset$, respectively.

A. Alexandru, G. Ciobanu, *Foundations of Finitely Supported Structures*,
https://doi.org/10.1007/978-3-030-52962-8_7

Let $U \in \wp_{fs}(X)$. The complement of U is $X \setminus U$. If S is a finite set of atoms supporting U, we claim that S supports $X \setminus U$. Indeed, let $\pi \in Fix(S)$, which means $\pi \cdot x \in U$ for all $x \in U$. We also have $\pi^{-1} \in Fix(S)$. Let $y \in X \setminus U$. If $\pi \cdot y \in U$, then $y \in \pi^{-1} \star U = U$. Therefore, $\pi \cdot y \in X \setminus U$, which means S supports $X \setminus U$. Therefore, $X \setminus U \in \wp_{fs}(X)$, and $(\wp_{fs}(X), \subseteq, \star)$ is an invariant Boolean lattice. □

Not every invariant complete lattice is a fully ZF complete lattice with respect to all atomic subsets, meaning that there may exist atomic (not finitely supported) subsets that have no least upper bounds. In this sense we present the following result.

Proposition 7.1 *The set $\wp_{fin}(A) \cup \wp_{confin}(A)$ is an invariant complete lattice, but it fails to be a fully ZF complete lattice (even if its construction is consistent with ZF).*

Proof We know that $\wp_{fin}(A) \cup \wp_{confin}(A) = \wp_{fs}(A)$, and so it is an invariant complete lattice according to Theorem 7.1. However, the union of a ZF simultaneously infinite and coinfinite family of singletons is not a finite subset of A, nor a cofinite subset of A, and so $\wp_{fin}(A) \cup \wp_{confin}(A)$ is not a fully ZF complete lattice with respect to all atomic subsets. □

The set $\wp_{fin}(A)$ is not an invariant complete lattice. This is because the (infinite) union of the terms from a finitely supported family of elements in $\wp_{fin}(A)$ is not necessarily a finite subset of A. For example, the union of the members of the equivariant family of those 1-sized subsets of A coincides with A, and so it is not finite. However, we prove that $\wp_{fin}(A)$ is an invariant cpo, although it is not necessarily a fully ZF cpo with respect to all atomic subsets.

Theorem 7.2 *Let $(X, \cdot)$ be an invariant set such that X does not contain an infinite uniformly supported subset. Then $(\wp_{fin}(X), \subseteq, \star)$ has the property that any finitely supported totally ordered subset in $\wp_{fin}(X)$ has a least upper bound in $\wp_{fin}(X)$.*

Proof Let $\mathscr{F}$ be a finitely supported totally ordered subset of $\wp_{fin}(X)$. Since $\mathscr{F}$ is totally ordered with respect to inclusion relation on $\wp_{fin}(X)$, then there do not exist two different finite subsets of X that belong to $\mathscr{F}$ and have the same cardinality (because different finite sets of the same cardinality are incomparable via $\subseteq$). Since $\mathscr{F}$ is finitely supported, then there exists a finite set $S \subseteq A$ such that $\mathscr{F}$ is left invariant under the effect of each permutation of atoms $\pi \in Fix(S)$ on $\mathscr{F}$, that is $\pi \star Y \in \mathscr{F}$ for each $Y \in \mathscr{F}$ and each $\pi \in Fix(S)$. However, for each $Y \in \mathscr{F}$ and each $\pi \in Fix(S)$ we have that $\pi \star Y$ is a finite subset of X, and because permutations of atoms are bijective functions, the cardinality of the finite set $\pi \star Y$ coincides with the cardinality of the finite set Y. More exactly, from Proposition 2.2, if $Y \in \mathscr{F}$ is of form $Y = \{y_1, \dots, y_m\}$, then $\pi \star Y = \{\pi \cdot y_1, \dots, \pi \cdot y_m\}$; since $y_i = y_j$ if and only if $\pi \cdot y_i = \pi \cdot y_j$, we have $|\pi \star Y| = |Y|$.

Since there do not exist two distinct elements in $\mathscr{F}$ having the same cardinality, we conclude that $\pi \star Y = Y$ for all $Y \in \mathscr{F}$ and all $\pi \in Fix(S)$. Thus, $\mathscr{F}$ is uniformly supported by S. We claim that $\mathscr{F}$ is finite. Suppose there exists an infinite injective subsequence $(X_i)_{i \in I} \subseteq \mathscr{F}$ (we used the representation $(X_i)_{i \in I}$ just for writing convention, without assuming that $i \mapsto X_i$ is finitely supported). We know that $supp(X_i) \subseteq S$

for all $i \in I$. Fix an arbitrary $j \in I$. We claim that X_j has the property that $supp(x) \subseteq S$ for all $x \in X_j$. Firstly, we note that X_j is a finite subset of X. Thus, $X_j = \{x_1, \ldots, x_n\}$ for some $n \in \mathbb{N}$, with $x_1, \ldots x_n \in X$. Let $T = supp(x_1) \cup \ldots \cup supp(x_n)$. According to Proposition 2.5, we have that $T = supp(X_j)$. Now, since $supp(X_j) \subseteq S$, we have $\bigcup_{x \in X_j} supp(x) \subseteq S$, and so $supp(x) \subseteq S$ for all $x \in X_j$. Since j has been arbitrarily chosen, it follows that every element from every set of form X_i is supported by S, and so $\bigcup_{i \in I} X_i$ is a uniformly supported subset of X (all its elements being supported by S). Furthermore, $\bigcup_{i \in I} X_i$ is infinite because the family $(X_i)_{i \in I}$ is infinite and injective. Otherwise, if $\bigcup_{i \in I} X_i$ was finite, the family $(X_i)_{i \in I}$ would be contained in the finite set $\wp(\bigcup_{i \in I} X_i)$, and so it could not be infinite and injective. We were able to construct an infinite uniformly supported subset of X, namely $\bigcup_{i \in I} X_i$, and this contradicts the hypothesis that X does not contain an infinite uniformly supported subset. Thus, $\mathscr{F}$ should be finite. Since any finitely supported totally ordered family $\mathscr{G} \subseteq \wp_{fin}(A)$ is finite, we have that $\bigcup_{X \in \mathscr{G}} X$ exists in $\wp_{fin}(X)$ (i.e. it is finite as a finite union of finite sets) and it is, obviously, the least upper bound of $\mathscr{G}$ in $\wp_{fin}(X)$. □

Corollary 7.1 *Let X be an invariant set such that X does not contain an infinite uniformly supported subset. Then we have that*

- $\wp_{fin}(X) = \{Z \subseteq X \mid Z \textit{finite}\}$ *is an invariant cpo;*
- $\wp_{fin}(A)$ *is an invariant cpo;*
- $\wp_{fin}(\wp_{fs}(A))$ *is an invariant cpo.*

7.2 Function Spaces (L-fuzzy sets)

In ZF, an L-fuzzy set is defined as a function from a ZF set to a ZF structure L (particularly, L can be a ZF complete lattice, a ZF inductive set, etc). A usual fuzzy set is an L-fuzzy set with $L = [0,1]$. We introduced and studied infinite, finitely supported usual fuzzy sets in [10]. Similarly, we can generalize the ZF concept of L-fuzzy sets (particularly when L is a complete lattice), and define the FSM notion of infinite, finitely supported L-fuzzy set as a finitely supported function from an invariant set to an FSM algebraic structure L (particularly L can be an invariant complete lattice). Formally, we express this by the following definition.

Definition 7.1 • An *FSM (usual) fuzzy set over the invariant set* $(U,\cdot)$ is a finitely supported function $\mu : U \to [0,1]$.
- Let L be an invariant algebraic structure. An *FSM L-fuzzy set over the invariant set* $(U,\cdot)$ is a finitely supported function $\mu : U \to L$.

If L is an invariant complete lattice, FSM L-fuzzy sets are characterized by the property presented below.

Theorem 7.3 *Let $(X,\cdot)$ be an invariant set and $(Y,\leq,\diamond)$ an invariant complete lattice. The family of those finitely supported functions $f: X \to Y$ (i.e. the family of all FSM Y-fuzzy sets over X) is an invariant complete lattice, with the order relation $\sqsubseteq$ defined by $f \sqsubseteq g$ if and only if $f(x) \leq g(x)$ for all $x \in X$.*

Proof Since Y is an invariant partially ordered set, we have that $\leq$ is an equivariant subset of the invariant set $Y \times Y$. We claim that $\sqsubseteq$ (which is constructed by involving only $\leq$) is also equivariant. Let $\pi \in S_A$, and $f,g: X \to Y$ be two finitely supported functions such that $f \sqsubseteq g$. Thus, $f(x) \leq g(x)$ for all $x \in X$, and so $f(\pi^{-1}\cdot x) \leq g(\pi^{-1}\cdot x)$ for all $x \in X$. Since $\leq$ is equivariant, we have $\pi \diamond f(\pi^{-1}\cdot x) \leq \pi \diamond g(\pi^{-1}\cdot x)$ for all $x \in X$, namely $(\pi \widetilde{\star} f)(x) \leq (\pi \widetilde{\star} g)(x)$ for all $x \in X$. It follows that $\pi \widetilde{\star} f \sqsubseteq \pi \widetilde{\star} g$, where $\widetilde{\star}$ represents the S_A-action on Y^X defined as in Proposition 2.7.

Let $\mathscr{F} = (f_i)_{i\in I}$ be a finitely supported family of finitely supported functions from X to Y. Let us fix an arbitrary $z \in X$. We claim that $\mathscr{F}' = (f_i(z))_{i\in I} \subseteq Y$ is finitely supported by $supp(\mathscr{F}) \cup supp(z)$, and so such a family has a least upper bound (because Y is an invariant complete lattice). Indeed, let us consider $\sigma \in Fix(supp(\mathscr{F}) \cup supp(z))$. Then $\sigma \widetilde{\star} f_k \in \mathscr{F}$ for all $k \in I$, and $\sigma \cdot z = z$. Let $f_j(z)$ be an arbitrary element of $\mathscr{F}'$. Then $f_j \in \mathscr{F}$, and so there exists a unique $i \in I$ such that $\sigma \widetilde{\star} f_j = f_i$, which means $f_j = \sigma^{-1} \widetilde{\star} f_i$. Thus, $f_j(z) = (\sigma^{-1} \widetilde{\star} f_i)(z) = \sigma^{-1} \diamond f_i((\sigma^{-1})^{-1}\cdot z) = \sigma^{-1} \diamond f_i(\sigma \cdot z) = \sigma^{-1} \diamond f_i(z)$. Therefore, $\sigma \diamond f_j(z) = f_i(z) \in \mathscr{F}'$.

We are able now to define $\bigsqcup_{i\in I} f_i : X \to Y$ by $(\bigsqcup_{i\in I} f_i)(x) = \underset{i\in I}{supremum}\{f_i(x) \mid i \in I\}$ for all $x \in X$. According to the above paragraph, $\bigsqcup_{i\in I} f_i$ is well-defined, and furthermore, it is the ZF least upper bound of $\mathscr{F}$. It remains to prove that $\bigsqcup_{i\in I} f_i$ is finitely supported. We claim that $supp(\mathscr{F})$ supports $\bigsqcup_{i\in I} f_i$. Let π be a permutation of atoms that fixes $supp(\mathscr{F})$ pointwise. According to Proposition 2.9, we should prove the relation $(\bigsqcup_{i\in I} f_i)(\pi\cdot x) = \pi \diamond (\bigsqcup_{i\in I} f_i)(x)$ for all $x \in X$, i.e. $\underset{i\in I}{supremum}\{f_i(\pi\cdot x) \mid i \in I\}$ $= \pi \diamond \underset{i\in I}{supremum}\{f_i(x) \mid i \in I\}$ for all $x \in X$. This follows because $supp(\mathscr{F})$ supports $\mathscr{F}$, and so every function belonging to $\mathscr{F}$ can be expressed as the effect of a unique function belonging to $\mathscr{F}$ under π, i.e. for each $j \in I$ there exists a unique $i \in I$ such that $f_j = \pi^{-1} \widetilde{\star} f_i$, or equivalently, for each $j \in I$ there exists a unique $i \in I$ such that $f_j(x) = (\pi^{-1} \widetilde{\star} f_i)(x) = \pi^{-1} \diamond f_i((\pi^{-1})^{-1}\cdot x) = \pi^{-1} \diamond f_i(\pi \cdot x)$ for all $x \in X$. Thus, for all $x \in X$ we have the following identity of sets: $\{f_i(\pi\cdot x) \mid i \in I\} = \{\pi \diamond f_i(x) \mid i \in I\} = \pi \star \{f_i(x) \mid i \in I\}$, where $\star$ represents the action on the powerset of Y. According to Proposition 6.2 (claiming that *supremum* is an equivariant mapping), $\underset{i\in I}{supremum}\{f_i(\pi\cdot x) \mid i \in I\} = \underset{i\in I}{supremum}(\pi \star \{f_i(x) \mid i \in I\}) =$ $\pi \diamond \underset{i\in I}{supremum}\{f_i(x) \mid i \in I\}$ for all $x \in X$. □

Example 7.1 1. If in the statement of the previous theorem Y is considered the compact real interval $[0,1]$ with the usual order, then the family of those finitely supported functions from an invariant set X to $[0,1]$ (which represents the family of those FSM fuzzy sets defined on X) is an invariant complete lattice [10].

2. The requirement that Y is invariant complete in Theorem 7.3 is mandatory. For example, let us fix an element $a \in A$. The family $(f_n)_{n\in\mathbb{N}}$ of functions from A to $\mathbb{N}$ defined by $f_i(b) = \begin{cases} i & \text{for } b = a \\ 0 & \text{for } b \in A \setminus \{a\} \end{cases}$ for all $i \in \mathbb{N}$ is finitely supported (each f_i is supported by the same set $\{a\}$), but it does not have a supremum modulo $\sqsubseteq$.

It is worth noting that FSM results are presented internally in the world of finitely supported structures, and so whenever a ZF structure appears in a certain result it has to be finitely supported. The construction of the family of finitely supported functions from X to Y makes sense in ZF, but this family is an invariant complete lattice and not a fully ZF complete lattice with respect to all atomic subsets.

Proposition 7.2 *There exists an invariant set $(X,\cdot)$ and an ordinary (non-atomic) ZF complete lattice $(Y,\leq)$ such that the family of those finitely supported functions $f : X \to Y$ is an invariant complete lattice, but it is not a fully ZF complete lattice with respect to all atomic subsets (although its construction is consistent with ZF).*

Proof Consider X to be the invariant set $(A,\cdot)$ of atoms with the canonical action of the group of permutations of atoms $(\pi,a) \mapsto \pi \cdot a \overset{def}{=} \pi(a)$. Consider a simultaneously infinite and coinfinite subset of atoms denoted by P that exists in ZF. Consider an arbitrary ZF complete lattice Y with at least two elements. For each $a \in A$ we consider $f_a : A \to Y$ defined by

$$f_a(b) = \begin{cases} supremum(Y) & \text{for } b = a \\ infimum(Y) & \text{for } b \in A \setminus \{a\} \end{cases}.$$

Any function f_a is supported by $supp(a) = \{a\}$. Let us consider now the infinite family $\mathscr{F}$ defined by $\mathscr{F} = \{f_a \mid a \in P\}$. According to the uniqueness of supremum of a subset of Y, the only possible least upper bound of $\mathscr{F}$ is the function $g : A \to Y$ defined by $g(x) = \underset{a\in P}{supremum}\{f_a(x) \mid a \in P\}$ for all $x \in A$.

It follows that $g(x) = \begin{cases} supremum(Y) & \text{for } x \in P \\ infimum(Y) & \text{for } x \in A \setminus P \end{cases}.$

For any permutation of atoms π, $g(\pi \cdot x) = g(x)$ for all $x \in A$ if and only if P is fixed by π (i.e. if and only if '$\pi \cdot x \in P$ whenever $x \in P$'). Thus, because P is not finitely supported, it follows that g is not finitely supported. This means that not every family of finitely supported functions from X to Y has a supremum; only those finitely supported families of finitely supported functions from X to Y have supremum. □

Theorem 7.4 *1. Let $(X,\cdot)$ be an invariant set and $(Y,\leq,\diamond)$ an invariant partially ordered set with the property that any finitely supported chain in Y has a least upper bound in Y. The family of those finitely supported functions $f : X \to Y$ (i.e. the family of all FSM Y-fuzzy sets over X) is an invariant partially ordered set with the order relation $\sqsubseteq$ defined by: $f \sqsubseteq g$ if and only if $f(x) \leq g(x)$ for all $x \in X$, and also has the property that any finitely supported chain in Y^X_{fs} has a least upper bound.*

2. Let $(X,\cdot)$ be an invariant set and $(Y,\leq,\diamond)$ an invariant cpo. The family of those finitely supported functions $f : X \to Y$ (i.e. the family of all FSM Y-fuzzy sets *over* X*) is an invariant cpo, with the order relation* $\sqsubseteq$ *defined by:* $f \sqsubseteq g$ *if and only if* $f(x) \leq g(x)$ *for all* $x \in X$*.*

Proof We follow the proof of Theorem 7.3. Since Y is an invariant partially ordered set, we have that $\sqsubseteq$ is equivariant.

1. Let $\mathscr{F} = (f_i)_{i \in I}$ be a finitely supported totally ordered family of finitely supported functions from X to Y. Fix some arbitrary $z \in X$. As in the proof of Theorem 7.3 we have that the family $\mathscr{F}' = (f_i(z))_{i \in I} \subseteq Y$ is finitely supported by $supp(\mathscr{F}) \cup supp(z)$. Furthermore, whenever $k, l \in I$ we have either $f_k \sqsubseteq f_l$ or $f_l \sqsubseteq f_k$, that is either $f_k(z) \leq f_l(z)$ or $f_l(z) \leq f_k(z)$. Thus, $\mathscr{F}'$ is totally ordered in Y, and so such a family has a least upper bound generically called *supremum*.

We define $\bigsqcup_{i \in I} f_i : X \to Y$ by $(\bigsqcup_{i \in I} f_i)(x) = supremum\{f_i(x) \mid i \in I\}$ for all $x \in X$. According to the above paragraph, $\bigsqcup_{i \in I} f_i$ is well-defined, and furthermore, it is the ZF least upper bound of $\mathscr{F}$. As in the proof of Theorem 7.3, it follows that $\bigsqcup_{i \in I} f_i$ is finitely supported by $supp(\mathscr{F})$ (Lemma 6.2 stating that *supremum* is an equivariant mapping can be directly reproved for strong invariant inductive sets, i.e. for invariant sets in which any chain has a least upper bound, just because $\leq$ is equivariant).

2. Let $\mathscr{F} = (f_i)_{i \in \mathbb{N}}$ be a finitely supported countable family of finitely supported functions from X to Y. According to Proposition 5.6, there exists a finite set S of atoms such that all the elements of $\mathscr{F}$ are supported by S. Fix some arbitrary $z \in X$. According to Proposition 2.9, the members of the family $\mathscr{F}' = (f_i(z))_{i \in \mathbb{N}} \subseteq Y$ are finitely supported by the same set $S \cup supp(z)$. Indeed, if $\sigma \in Fix(S \cup supp(z))$, then $\sigma \tilde{\star} f_i = f_i$ for all $i \in \mathbb{N}$ and $\sigma \cdot z = z$, and so $\sigma \diamond f_i(z) = f_i(\sigma \cdot z) = f_i(z)$ for all $i \in \mathbb{N}$. Thus, by Proposition 5.6, $\mathscr{F}'$ is a finitely supported countable family in Y. Since Y is an invariant cpo, we conclude that $\mathscr{F}'$ has a least upper bound in Y generically called *supremum*. It follows immediately that *supremum* is an equivariant function on its maximal domain of definition.

We define $\bigsqcup_{i \in \mathbb{N}} f_i : X \to Y$ by $(\bigsqcup_{i \in \mathbb{N}} f_i)(x) = supremum\{f_i(x) \mid i \in \mathbb{N}\}$ for all $x \in X$. According to the above paragraph, $\bigsqcup_{i \in \mathbb{N}} f_i$ is well-defined, and it is the ZF least upper bound of $\mathscr{F}$. As in (proof of) Theorem 7.3, with the mention that now $\pi \tilde{\star} f_i = f_i$ for all $\pi \in Fix(S)$ and all $i \in \mathbb{N}$, and so $f_i(\pi \cdot x) = \pi \diamond f_i(x)$ for all $\pi \in Fix(S)$ and all $i \in \mathbb{N}$ and $x \in X$, it follows from Proposition 2.9 that $\bigsqcup_{i \in \mathbb{N}} f_i$ is finitely supported by S. □

7.3 Subgroups of an Invariant Group

Definition 7.2 An *invariant group* is a triple $(G, \cdot, \diamond)$ such that the following conditions are satisfied:

- $(G, \cdot)$ is a group;
- $(G, \diamond)$ is a non-trivial invariant set;

- for each $\pi \in S_A$ and each $x, y \in G$ we have $\pi \diamond (x \cdot y) = (\pi \diamond x) \cdot (\pi \diamond y)$, which means that the internal law on G is equivariant.

Proposition 7.3 $(G, \cdot, \diamond)$ *be an invariant group. We have*

1. $\pi \diamond 1 = 1$ *for all* $\pi \in S_A$*, where* 1 *is the neutral element of* G.
2. $\pi \diamond x^{-1} = (\pi \diamond x)^{-1}$ *for all* $\pi \in S_A$ *and* $x \in G$.

Proof 1. Since $\pi \diamond 1 = \pi \diamond (1 \cdot 1) = (\pi \diamond 1) \cdot (\pi \diamond 1)$, it follows $\pi \diamond 1 = 1$.
2. We have $(\pi \diamond x) \cdot (\pi \diamond x^{-1}) = \pi \diamond (x \cdot x^{-1}) = \pi \diamond 1 = 1$, and similarly, $(\pi \diamond x^{-1}) \cdot (\pi \diamond x) = 1$. Therefore, $(\pi \diamond x)^{-1} = \pi \diamond x^{-1}$. □

Example 7.2

1. The group $(S_A, \circ, \cdot)$ of all (finite) permutations of A is an invariant group, where $\circ$ is the usual composition of permutations and $\cdot$ is the S_A-action on S_A defined as in Proposition 2.2(2). Since the composition law on S_A is associative, one can easily verify that $\pi \cdot (\sigma \circ \tau) = (\pi \cdot \sigma) \circ (\pi \cdot \tau)$ for all $\pi, \sigma, \tau \in S_A$.
2. Let $(X, \cdot)$ be an invariant set. The group of FSM one-to-one mappings of X onto X, $S_X = \{f : X \to X \mid f \text{ finitely supported and bijective}\}$ is an invariant group with the usual composition of functions.
   Indeed, X^X is an invariant set with the S_A-action $\star : S_A \times X^X \to X^X$ defined by $(\pi \star f)(x) = \pi \cdot (f(\pi^{-1} \cdot x))$. It is clear that $(S_X, \circ)$ is a group (where $\circ$ represents the usual composition of functions). We prove that for each $f, g \in S_X$ we have $\pi \star (f \circ g) = (\pi \star f) \circ (\pi \star g)$. Indeed, for each $x \in X$, we have $(\pi \star (f \circ g))(x) = \pi \cdot (f(g(\pi^{-1} \cdot x)))$. Also, if we denote $(\pi \star g)(x) = y$, we have $y = \pi \cdot (g(\pi^{-1} \cdot x))$ and $((\pi \star f) \circ (\pi \star g))(x) = (\pi \star f)(y) = \pi \cdot (f(\pi^{-1} \cdot y)) = \pi \cdot (f((\pi^{-1} \circ \pi) \cdot g(\pi^{-1} \cdot x))) = \pi \cdot (f(g(\pi^{-1} \cdot x)))$. Thus, $\pi \star (f \circ g) = (\pi \star f) \circ (\pi \star g)$.
   To prove that S_X is an invariant group it remains to prove that S_X is an invariant set. For this we must prove that $\star|_{S_X}$ (where $\star|_{S_X}$ represents the restriction of the function $\star$ to $S_A \times S_X$) is an S_A-action on S_X. All we have to do is to prove that the codomain of the action $\star|_{S_X}$ is indeed S_X. Thus, we should prove that $\pi \star f \in S_X$ for each $\pi \in S_A$ and each $f \in S_X$. According to Proposition 2.1, we have that $\pi \star f$ is finitely supported whenever $\pi \in S_A$ and f is finitely supported. It remains to prove that whenever $f : X \to X$ is bijective we have that $\pi \star f$ is also bijective for each $\pi \in S_A$. Let $\pi \in S_A$ and $f : X \to X$ be a bijective function. Suppose that $(\pi \star f)(x) = (\pi \star f)(y)$ for some $x, y \in X$. Then $\pi \cdot (f(\pi^{-1} \cdot x)) = \pi \cdot (f(\pi^{-1} \cdot y))$, and so $(\pi^{-1} \circ \pi) \cdot (f(\pi^{-1} \cdot x)) = (\pi^{-1} \circ \pi) \cdot (f(\pi^{-1} \cdot y))$. Thus, $f(\pi^{-1} \cdot x) = f(\pi^{-1} \cdot y)$, and so $\pi^{-1} \cdot x = \pi^{-1} \cdot y$. Therefore $x = y$, and $\pi \star f$ is one-to-one. Let us prove that $\pi \star f$ is onto. Let $y \in X$ be an arbitrary element. Since f is onto, we can find an element $z \in X$ such that $f(z) = \pi^{-1} \cdot y$. Let $x = \pi \cdot z$. We have $(\pi \star f)(x) = \pi \cdot (f(\pi^{-1} \cdot (\pi \cdot z))) = \pi \cdot (f(z)) = y$. Hence, $\pi \star f$ is onto.
3. - The free group $(F(X), \top, \tilde{\star})$ over an invariant set $(X, \diamond)$ (formed by those classes $[w]$ of words w, where two words are in the same class if one can be obtained from another by repeatedly cancelling or inserting terms of the form

$x^{-1}x$ or xx^{-1} for $x \in X$) is an invariant group, where $\widetilde{\star} : S_A \times F(X) \to F(X)$ is a group action defined by: $\pi \widetilde{\star} [x_1^{\varepsilon_1} x_2^{\varepsilon_2} \dots x_l^{\varepsilon_l}] = [(\pi \diamond x_1)^{\varepsilon_1} \dots (\pi \diamond x_l)^{\varepsilon_l}]$, and the internal law $\mathsf{T} : F(X) \times F(X) \to F(X)$ is defined by $[x_1^{\varepsilon_1} x_2^{\varepsilon_2} \dots x_n^{\varepsilon_n}] \mathsf{T} [y_1^{\delta_1} y_2^{\delta_2} \dots y_m^{\delta_m}] = [x_1^{\varepsilon_1} x_2^{\varepsilon_2} \dots x_n^{\varepsilon_n} y_1^{\delta_1} y_2^{\delta_2} \dots y_m^{\delta_m}]$.

- Let $(X, \diamond_X)$ be an invariant set. Let $i : X \to F(X)$ be the standard inclusion of X into $F(X)$ which maps each element $a \in X$ into the word $[a]$. If $(G, \cdot, \diamond_G)$ is an arbitrary invariant group and $\varphi : X \to G$ is an arbitrary finitely supported function, then there exists a unique finitely supported homomorphism of groups $\psi : F(X) \to G$ with $\psi \circ i = \varphi$. Moreover, if a finite set S supports φ, then the same set S supports ψ. Therefore, if φ is equivariant, then ψ is also equivariant.

Proof Firstly, we show that the statement of the theorem is well-formed in FSM. For this we have to prove that i is finitely supported. We prove that i is equivariant. This follows from Proposition 2.9, because $i(\pi \diamond_X x) = [\pi \diamond_X x] = \pi \widetilde{\star} [x] = \pi \widetilde{\star} i(x)$ for each $\pi \in S_A$ and each $x \in X$.

If $(G, \cdot, \diamond_G)$ is an invariant group, then clearly $(G, \cdot)$ is a group. From the general theory of groups, we can define a unique homomorphism of groups $\psi : F(X) \to G$ with $\psi \circ i = \varphi$. It remains to prove that ψ is indeed finitely supported. We prove that $S = supp(\varphi)$ supports ψ. Let $\pi \in Fix(S)$. We have $\varphi(\pi \diamond_X x) = \pi \diamond_G \varphi(x)$ for all $x \in X$. In the view of Proposition 2.8, for proving that ψ is finitely supported, we have to prove the relation $\psi(\pi \widetilde{\star} [x_1^{\varepsilon_1} x_2^{\varepsilon_2} \dots x_n^{\varepsilon_n}]) = \pi \diamond_G \psi([x_1^{\varepsilon_1} x_2^{\varepsilon_2} \dots x_n^{\varepsilon_n}])$ for all $[x_1^{\varepsilon_1} x_2^{\varepsilon_2} \dots x_n^{\varepsilon_n}] \in F(X)$. However, ψ is a group homomorphism between $F(X)$ and G, and $\psi \circ i = \varphi$. Thus, $\psi([x_1^{\varepsilon_1} x_2^{\varepsilon_2} \dots x_n^{\varepsilon_n}]) = \varphi(x_1)^{\varepsilon_1} \cdot \varphi(x_2)^{\varepsilon_2} \cdot \dots \cdot \varphi(x_n)^{\varepsilon_n}$. Since $(G, \cdot, \diamond_G)$ is an invariant group, we have that $\pi \diamond_G \psi([x_1^{\varepsilon_1} x_2^{\varepsilon_2} \dots x_n^{\varepsilon_n}]) = \pi \diamond_G (\varphi(x_1)^{\varepsilon_1} \cdot \varphi(x_2)^{\varepsilon_2} \cdot \dots \cdot \varphi(x_n)^{\varepsilon_n}) = (\pi \diamond_G \varphi(x_1))^{\varepsilon_1} \cdot (\pi \diamond_G \varphi(x_2))^{\varepsilon_2} \cdot \dots \cdot (\pi \diamond_G \varphi(x_n))^{\varepsilon_n} = \varphi(\pi \diamond_X x_1)^{\varepsilon_1} \cdot \varphi(\pi \diamond_X x_2)^{\varepsilon_2} \cdot \dots \cdot \varphi(\pi \diamond_X x_n)^{\varepsilon_n}$. However, we have that $\pi \widetilde{\star} [x_1^{\varepsilon_1} x_2^{\varepsilon_2} \dots x_n^{\varepsilon_n}] = [(\pi \diamond_X x_1)^{\varepsilon_1} \dots (\pi \diamond_X x_n)^{\varepsilon_n}]$, and so we obtain that $\psi(\pi \widetilde{\star} [x_1^{\varepsilon_1} x_2^{\varepsilon_2} \dots x_n^{\varepsilon_n}]) = \psi([(\pi \diamond_X x_1)^{\varepsilon_1} \dots (\pi \diamond_X x_n)^{\varepsilon_n}]) = \varphi(\pi \diamond_X x_1)^{\varepsilon_1} \cdot \varphi(\pi \diamond_X x_2)^{\varepsilon_2} \cdot \dots \cdot \varphi(\pi \diamond_X x_n)^{\varepsilon_n}$. Thus, $\psi(\pi \widetilde{\star} [x_1^{\varepsilon_1} x_2^{\varepsilon_2} \dots x_n^{\varepsilon_n}]) = \pi \diamond_G \psi([x_1^{\varepsilon_1} x_2^{\varepsilon_2} \dots x_n^{\varepsilon_n}])$ for all $\pi \in Fix(S)$, which means that S supports ψ. □

4. The set $\mathbb{Z}(A)$ of all FSM extended generalized multisets of atoms defined by $\mathbb{Z}(A) = \{f : A \to \mathbb{Z} \mid f \text{is finitely supported}\}$ is an invariant group. $\mathbb{Z}(A)$ is an invariant set formed by those finitely supported functions from A to $\mathbb{Z}$ with the S_A-action from Proposition 2.7. The equivariant group operation is the pointwise sum of FSM extended generalized multisets of atoms. According to Proposition 2.10, every $f \in \mathbb{Z}(A)$ associates the same multiplicity to all but finitely many atoms of A.
5. Generally, given an invariant set $(\Sigma, \cdot)$ (possible infinite), any function $f : \Sigma \to \mathbb{Z}$ with the property that $S_f \stackrel{def}{=} \{x \in \Sigma \mid f(x) \neq 0\}$ is finite is called an *extended generalized multiset over* Σ. The set of all extended generalized multisets over Σ is denoted by $\mathbb{Z}_{ext}(\Sigma)$.

- Let $(\Sigma, \cdot)$ be an invariant set. Then each function $f \in \mathbb{Z}_{ext}(\Sigma)$ is finitely supported with $supp(f) = supp(S_f)$

  Proof Let $f : \Sigma \to \mathbb{Z}$ be a function with the property that $S_f = \{a_1, \dots, a_k\}$. Let $S = supp(a_1) \cup \dots \cup supp(a_k)$. From Proposition 2.5, $S = supp(S_f)$. We have to prove that S supports f. Let $\pi \in Fix(S)$. We have that $\pi \in Fix(supp(a_i))$ for each $i \in \{1, \dots, k\}$. Therefore, $\pi \cdot a_i = a_i$ for each $i \in \{1, \dots, k\}$ because $supp(a_i)$ supports a_i for each $i \in \{1, \dots, k\}$. Let $x \in S_f$. We have $f(\pi \cdot x) = f(x) = \pi \diamond f(x)$ where by $\diamond$ we denoted the (single possible) trivial S_A-action on $\mathbb{Z}$. Let $x \notin S_f$. Since π fixes $supp(S_f)$ pointwise, we have that $\pi \cdot x \notin S_f$. Thus, $f(\pi \cdot x) = 0 = \pi \diamond 0 = \pi \diamond f(x)$. We obtained $f(\pi \cdot x) = \pi \diamond f(x)$ for all $x \in \Sigma$. Thus, according to Proposition 2.8, we have that f is finitely supported and $supp(f) \subseteq supp(S_f)$.
  It remains to prove that $supp(S_f) \subseteq supp(f)$. We prove that $supp(f)$ supports S_f. Let us fix a permutation of atoms $\pi \in Fix(supp(f))$. According to Proposition 2.8, we have that $f(\pi \cdot x) = f(x)$ for all $x \in \Sigma$. Thus, for each $x \in S_f$ we have $\pi \cdot x \in S_f$. Therefore, $\pi \star S_f \subseteq S_f$ for each $\pi \in Fix(supp(f))$, where $\star$ is the S_A-action on $\wp(\Sigma)$ defined as in Proposition 2.2. We remark that $\pi \in Fix(supp(f))$ if and only if $\pi^{-1} \in Fix(supp(f))$. Thus, for $\pi \in Fix(supp(f))$, we also have $\pi^{-1} \star S_f \subseteq S_f$, which means $\pi \star (\pi^{-1} \star S_f) \subseteq \pi \star S_f$, and $S_f \subseteq \pi \star S_f$. Therefore, $\pi \star S_f = S_f$ for each $\pi \in Fix(supp(f))$. Thus, $supp(f)$ supports S_f. Since the support of S_f is the least finite set supporting S_f, we obtain that $supp(S_f) \subseteq supp(f)$. □

- Let $(\Sigma, \cdot)$ be an invariant set. Then $\mathbb{Z}_{ext}(\Sigma)$ is a free Abelian invariant group.

  Proof Firstly, we remark that $(\mathbb{Z}_{ext}(\Sigma), +)$ is an Abelian group, where $f + g \colon \Sigma \to \mathbb{Z}$ is defined pointwise by $(f+g)(a) = f(a) + g(a)$ for all $a \in \Sigma$. Moreover, $\pi \star f \in \mathbb{Z}_{ext}(\Sigma)$ for all $f \in \mathbb{Z}_{ext}(\Sigma)$, where $\star$ is the standard S_A-action on $\mathbb{Z}^{\Sigma}$. Thus, $(\mathbb{Z}_{ext}(\Sigma), \star)$ is an invariant set with the S_A-action $\star \colon S_A \times \mathbb{Z}_{ext}(\Sigma) \to \mathbb{Z}_{ext}(\Sigma)$ defined by $(\pi \star f)(x) = f(\pi^{-1} \cdot x)$ for all $\pi \in S_A$, $f \in \mathbb{Z}_{ext}(\Sigma)$ and $x \in \Sigma$. Let $f, g \in \mathbb{Z}_{ext}(\Sigma)$. For each $x \in \Sigma$ we have $(\pi \star (f+g))(x) = (f+g)(\pi^{-1} \cdot x) = f(\pi^{-1} \cdot x) + g(\pi^{-1} \cdot x) = (\pi \star f)(x) + (\pi \star g)(x) = ((\pi \star f) + (\pi \star g))(x)$. Hence $\pi \star (f+g) = \pi \star f + \pi \star g$. Furthermore, every extended generalized multiset $f \in \mathbb{Z}_{ext}(\Sigma)$ can be expressed in a unique mode as $f = \sum\limits_{a \in S_f} f(a) \cdot \widetilde{a}$, where for each $a \in \Sigma$ the extended generalized multiset $\widetilde{a} \colon \Sigma \to \mathbb{N}$ is defined by $\widetilde{a}(b) = \begin{cases} 1, & \text{for } b = a; \\ 0, & \text{for } b \in \Sigma \setminus \{a\}. \end{cases}$ □

- Let $(\Sigma, \cdot)$ be an invariant set. Let $j \colon \Sigma \to \mathbb{Z}_{ext}(\Sigma)$ be the function which maps each $a \in \Sigma$ into $\widetilde{a} \in \widetilde{\Sigma}$. If $(G, +, \diamond)$ is an arbitrary Abelian invariant group and $\varphi \colon \Sigma \to G$ is an arbitrary finitely supported function, then there exists a unique finitely supported homomorphism of Abelian groups $\psi \colon \mathbb{Z}_{ext}(\Sigma) \to G$ with $\psi \circ j = \varphi$, i.e. $\psi(\widetilde{a}) = \varphi(a)$ for all $a \in \Sigma$. Moreover, if a finite set S supports φ, then the same set S supports ψ. Therefore, if φ is equivariant, then ψ is also equivariant.

Proof Firstly, we show that the statement of the theorem is well-formed in FSM, that is j is finitely supported. We actually prove that j is equivariant. Let $\pi \in S_A$ and $x \in \Sigma$ be arbitrary elements. For each $y \in \Sigma$ we have (by the definition of j) that $(j(\pi \cdot x))(y) = \begin{cases} 1, \text{ for } \pi \cdot x = y; \\ 0, \text{ for } \pi \cdot x \neq y. \end{cases}$ We also have $(\pi \star j(x))(y) = j(x)(\pi^{-1} \cdot y) = \begin{cases} 1, \text{ for } x = \pi^{-1} \cdot y; \\ 0, \text{ for } x \neq \pi^{-1} \cdot y. \end{cases} = \begin{cases} 1, \text{ for } \pi \cdot x = y; \\ 0, \text{ for } \pi \cdot x \neq y. \end{cases}$ Hence $j(\pi \cdot x) = \pi \star j(x)$ for each $\pi \in S_A$ and each $x \in \Sigma$, which means that j has empty support. As in the standard theory of free $\mathbb{Z}$-modules the homomorphism ψ is defined by $\psi(f) = \sum_{a \in S_f} f(a) \cdot \varphi(a)$ whenever $f = \sum_{a \in S_f} f(a) \cdot \widetilde{a}$. We must prove only that ψ is finitely supported, because the other properties of ψ required in the statement of the theorem have standard proofs as in the classical theory of modules (see Proposition 2.34 in [46]). We claim that $supp(\varphi)$ supports ψ. Let us take $\pi \in Fix(supp(\varphi))$. In the view of Proposition 2.8, it is sufficient to prove that $\psi(\pi \star g) = \pi \diamond \psi(g)$ for each $g \in \mathbb{Z}_{ext}(\Sigma)$. Let $f \in \mathbb{Z}_{ext}(\Sigma)$ be an arbitrary element. Then $f = \sum_{a \in S_f} f(a) \cdot \widetilde{a}$. Since j is equivariant and ψ is a group homomorphism, we obtain $\psi(\pi \star f) = \psi(\sum_{a \in S_f} f(a) \cdot (\pi \star \widetilde{a})) = \psi(\sum_{a \in S_f} f(a) \cdot (\widetilde{\pi \cdot a})) = \sum_{a \in S_f} f(a) \cdot \psi(\widetilde{\pi \cdot a}) = \sum_{a \in S_f} f(a) \cdot \varphi(\pi \cdot a)$. Also, $\pi \diamond \psi(f) = \pi \diamond (\sum_{a \in S_f} f(a) \cdot \varphi(a)) = \sum_{a \in S_f} f(a) \cdot (\pi \diamond \varphi(a)) = \sum_{a \in S_f} f(a) \cdot \varphi(\pi \cdot a)$; the second identity follows because $(G, +, \diamond)$ is an invariant group, and the third identity follows from Proposition 2.8 because π fixes $supp(\varphi)$ pointwise. Therefore, $supp(\varphi)$ supports ψ. □

Definition 7.3 Let $(G, \cdot, \diamond)$ be an invariant group. An *FSM subgroup* of G is a subgroup H of G which is finitely supported as an element of $\wp(G)$. We denote this by $H \leq_{fs} G$.

Example 7.3 1. Let $(G, \cdot, \diamond)$ be an invariant group. The centre of G, namely $Z(G) := \{g \in G \,|\, g \cdot x = x \cdot g \text{ for all } x \in G\}$ is an FSM (equivariant) subgroup of G, and it is itself an invariant group because it is empty supported as an element of $\wp(G)$.

Proof From the general theory, $Z(G)$ is a subgroup of G. It remains to prove that $\pi \diamond g \in Z(G)$ for all $\pi \in S_A$, $g \in Z(G)$. Let $\pi \in S_A$, $g \in Z(G)$ and $x \in G$. We have $(\pi \diamond g) \cdot x = (\pi \diamond g) \cdot ((\pi \circ \pi^{-1}) \diamond x) = (\pi \diamond g) \cdot (\pi \diamond (\pi^{-1} \diamond x)) = \pi \diamond (g \cdot (\pi^{-1} \diamond x)) = \pi \diamond ((\pi^{-1} \diamond x) \cdot g) = ((\pi \circ \pi^{-1}) \diamond x) \cdot (\pi \diamond g) = x \cdot (\pi \diamond g)$. Therefore, $\pi \diamond g \in Z(G)$. □

2. Let $(G, \cdot, \diamond)$ be an invariant group and F a finitely supported subset of G. Then, the subgroup $[F]$ generated by F is an FSM subgroup of G supported by $supp(F)$, but it is not itself an invariant group.

Proof We prove that $[F]$ is supported by $supp(F)$. Let $\pi \in Fix(supp(F))$. Let $x_1^{\varepsilon_1} \cdot x_2^{\varepsilon_2} \cdot \ldots \cdot x_n^{\varepsilon_n}$, $x_i \in F$, $\varepsilon_i = \pm 1$, $i = 1, \ldots, n$ be an arbitrary element from $[F]$.

Since $\pi \in Fix(supp(F))$, we have $\pi \diamond x_i \in F$ for all $i \in \{1,\dots,n\}$. Therefore, because the internal law on G is equivariant, we have $\pi \diamond (x_1^{\varepsilon_1} \cdot x_2^{\varepsilon_2} \cdot \dots \cdot x_n^{\varepsilon_n}) = (\pi \diamond x_1^{\varepsilon_1}) \cdot (\pi \diamond x_2^{\varepsilon_2}) \cdot \dots \cdot (\pi \diamond x_n^{\varepsilon_n}) = (\pi \diamond x_1)^{\varepsilon_1} \cdot (\pi \diamond x_2)^{\varepsilon_2} \cdot \dots \cdot (\pi \diamond x_n)^{\varepsilon_n} \in [F]$. Thus, $\pi \star [F] = [F]$, where $\star$ is the S_A-action on $\wp(G)$ defined as in Proposition 2.2, and so $supp(F)$ supports $[F]$. □

3. Let $(G, \cdot, \diamond)$ be an invariant group and $f : G \to G$ a finitely supported group automorphism. Then the set of all fixed points of f, $Fp(f) = \{g \in G \,|\, f(g) = g\}$ is an FSM subgroup of G supported by $supp(f)$.

   Proof According to Proposition 2.9, for any $\pi \in Fix(supp(f))$, whenever x is a fixed point of f, we have $f(\pi \diamond x) = \pi \diamond f(x) = \pi \diamond x$, and so $\pi \diamond x$ is another fixed point of f. Thus, $\pi \star Fp(f) = Fp(f)$ for all $\pi \in Fix(supp(f))$, which means $supp(f)$ supports $Fp(f)$. Obviously, $Fp(f)$ is a subgroup of G. □

4. Let $(G_1, \cdot, \diamond_1)$, $(G_2, \cdot, \diamond_2)$ be invariant groups and $f : G_1 \to G_2$ a finitely supported homomorphism. Then the kernel of f is an FSM normal subgroup of G_1, supported by $supp(f)$.

   Proof According to Proposition 2.9, for any $\pi \in Fix(supp(f))$, whenever $x \in Ker f$, we have $f(\pi \diamond x) = \pi \diamond f(x) = \pi \diamond e_2 = e_2$, and so $\pi \diamond x$ is in the kernel of f. Thus, $\pi \star Ker f = Ker f$ for all $\pi \in Fix(supp(f))$, which means $supp(f)$ supports $Ker f$. Obviously, $Ker f$ is a normal subgroup of G_1. □

5. Let $(G, \cdot, \diamond)$ be an invariant group and H an FSM subgroup of G. The centralizer of H in G, namely $C_G(H) = \{g \in G \,|\, g \cdot h = h \cdot g \text{ for all } h \in H\}$ is an FSM subgroup of G supported by $supp(H)$.

   Proof From the general group theory, $C_G(H)$ is a subgroup of G. We have to prove that $\pi \diamond g \in C_G(H)$ for all $\pi \in Fix(supp(H))$ and $g \in C_G(H)$. Let $\pi \in Fix(supp(H))$, $g \in C_G(H)$ and $h \in H$. Then $\pi^{-1} \diamond h \in H$. We have $(\pi \diamond g) \cdot h = (\pi \diamond g) \cdot ((\pi \circ \pi^{-1}) \diamond h) = (\pi \diamond g) \cdot (\pi \diamond (\pi^{-1} \diamond h)) = \pi \diamond (g \cdot (\pi^{-1} \diamond h)) \overset{\pi^{-1} \diamond h \in H}{=} \pi \diamond ((\pi^{-1} \diamond h) \cdot g) = ((\pi \circ \pi^{-1}) \diamond h) \cdot (\pi \diamond g) = h \cdot (\pi \diamond g)$. Thus, $\pi \diamond g \in C_G(H)$. □

6. Let $(G, \cdot, \diamond)$ be an invariant group and H an FSM subgroup of G. The normalizer of H in G, namely $N_G(H) = \{g \in G \,|\, g^{-1} \cdot H \cdot g = H\}$ is an FSM subgroup of G supported by $supp(H)$.

   Proof We know that $N_G(H)$ is a subgroup of G. It remains to prove that $\pi \diamond g \in N_G(H)$ for all $\pi \in Fix(supp(H))$ and $g \in N_G(H)$. Let $\pi \in Fix(supp(H))$ and $g \in N_G(H)$. Since H is supported by $supp(H)$ and π fixes $supp(H)$ pointwise, we obtain $\pi \diamond H = H$. We have $(\pi \diamond g)^{-1} \cdot H \cdot (\pi \diamond g) = (\pi \diamond g^{-1}) \cdot (\pi \diamond H) \cdot (\pi \diamond g) = \pi \diamond (g^{-1} \cdot H \cdot g) = \pi \diamond H = H$. Hence $\pi \diamond g \in N_G(H)$. □

7. Let $(G, \cdot, \diamond)$ be an invariant group and H, K FSM subgroups of G. The commutator subgroup$[H, K]$ generated by all the commutators $[h, k] = h^{-1} \cdot k^{-1} \cdot h \cdot k$ with $h \in H$ and $k \in K$ is an FSM subgroup of G supported by $supp(H) \cup supp(K)$.

Proof We have that $[H,K]$ is a subgroup of G. It remains to prove that $\pi \diamond g \in [H,K]$ for all $\pi \in Fix(supp(H) \cup supp(K))$ and $g \in [H,K]$. From the distributivity property of $\diamond$ and from the definition of the commutator subgroup, it is enough to prove that $\pi \diamond [h,k]$ is a commutator for all $\pi \in Fix(supp(H) \cup supp(K))$ and all $h \in H$, $k \in K$. Let $\pi \in Fix(supp(H) \cup supp(K))$, $h \in H$, $k \in K$. We have $\pi \diamond [h,k] = \pi \diamond (h^{-1} \cdot k^{-1} \cdot h \cdot k) = [\pi \diamond h, \pi \diamond k]$. This means $[H,K]$ is an FSM subgroup of G. □

8. Let $(G,\cdot,\diamond)$ be an invariant group and H,K are FSM subgroups of G with $HK = KH$. Then HK is an FSM subgroup of G supported by $supp(H) \cup supp(K)$.

   Proof From the general theory of groups, $HK = \{h \cdot k \,|\, h \in H,\ k \in K\}$ is a subgroup of G when the condition $HK = KH$ is satisfied. It remains to prove that $\pi \diamond g \in HK$ for all $\pi \in Fix(supp(H) \cup supp(K))$ and $g \in HK$. Let $\pi \in Fix(supp(H) \cup supp(K))$ and $g \in HK$. This means that $g = h \cdot k$ for some $h \in H$ and $k \in K$. From the definition of an invariant group, we have $\pi \diamond g = (\pi \diamond h) \cdot (\pi \diamond k)$. Since H and K are FSM subgroups of G, we have $\pi \diamond h \in H$ and $\pi \diamond k \in K$. Therefore, $\pi \diamond g \in HK$. □

9. Let $(G,\cdot,\diamond)$ be an invariant group and H,K FSM subgroups of G. Then $H \cap K$ is an FSM subgroup of G supported by $supp(H) \cup supp(K)$.

   Proof From the general theory, we know that $H \cap K = \{g \in G \,|\, g \in H \,and\, g \in K\}$ is a subgroup of G. It remains to prove that $\pi \diamond g \in H \cap K$ for all $\pi \in Fix(supp(H) \cup supp(K))$ and $g \in H \cap K$. Let $\pi \in Fix(supp(H) \cup supp(K))$ and $g \in H \cap K$. We obtain that $\pi \diamond g \in H$ because $\pi \in Fix(supp(H))$ and $g \in H$, and $\pi \diamond g \in K$ because $\pi \in Fix(supp(K))$ and $g \in K$. Therefore, $\pi \diamond g \in H \cap K$. □

Theorem 7.5 (Cayley Theorem)

Let $(G,\cdot,\diamond)$ be an invariant group. Then there exists an equivariant isomorphism from G to an equivariant subgroup of S_G, where S_G is the set of all finitely supported one-to-one mappings of G onto itself.

Proof For each $g \in G$ we consider the function $f_g : G \to G$ defined by $f_g(x) = g \cdot x$. Clearly, f_g is finitely supported by $supp(g)$ for all $g \in G$. Moreover, $f_g \in S_G$ for each $g \in G$. Let $H = \{f_g \,|\, g \in G\}$. Thus, H is a subgroup of S_G. We claim that H is equivariant. Indeed, if $h \in H$, we have that $h = f_g$ for some $g \in G$. Let $\pi \in S_A$. We have $(\pi \star h)(x) = \pi \diamond (f_g(\pi^{-1} \diamond x)) = \pi \diamond (g \cdot (\pi^{-1} \diamond x)) = (\pi \diamond g) \cdot x = f_{\pi \diamond g}(x)$. Hence $\pi \star h \in H$. Let $T : G \to S_G$ be the function defined by $T(g) = f_g$ for each $g \in G$. As in the standard proof of the Cayley theorem for groups, it can be proved (by direct calculation) that T is an injective group homomorphism whose image is H. It remains to prove that T is equivariant. It is sufficient to prove that $T(\pi \diamond g) = \pi \star T(g)$ for each $g \in G$ and $\pi \in S_A$ (where by $\star$ we denote the S_A-action on S_G defined in the proof of Example 7.2(2)). However, $T(\pi \diamond g) = f_{\pi \diamond g}$ and $\pi \star T(g) = \pi \star f_g$. We have just proved that $\pi \star f_g = f_{\pi \diamond g}$ for all $\pi \in S_A$. □

If $(G, \cdot, \diamond)$ is an invariant group, we denote by $\mathscr{L}(G)_{inv}$ the family of all finitely supported subgroups of G ordered by inclusion.

Theorem 7.6 *Let $(G, \cdot, \diamond)$ be an invariant group. Then the set $(\mathscr{L}(G)_{inv}, \subseteq, \star)$ is an invariant complete lattice, where $\subseteq$ represents the usual inclusion relation on $\wp(G)$ and $\star$ is the S_A-action on $\wp(G)$.*

Proof We know that $\star$ is the S_A-action on $\wp(G)$ defined as in Proposition 2.2. We claim that the restriction of $\star$ to $\mathscr{L}(G)_{inv}$ is an S_A-action on $\mathscr{L}(G)_{inv}$, that is the codomain of the restricted function $\star|_{\mathscr{L}(G)_{inv}}$ is also $\mathscr{L}(G)_{inv}$. We have to prove that for any $\pi \in S_A$ we have that $\pi \star H$ is a finitely supported subgroup of G whenever H is a finitely supported subgroup of G. Fix some $\pi \in S_A$ and $H \leq G$, H finitely supported as a subset of G. Let $\pi \diamond h_1$ and $\pi \diamond h_2$, $h_1, h_2 \in H$ be two arbitrary elements from $\pi \star H$. Since G is an invariant group (so $\cdot$ is equivariant) and because H is a subgroup of G, we have $(\pi \diamond h_1) \cdot (\pi \diamond h_2)^{-1} = (\pi \diamond h_1) \cdot (\pi \diamond h_2^{-1}) = \pi \diamond (h_1 \cdot h_2^{-1}) \in \pi \star H$. Since H is finitely supported as an element of the S_A-set $\wp(G)$, according to Proposition 2.1, we have that $\pi \star H$ is a finitely supported element in $\wp(G)$. Therefore, because $\pi \star H$ is also a subgroup of G, we have that $\pi \star H$ is a finitely supported subgroup of G. Thus, $(\mathscr{L}(G)_{inv}, \subseteq, \star)$ is an invariant set. The order relation $\subseteq$ on $\wp(G)$ is obviously an equivariant lattice order according to the definition of $\star$, and so $(\mathscr{L}(G)_{inv}, \subseteq, \star)$ is an invariant lattice.

Let $\mathscr{F} = (H_i)_{i \in I}$ be a finitely supported family of finitely supported subgroups of G. As in the proof of Theorem 7.1, we show that $\cup \mathscr{F} = \underset{i \in I}{\cup} H_i$ is supported by $supp(\mathscr{F})$. According to Example 7.3(2), we obtain that $[\underset{i \in I}{\cup} H_i]$ (which is the least upper bound of $\mathscr{F}$ in $\mathscr{L}(G)_{inv}$) is a finitely supported subgroup of G. Thus, any finitely supported family of finitely supported subgroups of G has a least upper bound in $\mathscr{L}(G)_{inv}$. □

Theorem 7.7 *Let $(G, \cdot, \diamond)$ and $(G', \cdot, \diamond)$ be invariant groups and $f : G \to G'$ a finitely supported homomorphism of groups. Then:*

1. $H \leq_{fs} G \Rightarrow f(H) \leq_{fs} G'$.
2. $H' \leq_{fs} G' \Rightarrow f^{-1}(H') \leq_{fs} G$.
3. $H' \lhd_{fs} G' \Rightarrow f^{-1}(H') \lhd_{fs} G$.

If f is surjective, then:

1'. $H \lhd_{fs} G \Rightarrow f(H) \lhd_{fs} G'$.
2'. $H \leq_{fs} G$ and $Ker f \subseteq H$ imply $H = f^{-1}(f(H))$.
3'. $H' \leq_{fs} G' \Rightarrow f(f^{-1}(H')) = H'$.
4'. There exists a finitely supported bijection between the finitely supported set of finitely supported subgroups of G containing $Ker f$ and the finitely supported (equivariant) set of finitely supported subgroups of G'.
5'. There exists a finitely supported bijection between the finitely supported set of finitely supported normal subgroups of G containing $Ker f$ and the finitely supported (equivariant) set of finitely supported normal subgroups of G'.

Proof The assertions of items 1, 2, 3, 1', 2' and 3' follow by direct calculation using the general theory (see Chapter 3 in [45]). All we have to prove is the property of being finitely supported for $f(H)$ and $f^{-1}(H')$, respectively. Let $H \leq_{fs} G$. We prove that $f(H) \leq G'$ is indeed an FSM subgroup supported by $supp(H) \cup supp(f)$, i.e. $\pi \diamond x \in f(H)$ for all $\pi \in Fix(supp(H) \cup supp(f))$ and $x \in f(H)$. Let $\pi \in Fix(supp(H) \cup supp(f))$ and $x \in f(H)$. There exists $h \in H$ such that $x = f(h)$. Since f is supported by $supp(f)$ and π fixes $supp(f)$ pointwise, by Proposition 2.9, we have $\pi \diamond x = \pi \diamond f(h) = f(\pi \diamond h) \in f(H)$. Hence $f(H)$ is an FSM subgroup of G'. Now let $H' \leq_{fs} G'$. We prove that $f^{-1}(H') \leq G$ is indeed an FSM subgroup supported by $supp(H') \cup supp(f)$. We must prove that $\pi \diamond x \in f^{-1}(H')$ for all $\pi \in Fix(supp(H') \cup supp(f))$ and $x \in f^{-1}(H')$. Let $\pi \in Fix(supp(H') \cup supp(f))$ and $x \in f^{-1}(H')$. We have $f(x) \in H'$. Since f is supported by $supp(f)$ and π fixes $supp(f)$ pointwise, by Proposition 2.9, we have $f(\pi \diamond x) = \pi \diamond f(x) \in H'$. Therefore, $\pi \diamond x \in f^{-1}(H')$, and $f^{-1}(H')$ is an FSM subgroup of G.

We prove item 4'. According to Proposition 2.1, Theorem 7.6 and Example 7.3(4), the set $A = \{H \,|\, H \leq_{fs} G, Ker f \subseteq H\}$ is a subset of G finitely supported by $supp(f)$. According to Proposition 2.1 and Theorem 7.6, the set $B = \{H' \,|\, H' \leq_{fs} G'\}$ is an equivariant subset of G. We define $\psi : A \to B$ by $\psi(H) = f(H)$ for all $H \in A$. From 1, it follows that ψ is well-defined. From 2', it follows that ψ is one-to-one. From 2 and 3', it follows that ψ is onto. We have to prove that ψ is supported by $supp(f)$. Let $\pi \in Fix(supp(f))$ and $H \in A$. Since π fixes $supp(f)$ pointwise and $supp(f)$ supports f, from Proposition 2.9 we have $\psi(\pi \star H) = f(\pi \star H) = \{f(\pi \diamond h) \,|\, h \in H\} = \{\pi \diamond f(h) \,|\, h \in H\} = \pi \star f(H) = \pi \star \psi(H)$, where by $\star$ we denoted the S_A-actions on $\wp(G)$ and $\wp(G')$, respectively. Hence, by Proposition 2.8, we have that ψ is finitely supported. Item 5' follows in a similar way. □

7.4 Fuzzy Subgroups of an Invariant Group

Definition 7.4 Let $(G, \cdot, \diamond)$ be an invariant group. An FSM fuzzy set μ over the invariant set G is called an *FSM fuzzy subgroup* of G if the following conditions are satisfied:

- $\mu(x \cdot y) \geq min\{\mu(x), \mu(y)\}$ for all $x, y \in G$;
- $\mu(x^{-1}) \geq \mu(x)$ for all $x \in G$.

An FSM fuzzy subgroup η of G that satisfies the additional condition $\eta(x \cdot y) = \eta(y \cdot x)$ for all $x, y \in G$ is called an *FSM fuzzy normal subgroup* of G.

Example 7.4 Let us consider the set A of atoms and consider $(F(A), \mathsf{T}, \widetilde{\star})$ the invariant free group over A described in Example 7.2(2). For an element in $F(A)$ of form $[w] = [x_1^{\varepsilon_1} x_2^{\varepsilon_2} \ldots x_l^{\varepsilon_l}]$, we define $s([w]) = \varepsilon_1 + \varepsilon_2 + \ldots + \varepsilon_l$. Whenever a word w is equivalent with w' modulo the reduction/inserting of terms of form xx^{-1} or $x^{-1}x$, i.e.

whenever $[w] = [w']$, we obviously have $s([w]) = s([w'])$, and so s is well-defined. It follows directly that

1. $s([w] \top [w']) = s([w]) + s([w'])$ for all $[w], [w'] \in F(A)$;
2. $s(\pi \widetilde{\star} [w]) = s([w])$ for all $\pi \in S_A$ and $[w] \in F(A)$, meaning that s is an equivariant function from $F(A)$ to $\mathbb{Z}$;
3. $s([w]^{-1}) = -s([w])$.

We proved that s is an equivariant group homomorphism between the invariant groups $F(A)$ and $\mathbb{Z}$. Let us consider $\mu : F(A) \to [0,1]$ defined by

$$\mu([w]) = \begin{cases} 0, & \text{if } s([w]) \text{ is odd in } \mathbb{Z}\,; \\ 1 - \frac{1}{n}, & \text{if } s([w]) = m \cdot 2^n \text{ with } m \text{ odd in } \mathbb{Z} \text{ and } n \in \mathbb{N}; \\ 1, & \text{if } s([w]) = 0\,. \end{cases}$$

Since every even integer k can be uniquely expressed as $k = m \cdot 2^n$ with m an odd integer and $n \in \mathbb{N}$, we have that μ is well-defined. Clearly, from Proposition 2.9 we have that μ is equivariant because, according to item 2 from above, we obtain $\mu(\pi \widetilde{\star} [w]) = \mu([w])$ for all $\pi \in S_A$ and $[w] \in F(A)$. Obviously, according to item 3 from above, $\mu([w]^{-1}) = \mu([w])$ for all $[w] \in F(A)$. It remains to prove that $\mu([w] \top [w']) \geq min\{\mu([w]), \mu([w'])\}$ for all $[w], [w'] \in F(A)$. Fix $[w], [w'] \in F(A)$. The non-trivial case to be analyzed is the case when both $s([w])$ and $s([w'])$ are non-zero even integers. Then there exist the unique expressions $s([w]) = m_1 \cdot 2^{n_1}$ and $s([w']) = m_2 \cdot 2^{n_2}$ with m_1, m_2 integer and odd. Assume $n_1 \leq n_2$ (the other case is similar). If $n_1 = n_2$, then $s([w] \top [w']) = s([w]) + s([w']) = (m_1 + m_2) \cdot 2^{n_1} = m \cdot 2^n$ with $m \in \mathbb{Z}$ odd and $n > n_1$ because $m_1 + m_2$ is non-zero and even (we considered only the case $m_1 \neq -m_2$ since the case $m_1 = -m_2$ leads to $s([w] \top [w']) = 0$, and so $\mu([w] \top [w']) = 1$ and the result follows trivially). Thus, $\mu([w] \top [w']) = 1 - \frac{1}{n} > 1 - \frac{1}{n_1} = min\{\mu([w]), \mu([w'])\}$. Suppose now $n_1 < n_2$. Then $s([w] \top [w']) = s([w]) + s([w']) = (m_1 + m_2 \cdot 2^{n_2 - n_1}) \cdot 2^{n_1}$ with $m_1 + m_2 \cdot 2^{n_2 - n_1}$ integer and odd. Thus, $\mu([w] \top [w']) = 1 - \frac{1}{n_1} = min\{1 - \frac{1}{n_1}, 1 - \frac{1}{n_2}\}$. We have that μ is an FSM fuzzy subgroup of $F(A)$.

Theorem 7.8 *Let $(G, \cdot, \diamond)$ be an invariant group. The set $FL_{inv}(G)$ consisting of all FSM fuzzy subgroups of G forms an invariant complete lattice with respect to fuzzy sets inclusion.*

Proof We firstly prove that $FL_{inv}(G)$ is an invariant poset. Clearly, $FL_{inv}(G)$ is a subset of the invariant set formed by those finitely supported functions from G to $[0,1]$. We have to prove that $FL_{inv}(G)$ is itself invariant, that is $\pi \widetilde{\star} \mu$ is an FSM fuzzy subgroup of G for all $\pi \in S_A$ and $\mu \in FL_{inv}(G)$, where $\widetilde{\star}$ is the S_A-action on $[0,1]^G$ defined in Proposition 2.7 (we denote this action by $\widetilde{\star}$ in order to avoid the confusion with the action $\star$ defined on $\wp(G)$). Let us fix $\pi \in S_A$ and $\mu \in FL_{inv}(G)$. We have $\mu(x \cdot y) \geq min\{\mu(x), \mu(y)\}$ for all $x, y \in G$, and $\mu(x^{-1}) \geq \mu(x)$ for all $x \in G$. According to Proposition 2.1, $\pi \widetilde{\star} \mu$ is a finitely supported function from G to $[0,1]$. According to Proposition 2.7 and since $[0,1]$ is a trivial invariant set (i.e. $[0,1]$ is an S_A-set equipped with the trivial S_A-action $(\sigma, x) \mapsto x$ for all $(\sigma, x) \in S_A \times [0,1]$), we have

$(\pi\widetilde{\star}\mu)(x\cdot y)=\mu(\pi^{-1}\diamond(x\cdot y))=\mu((\pi^{-1}\diamond x)\cdot(\pi^{-1}\diamond y))\geq min\{\mu(\pi^{-1}\diamond x),\mu(\pi^{-1}\diamond y)\}=min\{(\pi\widetilde{\star}\mu)(x),(\pi\widetilde{\star}\mu)(y)\}$ for all $x,y\in G$; the second identity holds because G is an invariant group, and so the internal law on G is equivariant. Moreover, $(\pi\widetilde{\star}\mu)(x^{-1})=\mu(\pi^{-1}\diamond(x^{-1}))=\mu((\pi^{-1}\diamond x)^{-1})\geq\mu(\pi^{-1}\diamond x)=(\pi\widetilde{\star}\mu)(x)$; the second identity holds from Proposition 7.3(2). This means that $(FL_{inv}(G),\widetilde{\star})$ is an invariant set.

Now we prove that $(FL_{inv}(G),\sqsubseteq)$ is an invariant poset where $\sqsubseteq$ is the classical inclusion order on the family of all FSM fuzzy subgroups of G, defined by $\mu\sqsubseteq\eta$ if and only if $\mu(x)\leq\eta(x)$ for all $x\in G$. We have to prove that $\sqsubseteq$ is equivariant. Indeed, let $\pi\in S_A$, and μ,η be two FSM fuzzy subgroups of G such that $\mu\sqsubseteq\eta$. Since $\mu(x)\leq\eta(x)$ for all $x\in G$, we have $\mu(\pi^{-1}\cdot x)\leq\eta(\pi^{-1}\cdot x)$ for all $x\in G$, namely $(\pi\widetilde{\star}\mu)(x)\leq(\pi\widetilde{\star}\eta)(x)$ for all $x\in G$. It follows that $\pi\widetilde{\star}\mu\sqsubseteq\pi\widetilde{\star}\eta$.

For each $\alpha\in[0,1]$ and $\mu\in[0,1]^G_{fs}$, we define $G^\mu_\alpha=\{x\in G\mid\mu(x)\geq\alpha\}$. We prove that each G^μ_α is a finitely supported subset of G, and furthermore, $supp(G^\mu_\alpha)\subseteq supp(\mu)$, for all $\alpha\in[0,1]$. Indeed, let us fix $\alpha\in[0,1]$ and $\mu\in FL_{inv}(G)$. Recall that $[0,1]$ is a trivial invariant set. Let $\pi\in Fix(supp(\mu))$. According to Proposition 2.8, we have $\mu(\pi\diamond x)=\mu(x)$ for all $x\in G$. Thus, for each $x\in G^\mu_\alpha$ and each $\pi\in Fix(supp(\mu))$ we have $\pi\diamond x\in G^\mu_\alpha$. Therefore, $\pi\star G^\mu_\alpha\subseteq G^\mu_\alpha$ for each $\pi\in Fix(supp(\mu))$, where $\star$ is the S_A-action on $\wp(G)$ defined as in Proposition 2.2. By contradiction, let us assume that there is a $\pi\in Fix(supp(\mu))$ such that $\pi\star G^\mu_\alpha\subsetneq G^\mu_\alpha$. By induction, we get $\pi^n\star G^\mu_\alpha\subsetneq G^\mu_\alpha$ for all $n\geq 1$. However, π is a (finite) permutation, and so there exists $k\in\mathbb{N}$ such that $\pi^k=Id$. We obtain $G^\mu_\alpha\subsetneq G^\mu_\alpha$, a contradiction. Therefore, $\pi\star G^\mu_\alpha=G^\mu_\alpha$ for all $\pi\in Fix(supp(\mu))$, and so G^μ_α is supported by $supp(\mu)$. Since the support of G^μ_α is the least finite set supporting G^μ_α, we obtain that $supp(G^\mu_\alpha)\subseteq supp(\mu)$.

Let $[G^\mu_\alpha]$ be the subgroup of G generated by G^μ_α, i.e. the smallest subgroup of G containing G^μ_α. As in Example 7.3(2), we show that each subgroup $[G^\mu_\alpha]$ generated by G^μ_α is finitely supported by $supp(\mu)$.

For any finitely supported function $\nu:G\to[0,1]$, we define the function $\nu^*:G\to[0,1]$ by $\nu^*(x)=supremum\{\alpha\in[0,1]\,|\,x\in[G^\nu_\alpha]\}$ for any $x\in G$. We claim that $\nu^\star$ is supported by $supp(\nu)$. Let $\pi\in Fix(supp(\nu))$. We have $\nu^*(\pi\diamond x)=supremum\{\alpha\in[0,1]\mid\pi\diamond x\in[G^\nu_\alpha]\}=supremum\{\alpha\in[0,1]\mid x\in\pi^{-1}\star[G^\nu_\alpha]\}$. However, $\pi^{-1}\in Fix(supp(\nu))$ and $supp([G^\nu_\alpha])\subseteq supp(\nu)$ for all $\alpha\in[0,1]$ (meaning that there exists a set of atoms not depending on α which supports all $[G^\nu_\alpha]$), and so $\pi^{-1}\star[G^\nu_\alpha]=[G^\nu_\alpha]$. Therefore, $\nu^*(\pi\diamond x)=supremum\{\alpha\in[0,1]\,|\,x\in[G^\nu_\alpha]\}=\nu^*(x)$ for all $x\in G$. Thus, ν^* is finitely supported. Furthermore, as in the standard fuzzy groups theory we have that ν^* is a fuzzy subgroup of G [1], and so it is an FSM fuzzy subgroup of G.

In order to prove that $FL_{inv}(G)$ is an invariant complete lattice it remains to establish that any finitely supported family of elements from $FL_{inv}(G)$ has a least upper bound. Let us consider now $\mathscr{F}=(\mu_i)_{i\in I}$ a finitely supported family of elements from $FL_{inv}(G)$. We define $\bigsqcup_{i\in I}\mu_i:G\to[0,1]$ by $\bigsqcup_{i\in I}\mu_i(x)=\underset{i\in I}{supremum}\{\mu_i(x)\mid i\in I\}$ for all $x\in G$. Since $[0,1]$ is a ZF complete lattice, from Theorem 7.3 we have that $supp(\mathscr{F})$ supports $\bigsqcup_{i\in I}\mu_i$. Furthermore, we have that $(\bigsqcup_{i\in I}\mu_i)^*$ is finitely supported by

$supp(\mathscr{F})$. Moreover, as in the standard fuzzy groups theory we have that $(\bigsqcup_{i\in I}\mu_i)^*$ is the least upper bound of $\mathscr{F}$ in $FL(G)$ with respect to the order relation $\sqsubseteq$ [1]. Since $(\bigsqcup_{i\in I}\mu_i)^*$ is also finitely supported, it follows that $(\bigsqcup_{i\in I}\mu_i)^*$ is the least upper bound of $\mathscr{F}$ in $FL_{inv}(G)$ with respect to the equivariant order relation $\sqsubseteq$. □

Theorem 7.9 *Let $(G,\cdot,\diamond)$ be an invariant group. The set $FN_{inv}(G)$ consisting of all FSM fuzzy normal subgroups of G forms an invariant modular lattice with respect to fuzzy sets inclusion.*

Proof We prove that $FN_{inv}(G)$ is an invariant subset of the invariant set $FL_{inv}(G)$, that is $\pi\tilde{\star}\mu$ is an FSM fuzzy normal subgroup of G for all $\pi\in S_A$ and $\mu\in FN_{inv}(G)$, where $\tilde{\star}$ is the S_A-action on $FL_{inv}(G)$ defined as in Proposition 2.7. Fix $\pi\in S_A$ and $\mu\in FN_{inv}(G)$. We have $\mu(x\cdot y)=\mu(y\cdot x)$ for all $x,y\in G$. According to Proposition 2.1, $\pi\tilde{\star}\mu$ is a finitely supported function from G to $[0,1]$. According to Proposition 2.7, because $[0,1]$ is a trivial invariant set, and because the internal law in G is equivariant, we have $(\pi\tilde{\star}\mu)(x\cdot y)=\mu(\pi^{-1}\diamond(x\cdot y))=\mu((\pi^{-1}\diamond x)\cdot(\pi^{-1}\diamond y))=\mu((\pi^{-1}\diamond y)\cdot(\pi^{-1}\diamond x))=\mu(\pi^{-1}\diamond(y\cdot x))=(\pi\tilde{\star}\mu)(y\cdot x)$ for all $x,y\in G$. This means $\pi\tilde{\star}\mu$ is an FSM fuzzy normal subgroup of G, and so $FN_{inv}(G)$ is an invariant subset of $FL_{inv}(G)$.

As in the proof of Theorem 7.8, it follows that the inclusion order on $FN_{inv}(G)$ is equivariant. Furthermore, for any finitely supported functions $\mu,\eta:G\to[0,1]$ we have that the functions $\mu\cup\eta:G\to[0,1]$ defined by $(\mu\cup\eta)(x)=max[\mu(x),\eta(x)]$ for all $x\in G$ and $\mu\cap\eta:G\to[0,1]$ defined by $(\mu\cap\eta)(x)=min[\mu(x),\eta(x)]$ for all $x\in G$ are both finitely supported by $supp(\mu)\cup supp(\eta)$. As in the proof of Theorem 7.8, we also have that $(\mu\cup\eta)^*$ is supported by $supp(\mu)\cup supp(\eta)$. Similar to the standard ZF theory of fuzzy sets, we have that $(\mu\cup\eta)^*$ and $\mu\cap\eta$ are the least upper bound and the greatest lower bound of μ and η in $FN_{inv}(G)$, respectively. This means that $(\mu\cup\eta)^*$ and $\mu\cap\eta$ are the least upper bound and the greatest lower bound of μ and η in $FN_{inv}(G)$, respectively. Therefore, $FN_{inv}(G)$ is an invariant lattice. The modularity property of $FN_{inv}(G)$ holds as in the ZF (see [2]) due to the special property of fuzzy normal subgroups described in Definition 7.4. □

Chapter 8
Galois Connections in Finitely Supported Mathematics

Abstract We introduce and study Galois connections between finitely supported ordered structures. Particularly, we present properties of finitely supported Galois connections between invariant complete lattices. As an application, we investigate upper and lower approximations of finitely supported sets using the approximation techniques from the theory of rough sets translated into the framework of atomic sets with finite supports.

8.1 Basic Definitions

Galois connections are pairs of mappings which allow us to move back and forth between two different structures. After an element is mapped to the other structure and back, a certain stability is reached in such a way that further mappings give the same results. Furthermore, the image sets of the mappings forming the Galois connection are isomorphic. The ZF concept of Galois connection was firstly translated into FSM in [7]. However, in [7] we considered and studied only the invariant Galois connections by employing only equivariant adjunctions. In the view of the S-finite support principle from FSM which generalizes the classical equivariance principle of Pitts, we can reformulate the related results from [7] by generalizing 'equivariant' to 'finitely supported'.

Definition 8.1 Let $(P, \sqsubseteq_P, \cdot_P)$ and $(Q, \sqsubseteq_Q, \cdot_Q)$ be two invariant posets, and $f : P \to Q$, $g : Q \to P$ two functions. The pair (f, g) is a *finitely supported Galois connection* between P and Q if the following conditions are satisfied:

- f and g are finitely supported functions;
- for all $p \in P$ and $q \in Q$ we have that $f(p) \sqsubseteq_Q q$ if and only if $p \sqsubseteq_P g(q)$.

The function g is called the finitely supported adjoint (of f) and f is called the finitely supported co-adjoint (of g). The following FSM characterization of Galois connections can be directly proved as in the ZF framework.

A. Alexandru, G. Ciobanu, *Foundations of Finitely Supported Structures*,
https://doi.org/10.1007/978-3-030-52962-8_8

Proposition 8.1 *Let $(P, \sqsubseteq_P, \cdot_P)$ and $(Q, \sqsubseteq_Q, \cdot_Q)$ be two invariant posets, and $f : P \to Q$, $g : Q \to P$ two functions. The pair (f, g) is a finitely supported Galois connection between P and Q if and only if the following conditions are satisfied:*

- *f and g are finitely supported monotone functions;*
- *for all $p \in P$ and $q \in Q$ we have that $f(g(q)) \sqsubseteq_Q q$ and $p \sqsubseteq_P g(f(p))$.*

Example 8.1 Let $(U, \cdot)$ be an invariant set and ε a finitely supported equivalence relation on U. We denote by $[x]_\varepsilon$ the equivalence class of an element $x \in U$, i.e. $[x]_\varepsilon = \{y \,|\, y\varepsilon x\}$. We define the following functions (also named FSM Pawlak approximations of ε):

- $\underline{\varepsilon} : \wp_{fs}(U) \to \wp_{fs}(U)$, $\underline{\varepsilon}(X) = \{x \in U \,|\, [x]_\varepsilon \subseteq X\}$;
- $\overline{\varepsilon} : \wp_{fs}(U) \to \wp_{fs}(U)$, $\overline{\varepsilon}(X) = \{x \in U \,|\, [x]_\varepsilon \cap X \neq \emptyset\}$.

The pair $(\overline{\varepsilon}, \underline{\varepsilon})$ is a finitely supported Galois connection on $\wp_{fs}(U)$.

Proof We know that $(\wp_{fs}(U), \star)$ is an invariant set with the notations in Proposition 2.2. Let $\pi \in Fix(supp(\varepsilon))$ and $x \in U$. We have $\pi \star [x]_\varepsilon = \{\pi \cdot y \,|\, y\varepsilon x\}$ and $[\pi \cdot x]_\varepsilon = \{z \,|\, z\varepsilon(\pi \cdot x)\}$. Let $t \in \pi \star [x]_\varepsilon$. Then $t = \pi \cdot y$, for some $y\varepsilon x$. Since ε is finitely supported and π fixes its support pointwise, we have $(\pi \cdot y)\varepsilon(\pi \cdot x)$, and so $t \in [\pi \cdot x]_\varepsilon$. Conversely, if $t \in [\pi \cdot x]_\varepsilon$, then $t\varepsilon(\pi \cdot x)$. Since ε is finitely supported and π fixes its support pointwise, we also have $(\pi^{-1} \cdot t)\varepsilon x$, that is $t = \pi \cdot (\pi^{-1} \cdot t) \in \pi \star [x]_\varepsilon$. Therefore, $\pi \star [x]_\varepsilon = [\pi \cdot x]_\varepsilon$ for all $\pi \in Fix(supp(\varepsilon))$ and all $x \in U$.

In order to prove that $\underline{\varepsilon}$ and $\overline{\varepsilon}$ are well-defined, we have to prove that the sets $C_X = \{x \in U \,|\, [x]_\varepsilon \subseteq X\}$ and $D_X = \{x \in U \,|\, [x]_\varepsilon \cap X \neq \emptyset\}$ are finitely supported for each $X \in \wp_{fs}(U)$. This follows from S-finite support principle. Below we provide the full calculation. Fix some $X \in \wp_{fs}(U)$. We claim that C_X and D_X are supported by $supp(X) \cup supp(\varepsilon)$. Let $\pi \in Fix(supp(X) \cup supp(\varepsilon))$. Thus, $\pi \star X = X$. Let $c \in C_X$, that is $[c]_\varepsilon \subseteq X$. Since $\pi \in Fix(supp(\varepsilon))$, we have $[\pi \cdot c]_\varepsilon = \pi \star [c]_\varepsilon \subseteq \pi \star X = X$. Therefore, $\pi \cdot c \in C_X$, and so C_X is finitely supported. Let $d \in D_X$, that is $[d]_\varepsilon \cap X \neq \emptyset$. There exists $d_1 \in [d]_\varepsilon$ and $d_1 \in X$. Since π fixes both $supp(X)$ and $supp(\varepsilon)$, we have $\pi \cdot d_1 \in \pi \star [d]_\varepsilon = [\pi \cdot d]_\varepsilon$ and $\pi \cdot d_1 \in X$ Therefore, $\pi \cdot d_1 \in [\pi \cdot d]_\varepsilon \cap X$, and so $\pi \cdot d \in D_X$ and D_X is finitely supported.

It remains to prove that $\underline{\varepsilon}$ and $\overline{\varepsilon}$ are finitely supported (by $supp(\varepsilon)$). This follows directly from the S-finite support principle, but we also provide an alternative direct proof. Let $\pi \in Fix(supp(\varepsilon))$ and $X \in \wp_{fs}(U)$. We have to prove that $\underline{\varepsilon}(\pi \star X) = \pi \star \underline{\varepsilon}(X)$ and $\overline{\varepsilon}(\pi \star X) = \pi \star \overline{\varepsilon}(X)$. Let $y \in \underline{\varepsilon}(\pi \star X)$, that is $[y]_\varepsilon \subseteq \pi \star X$. We have $y = \pi \cdot (\pi^{-1} \cdot y)$, where $[\pi^{-1} \cdot y]_\varepsilon = \pi^{-1} \star [y]_\varepsilon \subseteq \pi^{-1} \star (\pi \star X) = X$. Therefore, $y \in \pi \star \underline{\varepsilon}(X)$. Conversely, let $y \in \pi \star \underline{\varepsilon}(X)$, that is $y = \pi \cdot x$ where $[x]_\varepsilon \subseteq X$. We have $[y]_\varepsilon = [\pi \cdot x]_\varepsilon = \pi \star [x]_\varepsilon \subseteq \pi \star X$, and so $y \in \underline{\varepsilon}(\pi \star X)$.

Now, let $y \in \overline{\varepsilon}(\pi \star X)$, that is $[y]_\varepsilon \cap (\pi \star X) \neq \emptyset$. Let $z \in [y]_\varepsilon \cap (\pi \star X)$. We have $\pi^{-1} \cdot z \in \pi^{-1} \star [y]_\varepsilon = [\pi^{-1} \cdot y]_\varepsilon$ and $\pi^{-1} \cdot z \in \pi^{-1} \star (\pi \star X) = X$. Therefore, $[\pi^{-1} \cdot y]_\varepsilon \cap X \neq \emptyset$. Since $y = \pi \cdot (\pi^{-1} \cdot y)$, we have $y \in \pi \star \overline{\varepsilon}(X)$. Conversely, let $y \in \pi \star \overline{\varepsilon}(X)$. We have $y = \pi \cdot x$ where $[x]_\varepsilon \cap X \neq \emptyset$. If $z \in [x]_\varepsilon \cap X$, then $\pi \cdot z \in [y]_\varepsilon \cap (\pi \star X)$. Therefore, $[y]_\varepsilon \cap (\pi \star X) \neq \emptyset$, and so $y \in \overline{\varepsilon}(\pi \star X)$.

Now, fix some $X \in \wp_{fs}(U)$. Let x be an arbitrary element of X. Let $y \in [x]_\varepsilon$. Since ε is an equivalence relation, we also have $x \in [y]_\varepsilon$. However, $x \in X$, and so

$[y]_\varepsilon \cap X \neq \emptyset$. Therefore, $[x]_\varepsilon \subseteq \overline{\varepsilon}(X)$. Since x is an arbitrary element from X, we obtain $X \subseteq \underline{\varepsilon}(\overline{\varepsilon}(X))$. Now, let x be an arbitrary element from $\overline{\varepsilon}(\underline{\varepsilon}(X))$, that is $[x]_\varepsilon \cap \underline{\varepsilon}(X) \neq \emptyset$. Let $y \in [x]_\varepsilon \cap \underline{\varepsilon}(X)$. This means $x \in [y]_\varepsilon$, and $[y]_\varepsilon \subseteq X$. Therefore, $x \in X$, and so $\overline{\varepsilon}(\underline{\varepsilon}(X)) \subseteq X$. According to Proposition 8.1, we have that $(\overline{\varepsilon}, \underline{\varepsilon})$ is a finitely supported Galois connection on $\wp_{fs}(U)$. □

8.2 Basic Properties

Proposition 8.2 *Let $(P, \sqsubseteq_P, \cdot_P)$ and $(Q, \sqsubseteq_Q, \cdot_Q)$ be two invariant posets and (f, g) a finitely supported Galois connection between P and Q. The function f is a finitely supported complete join-morphism and g is a finitely supported complete meet-morphism.*

Proof We prove only that f is a finitely supported complete join-morphism (the other part can be proved in a similar way). Let $X \subseteq P$ be finitely supported such that $\sqcup X$ exists in P. Since there exists a finite set S supporting X, we obtain that the set $f(X) = \{f(p) \,|\, p \in X\}$ is also supported by $S \cup supp(f)$. Indeed, for each $p \in X$ and each $\pi \in Fix(supp(f))$ we have $\pi \cdot_Q f(p) = f(\pi \cdot_P p)$. Let $\pi \in Fix(S \cup supp(f))$ and $p \in X$. Since π fixes S pointwise and S supports X, we have $\pi \cdot_P p \in X$. Therefore, $\pi \cdot_Q f(p) \in f(X)$ for all $\pi \in Fix(S \cup supp(f))$ and $x \in X$. This means that $S \cup supp(f)$ supports $f(X)$. The rest of the proof is standard and does not involve specific FSM results - see Proposition 82(e) from [36]. □

Proposition 8.3 *Let $(L, \sqsubseteq_L, \cdot_L)$ and $(K, \sqsubseteq_K, \cdot_K)$ be two invariant complete lattices.*

- *A finitely supported function $f : L \to K$ has a finitely supported adjoint if and only if f is a finitely supported complete join-morphism.*
- *A finitely supported function $g : K \to L$ has a finitely supported co-adjoint if and only if g is a finitely supported complete meet-morphism.*

Proof According to Proposition 8.2, whenever f has a finitely supported adjoint we have that f is a finitely supported complete join-morphism. Conversely, let f be a finitely supported complete join-morphism. For each $k \in K$ we consider the set $X_k = \{l \in L \,|\, f(l) \sqsubseteq_K k\}$. We claim that $supp(k) \cup supp(f)$ supports X_k. Let $\pi \in Fix(supp(k) \cup supp(f))$, and $l \in X_k$ be arbitrarily chosen. Then $f(l) \sqsubseteq_K k$. Since $\sqsubseteq$ is equivariant, we also have $\pi \cdot_K f(l) \sqsubseteq_K \pi \cdot_K k$. Since f is supported by $supp(f)$ and π fixes $supp(k)$ and $supp(f)$ pointwise, from Proposition 2.9 we have that $f(\pi \cdot_L l) = \pi \cdot_K f(l) \sqsubseteq_K \pi \cdot k = k$. Therefore, $\pi \star X_k \subseteq X_k$, and so (because π is an element with finite order in the group S_A), $X_k = \pi \star X_k$. Since X_k is finitely supported, there exists its supremum $\sqcup X_k$. We define $g : K \to L$ by $g(k) = \sqcup X_k$ for each $k \in K$. Since $\sqsubseteq_K$ is equivariant, we have for each $\pi \in Fix(supp(f))$ that $\{l \in L \,|\, f(l) \sqsubseteq_K \pi \cdot_K k\} = \{l \in L \,|\, f(\pi^{-1} \cdot_L l) \sqsubseteq_K k\} = \{\pi \cdot_L l \in L \,|\, f(l) \sqsubseteq_K k\} = \pi \star \{l \in L \,|\, f(l) \sqsubseteq_K k\}$. Furthermore, $g(\pi \cdot_K k) = \sqcup\{l \in L \,|\, f(l) \sqsubseteq_K \pi \cdot_K k\}$ and $\pi \cdot_L g(k) = \pi \cdot_L \sqcup\{l \in L \,|\, f(l) \sqsubseteq_K k\} = \sqcup(\pi \star \{l \in L \,|\, f(l) \sqsubseteq_K k\})$. Therefore, because we obtained $g(\pi \cdot_K k) = \pi \cdot_L g(k)$

for all $k \in K$ and all $\pi \in Fix(supp(f))$, according to Proposition 2.9, we have that g is supported by $supp(f)$.

We have to check the rest of the conditions in Definition 8.1. Let us consider $l \in L$ and $k \in K$. If $f(l) \sqsubseteq_K k$, then $l \in X_k$, which means $l \sqsubseteq_L \sqcup X_k = g(k)$. Conversely, assume $l \sqsubseteq_L g(k) = \sqcup\{l \in L \,|\, f(l) \sqsubseteq_K k\}$. Since $X_k = \{l \in L \,|\, f(l) \sqsubseteq_K k\}$ is supported by $supp(k) \cup supp(f)$, according to Proposition 2.9, we obtain that $\{f(l) \in L \,|\, f(l) \sqsubseteq_K k\} = f(X_k)$ is also supported by $supp(k) \cup supp(f)$. Therefore, there exists $\sqcup\{f(l) \in L \,|\, f(l) \sqsubseteq_K k\}$. Since f is a monotone complete join-morphism, we have $f(l) \sqsubseteq_K \sqcup\{f(l) \in L \,|\, f(l) \sqsubseteq_K k\} \sqsubseteq_K k$. □

In the following we study the finitely supported conjugate functions on an invariant Boolean lattice, and we establish a connection between the finitely supported Galois connections and finitely supported conjugate function pairs.

Definition 8.2 Let $(L, \sqsubseteq, \cdot)$ be an invariant Boolean lattice, and $f, g : L \to L$ two functions. We say that g is a *finitely supported conjugate* of f if the following conditions are satisfied:

- f and g are finitely supported;
- for all $x, y \in L$ we have $x \wedge f(y) = 0$ if and only if $y \wedge g(x) = 0$.

Clearly, if g is a finitely supported conjugate of f, then f is also a finitely supported conjugate of g. If a function f is the finitely supported conjugate of itself, we say that f is a *finitely supported self-conjugate*.

Proposition 8.4 *Let $(L, \sqsubseteq, \cdot)$ be an invariant complete Boolean lattice, and $f : L \to L$ a function on L. Then f has a finitely supported conjugate if and only if f is a finitely supported complete join-morphism.*

Proof Let $f : L \to L$ be a function on L which has a finitely supported conjugate. A simple calculation which does not involve FSM results show us that f is monotone. Let $X \subseteq L$ be a finitely supported subset. According to the proof of Proposition 8.2, because f is finitely supported, $f(X)$ is also finitely supported and $\sqcup X$, $\sqcup f(X)$ exist. The relation $f(\sqcup X) = \sqcup f(X)$ follows directly, using the technical details presented in Proposition 86 from [36].

Conversely, let us suppose that f is a finitely supported complete join-morphism. For each $y \in L$ we consider the set $X_y = \{x \,|\, f(x) \sqsubseteq y'\}$. We claim that $supp(f) \cup supp(y') = supp(f) \cup supp(y)$ supports X_y. Let $\pi \in Fix(supp(f) \cup supp(y'))$, and $x \in X_y$ be arbitrarily chosen. Then $f(x) \sqsubseteq y'$. Since $\sqsubseteq$ is equivariant, we also have $\pi \cdot f(x) \sqsubseteq \pi \cdot y'$. Thus, $f(\pi \cdot x) = \pi \cdot f(x) \sqsubseteq \pi \cdot y' = y'$. Therefore, $\pi \star X_y \subseteq X_y$ whenever $\pi \in Fix(supp(f) \cup supp(y'))$, and so $X_y = \pi \star X_y$. Since X_y is finitely supported, there exists its supremum $\sqcup X_y$. We define $g : L \to L$ by $g(y) = (\sqcup X_y)'$ for each $y \in L$. According to Proposition 2.9 we have for each $\pi \in Fix(supp(f))$ that $f(\pi^{-1} \cdot x) \sqsubseteq y' \Leftrightarrow \pi^{-1} \cdot f(x) \sqsubseteq y' \Leftrightarrow f(x) \sqsubseteq \pi \cdot y'$. Therefore, $\{x \in L \,|\, f(x) \sqsubseteq \pi \cdot y'\}$ $= \{x \in L \,|\, f(\pi^{-1} \cdot x) \sqsubseteq y'\} = \{\pi \cdot x \in L \,|\, f(x) \sqsubseteq y'\} = \pi \star \{x \in L \,|\, f(x) \sqsubseteq y'\}$. Thus, $g(\pi \cdot y) = (\sqcup\{x \in L \,|\, f(x) \sqsubseteq (\pi \cdot y)'\})' = (\sqcup\{x \in L \,|\, f(x) \sqsubseteq \pi \cdot y'\})'$ for each $y \in L$. According to Proposition 6.3 and Proposition 6.2, we have $\pi \cdot g(y) = \pi \cdot (\sqcup\{x \in L \,|\, f(x) \sqsubseteq y'\})' = (\pi \cdot \sqcup\{x \in L \,|\, f(x) \sqsubseteq y'\})' = (\sqcup(\pi \star \{x \in L \,|\, f(x) \sqsubseteq y'\}))'$ for each

$y \in L$. Therefore, g is supported by $supp(f)$. According to the definition of g and Proposition 6.4, we can say that $g(y) = \sqcap\{x' \,|\, f(x) \wedge y = 0\}$. According to Proposition 6.3 and Proposition 6.1, and because f is finitely supported, we have that $\{x' \,|\, f(x) \wedge y = 0\}$ is supported by $supp(y) \cup supp(f)$ for each $y \in L$. Therefore, its infimum exists according to Theorem 6.1. Let us consider $x, y \in L$. If $f(x) \wedge y = 0$, then $g(y) \sqsubseteq x'$, i.e. $g(y) \wedge x = 0$. On the other hand, because X_y is finitely supported we have that $f(X_y)$ is also finitely supported. Therefore, $f(g(y)')$=$f(\sqcup\{x \,|\, f(x) \sqsubseteq y'\})$=$\sqcup\{f(x) \,|\, f(x) \sqsubseteq y'\} \sqsubseteq y'$. If x and y are such that $g(y) \wedge x = 0$, then $x \sqsubseteq g(y)'$. Thus, $f(x) \sqsubseteq f(g(y)') \sqsubseteq y'$, which means $f(x) \wedge y = 0$. □

Definition 8.3 Let $(L, \sqsubseteq, \cdot)$ be an invariant complete Boolean lattice, and $f, g : L \to L$ two functions. We say that g is the *finitely supported dual* of f if the following conditions are satisfied:

- f and g are finitely supported;
- for any $x \in L$ we have $f(x') = g(x)'$.

For any finitely supported function f over an invariant complete Boolean lattice $(L, \sqsubseteq, \cdot)$, there exists a function $g : L \to L$ defined by $g(x) = f(x')'$ for all $x \in L$. According to Proposition 6.3, we have $g(\pi \cdot x) = f((\pi \cdot x)')' = f(\pi \cdot x')'$. On the other hand, because f is finitely supported, we obtain $\pi \cdot g(x) = \pi \cdot f(x')' = (\pi \cdot f(x'))' = f(\pi \cdot x')'$ for all $\pi \in Fix(supp(f))$ and all $x \in L$. Therefore, g is supported by $supp(f)$, and g is the finitely supported dual of f. The finitely supported dual of f is denoted by f^{δ}.

Proposition 8.5 *Let $(L, \sqsubseteq, \cdot)$ be an invariant complete Boolean lattice. A function $f : L \to L$ is a finitely supported complete join-morphism if and only if f^{δ} is a finitely supported complete meet-morphism.*

Proof According to Proposition 6.3, whenever X is a subset of L supported by a finite set S we have that $\{x' \,|\, x \in X\}$ is supported by S. Also, because f is finitely supported, whenever X is a subset of L supported by a finite set S we have that $f(X)$ is supported by $S \cup supp(f)$. Suppose that $f(\sqcup Y) = \sqcup f(Y)$ for all finitely supported subsets Y of L. Then, according to Proposition 6.4, we have $f^{\delta}(\sqcap Y)$=$(f((\sqcap Y)'))'$ =$(f(\sqcup\{x' \,|\, x \in Y\}))'$=$(\sqcup\{f(x') \,|\, x \in Y\})'$ =$\sqcap\{f(x')' \,|\, x \in Y\}$=$\sqcap\{f^{\delta}(x) \,|\, x \in Y\}$. Thus, f^{δ} is a finitely supported (supported by $supp(f)$) complete meet-morphism. The converse holds similarly. □

In the ZF framework, according to [36], if L is a complete Boolean lattice and $f : L \to L$ is a complete join-morphism, then f has a unique adjoint. Therefore, if L is an invariant complete Boolean lattice and $f : L \to L$ is a finitely supported complete join-morphism, then f has at most one finitely supported adjoint because each finitely supported adjoint of f has to satisfy the second condition in Definition 8.1 which is in fact the ZF condition of being a ZF adjoint of f. Moreover, according to Proposition 8.3, if L is an invariant complete Boolean lattice and $f : L \to L$ is a finitely supported complete join-morphism, then f has at least one finitely supported adjoint. Therefore, if L is an invariant complete Boolean lattice and $f : L \to L$

is finitely supported complete join-morphism, then f has a unique finitely supported adjoint which can be defined in the following way.

Theorem 8.1 *Let $(L, \sqsubseteq, \cdot)$ be an invariant complete Boolean lattice.*

- *For any finitely supported complete join-morphism $f : L \to L$, its finitely supported adjoint is the finitely supported dual of the finitely supported conjugate of f.*
- *For any finitely supported complete meet-morphism $g : L \to L$, its finitely supported co-adjoint is the finitely supported conjugate of the finitely supported dual of g.*

Proof Let $f : L \to L$ be a finitely supported complete join-morphism. According to Proposition 8.3, it has a finitely supported adjoint $f^{adj} : L \to L$ defined by $f^{adj}(x) = \sqcup\{y \,|\, f(y) \sqsubseteq x\}$ for all $x \in L$. According to Proposition 8.4 and Proposition 6.4, the finitely supported conjugate of f is the function $g : L \to L$ defined by $g(x) = (\sqcup\{y \,|\, f(y) \sqsubseteq x'\})' = \sqcap\{y' \,|\, f(y) \sqsubseteq x'\}$ for all $x \in L$. The finitely supported dual of g is $g^{\delta} : L \to L$ defined by $g^{\delta}(x) = g(x')'$ for all $x \in L$. A simple calculation which does not involve specific FSM results shows us that $g^{\delta} = f^{adj}$. The second part of the theorem can be proved in a similar way. □

Theorem 8.2 *Let $(L, \sqsubseteq, \cdot)$ be an invariant complete Boolean lattice and (f, f^{adj}) a finitely supported Galois connection on L. If f is finitely supported self-conjugate and $x \sqsubseteq f(x)$ for all $x \in L$, then the set of all fixed points of f forms a finitely supported complete Boolean sublattice of L.*

Proof We denote by P the set of all fixed points of f. Since f has an adjoint, it is monotone (order preserving). Since f is finitely supported, by Theorem 6.2, we know that $(P, \cdot)$ is a finitely supported complete sublattice of L. According to Proposition 8.1, since f is finitely supported self-conjugate, its finitely supported adjoint coincides with its finitely supported dual. Therefore, $f^{adj} = f^{\delta}$. If $x \in P$, then $f^{\delta}(x') = f(x)' = x'$. Moreover, by Proposition 8.1, we have $x' \sqsupseteq f(f^{\delta}(x')) = f(x')$. However, $x' \sqsubseteq f(x')$, that is $x' \in P$. □

Example 8.2 Let $(U, \cdot)$ be an invariant set and ε a finitely supported equivalence relation on U. The set of all fixed points of $\overline{\varepsilon}$ (defined in Example 8.1) forms a finitely supported complete Boolean sublattice of $\wp_{fs}(U)$.

Proof We know that $(\wp_{fs}(U), \subseteq, \star)$ is an invariant complete Boolean lattice. According to Example 8.1, we have that $X \subseteq \overline{\varepsilon}(X)$ for all $X \in \wp_{fs}(U)$. Also, from Example 8.1, we can prove that $\overline{\varepsilon}^{\delta}(X) = U \setminus (\overline{\varepsilon}(U \setminus X)) = \underline{\varepsilon}(X)$ for all $X \in \wp_{fs}(X)$, that is $\overline{\varepsilon}^{\delta} = \underline{\varepsilon}$. Since $(\overline{\varepsilon}, \underline{\varepsilon})$ is a finitely supported Galois connection on $\wp_{fs}(U)$, according to Theorem 8.1 we have that the finitely supported conjugate of $\overline{\varepsilon}$ coincides with $\overline{\varepsilon}$, that is $\overline{\varepsilon}$ is finitely supported self-conjugate. According to Theorem 8.2, the set of all fixed points of $\overline{\varepsilon}$ forms a finitely supported complete Boolean sublattice of $\wp_{fs}(U)$. □

Chapter 9
Several Forms of Infinity for Finitely Supported Structures

Abstract The theory of finitely supported sets allows the study of structures which are very large, but contain enough symmetries such that they can be concisely represented and manipulated. The equivalence of various definitions of infinity is provable in Zermelo-Fraenkel set theory under the consideration of the axiom of choice. Since in the theory of atomic finitely supported structures the axiom of choice fails, we study various FSM forms of infinity (of Tarski type, of Dedekind type, of Mostowski type, of Kuratowski type, and so on), and provide several relationships between them. We emphasize the connections and differences between various definitions of infinity internally in FSM. By presenting examples of atomic sets that satisfy a certain form of infinity, while they do not satisfy other forms of infinity, we are able to conclude that the FSM definitions of infinity we introduce are pairwise non-equivalent. Especially, we are interested on FSM uniformly infinite sets that are finitely supported sets containing infinite, uniformly supported subsets. Uniformly supported sets are particularly of interest because they involve boundedness properties of supports, meaning that the support of each element in a uniformly supported set is contained in the same finite set of atoms; in this way, all the individuals in an infinite uniformly supported family can be characterized by involving only the same finitely many characteristics. We prove that the set of all finitely supported functions from A to the FSM powerset of A (and consequently, the set of all finitely supported functions from A to A, as well as the set of all finitely supported functions from A to the set of all finite injective tuples of atoms) is not FSM uniformly infinite (i.e. it does not contain an infinite uniformly supported subset). In this way several fixed point properties can be obtained. Connections between FSM uniformly finiteness and injectivity/surjectivity of self-mappings on FSM sets are presented. In particular, we prove that for finitely supported self-mappings defined on A, and on the finite powerset of A, respectively, the injectivity is equivalent with the surjectivity. A pictorial summary catching all the relations between several forms of infinity is emphasized together with a table presenting all the forms of infinity satisfied by fundamental atomic sets. We also discuss the concept of countability in FSM, and present a connection between countable union theorems and countable choice principles. This chapter is based on [12].

A. Alexandru, G. Ciobanu, *Foundations of Finitely Supported Structures*,
https://doi.org/10.1007/978-3-030-52962-8_9

9.1 Relationships Between Various Notions of Infinity in FSM

Definition 9.1 Let X be a finitely supported subset of an invariant set Y.

1. X is called *FSM classical infinite* if X does not correspond one-to-one and onto to a finite ordinal (i.e. to an ordinal $n < \omega$), that is if X cannot be represented as $\{x_1, \dots, x_n\}$ for some $n \in \mathbb{N}$. We simply call an FSM usual infinite set as *infinite*, and an FSM non-usual infinite set as *finite*.
2. X is *FSM Kuratowski infinite* if there is a finitely supported directed family $\mathscr{F}$ of finitely supported subsets of Y with the property that X is contained in the union of the members of $\mathscr{F}$, but there does not exist $Z \in \mathscr{F}$ such that $X \subseteq Z$.
3. X is called *FSM Tarski I infinite* (TI inf) if there exists a finitely supported one-to-one mapping of X onto $X \times X$.
4. X is called *FSM Tarski II infinite* (TII inf) if there exists a finitely supported family of finitely supported subsets of X, totally ordered by inclusion, having no maximal element.
5. X is called *FSM Tarski III infinite* (TIII inf) if there exists a finitely supported one-to-one mapping of X onto $X + X$ (where $X + X$ is the disjoint union of X with X), i.e. if $|X| = 2|X|$.
6. X is called *FSM Mostowski infinite* (M inf) if there exists an infinite, finitely supported, totally ordered subset of X.
7. X is called *FSM Dedekind infinite* (D inf) if there exists a finitely supported one-to-one mapping of X onto a finitely supported proper subset of X.
8. X is called *FSM ascending infinite* (A inf) if there is a finitely supported increasing countable chain of finitely supported sets $X_0 \subseteq X_1 \subseteq \dots \subseteq X_n \subseteq \dots$ with $X \subseteq \cup X_n$, but there does not exist $n \in \mathbb{N}$ such that $X \subseteq X_n$;
9. X is called *FSM non-amorphous* (N-am) if X contains two disjoint, infinite, finitely supported supported subsets.

Note that in the definition of FSM Tarski II infinity for a certain X, the existence of a totally ordered, finitely supported family of finitely supported subsets of X is required, while in the definition of FSM ascending infinity for X, the related family of finitely supported subsets of X has to be FSM countable (i.e. the mapping $n \mapsto X_n$ should be finitely supported, which means the family $(X_n)_{n \in \mathbb{N}}$ should be uniformly supported). It is immediate that if X is FSM ascending infinite, then it is also FSM Tarski II infinite, while the reverse is not necessarily valid in the absence of axiom of choice over (non-atomic) ZF sets (see Remark 9.3). Actually, we can prove that a totally ordered, finitely supported family of finitely supported subsets of X should be uniformly supported (see the proof of Theorem 9.7), but for concluding that such a family has a countable uniformly supported sub-family we need to assume a form of choice over non-atomic sets.

Theorem 9.1 *Let X be a finitely supported subset of an invariant set Y.*

1. X is FSM classical infinite if and only if X is FSM Kuratowski infinite.

2. *X is FSM classical infinite if and only if there exists a non-empty, finitely supported family of finitely supported subsets of X having no maximal element under inclusion.*

Proof 1. Let us suppose that X is FSM classical infinite. Let $\mathscr{F}$ be the family of all FSM classical non-infinite (FSM classical finite) subsets of X ordered by inclusion. Since X is finitely supported, it follows that $\mathscr{F}$ is supported by $supp(X)$. Moreover, since all the elements of $\mathscr{F}$ are finite sets, it follows that all the elements of $\mathscr{F}$ are finitely supported. Clearly, $\mathscr{F}$ is directed, and X is the union of the members of $\mathscr{F}$. Suppose by contradiction that X is not FSM Kuratowski infinite. Then there exists $Z \in \mathscr{F}$ such that $X \subseteq Z$. Therefore, X should by FSM classical finite which is a contradiction with our original assumption.

Conversely, assume that X is FSM Kuratowski infinite. Suppose, by contradiction that X is FSM classical finite, i.e. $X = \{x_1, \ldots, x_n\}$. Let $\mathscr{F}$ be a directed family such that X is contained in the union of the members of $\mathscr{F}$ (at least one such a family exists, for example $\wp_{fs}(X)$). Then for each $i \in \{1, \ldots, n\}$ there exists $F_i \in \mathscr{F}$ such that $x_i \in F_i$. Since $\mathscr{F}$ is directed, there is $Z \in \mathscr{F}$ such that $F_i \subseteq Z$ for all $i \in \{1, \ldots, n\}$, and so $X \subseteq Z$ with $Z \in \mathscr{F}$, which is a contradiction.

2. Assume that X is FSM classical infinite. By contradiction, suppose that every non-empty, finitely supported family of finitely supported subsets of X has a maximal element under inclusion. Particularly, $\wp_{fin}(X)$ is an equivariant family of finite (and so finitely supported) subsets of X, and so it should have a maximal element $Z \in \wp_{fin}(X)$. Assume $Z = \{z_1, \ldots, z_n\}$. Since X is infinite, there is an element $z \in X$ different from all $z_1, \ldots, z_n$. However, in this case we have $\{z_1, \ldots, z_n, z\} \in \wp_{fin}(X)$ and $\{z_1, \ldots, z_n\} \subsetneq \{z_1, \ldots, z_n, z\}$ contradicting the maximality of Z.

Conversely, assume there exists a non-empty, finitely supported family of finitely supported subsets of X having no a maximal element under inclusion. Suppose by contradiction that X is finite. Then X has only finitely many subsets, and all its subsets are finite. Let $\mathscr{F}$ be an arbitrary family of subsets of X. Then $\mathscr{F}$ is a finite family of finite sets, and so the elements of $\mathscr{F}$ of greatest cardinality are maximal elements of $\mathscr{F}$ with respect to inclusion, contradicting the assumption. □

Theorem 9.2 *The following properties of FSM Dedekind infinite sets hold.*

1. *Let X be a finitely supported subset of an invariant set Y (particularly, X is an FM set). Then X is FSM Dedekind infinite if and only if there exists a finitely supported one-to-one mapping $f : \mathbb{N} \to X$. As a consequence, an FSM superset of an FSM Dedekind infinite set is FSM Dedekind infinite, and an FSM subset of an FSM set that is not Dedekind infinite is also not FSM Dedekind infinite.*
2. *Let X be an infinite, finitely supported subset of an invariant set Y (particularly, X is an infinite FM set). Then the set $\wp_{fs}(\wp_{fin}(X))$ is FSM Dedekind infinite.*
3. *Let X be an infinite, finitely supported subset of an invariant set Y (particularly, X is an infinite FM set). Then the set $\wp_{fs}(\wp_{fs}(X))$ is FSM Dedekind infinite.*
4. *Let X be a finitely supported subset of an invariant set Y (particularly, X is an FM set) such that X does not contain an infinite subset Z with the property that all the elements of Z are supported by the same set of atoms. Then X is not FSM Dedekind infinite.*

5. *Let X be a finitely supported subset of an invariant set Y (particularly, X is an FM set) such that X does not contain an infinite subset Z with the property that all the elements of Z are supported by the same set of atoms. Then $\wp_{fin}(X)$ is not FSM Dedekind infinite.*
6. *Let X and Y be two finitely supported subsets of an invariant set Z (particularly, X and Y are FM sets). If neither X nor Y is FSM Dedekind infinite, then $X \times Y$ is not FSM Dedekind infinite.*
7. *Let X and Y be two finitely supported subsets of an invariant set Z (particularly, X and Y are FM sets). If neither X nor Y is FSM Dedekind infinite, then $X+Y$ is not FSM Dedekind infinite.*
8. *Let X be a finitely supported subset of an invariant set Y. Then $\wp_{fs}(X)$ is FSM Dedekind infinite if and only if X is FSM ascending infinite.*
9. *Let X be a finitely supported subset of an invariant set Y. If X is FSM Dedekind infinite, then X is FSM ascending infinite. The reverse implication is not valid.*

Proof 1. Let us suppose that $(X,\cdot)$ is FSM Dedekind infinite, and $g : X \to X$ is an injection supported by the finite set $S \subsetneq A$ with the property that $Im(g) \subsetneq X$. This means that there exists $supp(g) \subseteq S$ and there exists $x_0 \in X$ such that $x_0 \notin Im(g)$. We can form a sequence of elements from X which has the first term x_0 and the general term $x_{n+1} = g(x_n)$ for all $n \in \mathbb{N}$. Since $x_0 \notin Im(g)$ it follows that $x_0 \neq g(x_0)$. Since g is injective and $x_0 \notin Im(g)$, by induction we obtain that $g^n(x_0) \neq g^m(x_0)$ for all $n,m \in \mathbb{N}$ with $n \neq m$. Furthermore, x_{n+1} is supported by $supp(g) \cup supp(x_n)$ for all $n \in \mathbb{N}$. Indeed, let $\pi \in Fix(supp(g) \cup supp(x_n))$. According to Proposition 2.9, $\pi \cdot x_{n+1} = \pi \cdot g(x_n) = g(\pi \cdot x_n) = g(x_n) = x_{n+1}$. Since $supp(x_{n+1})$ is the least set supporting x_{n+1}, we obtain $supp(x_{n+1}) \subseteq supp(g) \cup supp(x_n)$ for all $n \in \mathbb{N}$. By finite recursion, we have $supp(x_n) \subseteq supp(g) \cup supp(x_0)$ for all $n \in \mathbb{N}$. Since all x_n are supported by the same set of atoms $supp(g) \cup supp(x_0)$, we have that the function $f : \mathbb{N} \to X$, defined by $f(n) = x_n$, is also finitely supported (by the set $supp(g) \cup supp(x_0) \cup supp(X)$ not depending on n). Indeed, for any $\pi \in Fix(supp(g) \cup supp(x_0) \cup supp(X))$ we have $f(\pi \diamond n) = f(n) = x_n = \pi \cdot x_n = \pi \cdot f(n)$ for all $n \in \mathbb{N}$, where by $\diamond$ we denoted the trivial S_A-action on $\mathbb{N}$. Furthermore, because π fixes $supp(X)$ pointwise we have $\pi \cdot f(n) \in X$ for all $n \in \mathbb{N}$. From Proposition 2.9 we have that f is finitely supported. Obviously, f is also injective.

Conversely, suppose there is a finitely supported injective mapping $f : \mathbb{N} \to X$. According to Proposition 2.9, it follows that for any $\pi \in Fix(supp(f))$ we have $\pi \cdot f(n) = f(\pi \diamond n) = f(n)$ and $\pi \cdot f(n) \in X$ for all $n \in \mathbb{N}$. Let us define $g : X \to X$ by

$$g(x) = \begin{cases} f(n+1), & \text{if there exists } n \in \mathbb{N} \text{ with } x = f(n); \\ x, & \text{if } x \notin Im(f)\,. \end{cases}$$

We claim that g is supported by $supp(f) \cup supp(X)$. Indeed, let us consider $\pi \in Fix(supp(f) \cup supp(X))$ and $x \in X$. If there exists $n \in \mathbb{N}$ such that $x = f(n)$, we have that $\pi \cdot x = \pi \cdot f(n) = f(n)$, and so $g(\pi \cdot x) = g(f(n)) = f(n+1) = \pi \cdot f(n+1) = \pi \cdot g(x)$. If $x \notin Im(f)$, we prove by contradiction that $\pi \cdot x \notin Im(f)$. Indeed, suppose that $\pi \cdot x \in Im(f)$. Then there is $y \in \mathbb{N}$ such that $\pi \cdot x = f(y)$, or equivalently $x = \pi^{-1} \cdot f(y)$. However, since $\pi \in Fix(supp(f))$, from Proposition 2.9 we have $\pi^{-1} \cdot f(y) = f(\pi^{-1} \diamond y)$, and so we get $x = f(\pi^{-1} \diamond y) = f(y) \in Im(f)$ which contradicts the assumption that $x \notin Im(f)$. Thus, $\pi \cdot x \notin Im(f)$, and so $g(\pi \cdot x) = \pi \cdot x = \pi \cdot g(x)$. We obtained that $g(\pi \cdot x) = \pi \cdot x = \pi \cdot g(x)$ for all $x \in X$ and all $\pi \in Fix(supp(f) \cup supp(X))$. Furthermore, $\pi \cdot g(x) \in \pi \star X = X$ (where by $\star$ we denoted the S_A-action on $\wp_{fs}(Y)$), and so g is finitely supported. Since f is injective, it follows immediately that g is injective. Furthermore, $Im(g) = X \setminus \{f(0)\}$ which is a proper subset of X, finitely supported by $supp(f(0)) \cup supp(X) = supp(f) \cup supp(X)$.

2. The family $\wp_{fin}(X)$ represents the family of those finite subsets of X (these subsets of X are finitely supported as subsets of the invariant set Y in the sense of Definition 2.4). Obviously, $\wp_{fin}(X)$ is a finitely supported subset of the invariant set $\wp_{fs}(Y)$, supported by $supp(X)$. This is because whenever Z is an element of $\wp_{fin}(X)$ (i.e. whenever Z is a finite subset of X) and π fixes $supp(X)$ pointwise, we have that $\pi \star Z$ is also a finite subset of X. The family $\wp_{fs}(\wp_{fin}(X))$ represents the family of those subsets of $\wp_{fin}(X)$ which are finitely supported as subsets of the invariant set $\wp_{fs}(Y)$ in the sense of Definition 2.4. As above, according to Proposition 2.1, we have that $\wp_{fs}(\wp_{fin}(X))$ is a finitely supported subset of the invariant set $\wp_{fs}(\wp_{fs}(Y))$, supported by $supp(\wp_{fin}(X)) \subseteq supp(X)$.
Let X_i be the set of all i-sized subsets from X, i.e. $X_i = \{Z \subseteq X \,|\, |Z| = i\}$. Since X is infinite, it follows that each $X_i, i \geq 1$ is non-empty. Obviously, we have that any i-sized subset $\{x_1, \dots, x_i\}$ of X is finitely supported (as a subset of Y) by $supp(x_1) \cup \dots \cup supp(x_i)$. Therefore, $X_i \subseteq \wp_{fin}(X)$ and $X_i \subseteq \wp_{fs}(Y)$ for all $i \in \mathbb{N}$. Since $\cdot$ is a group action, the image of an i-sized subset of X under an arbitrary permutation of atoms is an i-sized subset of Y. However, any permutation of atoms that fixes $supp(X)$ pointwise also leaves X invariant, and so for any permutation $\pi \in Fix(supp(X))$ we have that $\pi \star Z$ is an i-sized subset of X whenever Z is an i-sized subset of X. Thus, each X_i is a subset of $\wp_{fin}(X)$ finitely supported by $supp(X)$, and so $X_i \in \wp_{fs}(\wp_{fin}(X))$.
We define $f : \mathbb{N} \to \wp_{fs}(\wp_{fin}(X))$ by $f(n) = X_n$. We claim that $supp(X)$ supports f. Indeed, let $\pi \in Fix(supp(X))$. Since $supp(X)$ supports X_n for all $n \in \mathbb{N}$, we have $\pi \star f(n) = \pi \star X_n = X_n = f(n) = f(\pi \diamond n)$ (where $\diamond$ is the trivial S_A-action on $\mathbb{N}$) and $\pi \star f(n) = \pi \star X_n = X_n \in \wp_{fs}(\wp_{fin}(X))$ for all $n \in \mathbb{N}$. According to Proposition 2.9, we have that f is finitely supported. Furthermore, f is injective, and by item 1, we have that $\wp_{fs}(\wp_{fin}(X))$ is Dedekind infinite in FSM.
3. The proof is actually the same as in the above item since every $X_i \in \wp_{fs}(\wp_{fs}(A))$.
4. If there does not exist a uniformly supported subset of X, then there does not exist a finitely supported injective mapping $f : \mathbb{N} \to X$, and so f cannot be FSM Dedekind infinite.
5. We prove the following lemma:

Lemma 9.1 *Let X be a finitely supported subset of an invariant set Y such that X does not contain an infinite uniformly supported subset. Then the set $\wp_{fin}(X) = \{Z \subseteq X \,|\, Z \text{finite}\}$ does not contain an infinite uniformly supported subset.* □

Proof of Lemma 9.1. Suppose by contradiction that the set $\wp_{fin}(X)$ contains an infinite subset $\mathscr{F}$ such that all the elements of $\mathscr{F}$ are different and supported by the same finite set S. Just for writing convention (without assuming that $i \mapsto X_i$ is finitely supported) we denote $\mathscr{F}$ as $\mathscr{F} = (X_i)_{i\in I} \subseteq \wp_{fin}(X)$ with the properties that $X_i \neq X_j$ whenever $i \neq j$ and $supp(X_i) \subseteq S$ for all $i \in I$. Fix an arbitrary $j \in I$. However, from Proposition 2.5, because $supp(X_j) = \underset{x\in X_j}{\cup} supp(x)$, we have that X_j has the property that $supp(x) \subseteq S$ for all $x \in X_j$. Since j has been arbitrarily chosen from I, it follows that every element from every set of form X_i is supported by S, and so $\underset{i}{\cup} X_i$ is a uniformly supported subset of X (all its elements being supported by S). Furthermore, $\underset{i\in I}{\cup} X_i$ is infinite because the family $(X_i)_{i\in I}$ is infinite and $X_i \neq X_j$ whenever $i \neq j$. Otherwise, if $\underset{i}{\cup} X_i$ was finite, the family $(X_i)_{i\in I}$ would be contained in the finite set $\wp(\underset{i}{\cup} X_i)$, and so it could not be infinite with the property that $X_i \neq X_j$ whenever $i \neq j$. We were able to construct an infinite uniformly supported subset of X, namely $\underset{i}{\cup} X_i$, and this contradicts the hypothesis that X does not contain an infinite uniformly supported subset.

Proof of item 5. According to the above lemma, if X does not contain an infinite uniformly supported subset, then $\wp_{fin}(X)$ does not contain an infinite uniformly supported subset. Suppose by contradiction that $\wp_{fin}(X)$ is FSM Dedekind infinite. According to item 1, there exists a finitely supported injective mapping $f: \mathbb{N} \to \wp_{fin}(X)$. Thus, because $\mathbb{N}$ is a trivial invariant set, according to Proposition 2.9, there exists an infinite injective (countable) sequence $f(\mathbb{N}) = (X_i)_{i\in\mathbb{N}} \subseteq \wp_{fin}(X)$ having the property $supp(X_i) \subseteq supp(f)$ for all $i \in \mathbb{N}$. We obtained that $\wp_{fin}(X)$ contains an infinite uniformly supported subset $(X_i)_{i\in\mathbb{N}}$, which is a contradiction.

6. Suppose by contradiction that $X \times Y$ is FSM Dedekind infinite. According to item 1, there exists a finitely supported injective mapping $f: \mathbb{N} \to X \times Y$ Thus, according to Proposition 2.9, there exists an infinite injective sequence $f(\mathbb{N}) = ((x_i,y_i))_{i\in\mathbb{N}} \subseteq X \times Y$ with the property that $supp((x_i,y_i)) \subseteq supp(f)$ for all $i \in \mathbb{N}$ (1). Fix some $j \in \mathbb{N}$. We claim that $supp((x_j,y_j)) = supp(x_j) \cup supp(y_j)$. Let $U = (x_j,y_j)$, and $S = supp(x_j) \cup supp(y_j)$. Obviously, S supports U. Indeed, let us consider $\pi \in Fix(S)$. We have that $\pi \in Fix(supp(x_j))$ and also $\pi \in Fix(supp(y_j))$ Therefore, $\pi \cdot x_j = x_j$ and $\pi \cdot y_j = y_j$, and so $\pi \otimes (x_j,y_j) = (\pi \cdot x_j, \pi \cdot y_j) = (x_j,y_j)$, where $\otimes$ represents the S_A action on $X \times Y$ described in Proposition 2.2(5). Thus, $supp(U) \subseteq S$. It remains to prove that $S \subseteq supp(U)$. Fix $\pi \in Fix(supp(U))$. Since $supp(U)$ supports U, we have $\pi \otimes (x_j,y_j) = (x_j,y_j)$, and so $(\pi \cdot x_j, \pi \cdot y_j) = (x_j,y_j)$, from which we get $\pi \cdot x_j = x_j$ and $\pi \cdot y_j = y_j$. Thus, $supp(x_j) \subseteq supp(U)$ and $supp(y_j) \subseteq supp(U)$. Hence $S = supp(x_j) \cup supp(y_j) \subseteq supp(U)$.
According to relation (1) we obtain, $supp(x_i) \cup supp(y_i) \subseteq supp(f)$ for all $i \in \mathbb{N}$. Thus, $supp(x_i) \subseteq supp(f)$ and $supp(y_i) \subseteq supp(f)$ for all $i \in \mathbb{N}$ (2). Since the sequence $((x_i,y_i))_{i\in\mathbb{N}}$ is infinite and injective, then at least one of the sequences $(x_i)_{i\in\mathbb{N}}$ and $(y_i)_{i\in\mathbb{N}}$ is infinite. Assume that $(x_i)_{i\in\mathbb{N}}$ is infinite. Then there exists an infinite subset B of $\mathbb{N}$ such that $(x_i)_{i\in B}$ is injective, and so there exists an

injection $u : B \to X$ defined by $u(i) = x_i$ for all $i \in B$ which is supported by $supp(f)$ (according to relation (2) and Proposition 2.9). However, since B is an infinite subset of $\mathbb{N}$, there exists a ZF bijection $h : \mathbb{N} \to B$. The construction of h requires only the fact that $\mathbb{N}$ is well-ordered which is obtained from the Peano construction of $\mathbb{N}$ and does not involve a form of the axiom of choice. Since both B and $\mathbb{N}$ are trivial invariant sets (see Proposition 2.2(3)), it follows that h is equivariant. Thus, $u \circ h$ is an injection from $\mathbb{N}$ to X which is finitely supported by $supp(u) \subseteq supp(f)$. This contradicts the assumption that X is not FSM Dedekind infinite.

Remark 9.1 Similarly, using the relation $supp(x) \cup supp(y) = supp((x,y))$ for all $x \in X$ and $y \in Y$ derived from Proposition 2.5, it can be proved that $X \times Y$ does not contain an infinite uniformly supported subset if neither X nor Y contain an infinite uniformly supported subset. □

7. Suppose by contradiction that $X+Y$ is FSM Dedekind infinite. According to item 1, there exists a finitely supported injective mapping $f : \mathbb{N} \to X+Y$. Thus, there exists an infinite injective sequence $(z_i)_{i \in \mathbb{N}} \subseteq X+Y$ such that $supp(z_i) \subseteq supp(f)$ for all $i \in \mathbb{N}$. According to the construction of the disjoint union of two S_A-sets (see Proposition 2.2(6)), as in the proof of item 6, there should exist an infinite subsequence of $(z_i)_i$ of form $((0,x_j))_{x_j \in X}$ which is uniformly supported by $supp(f)$, or an infinite sequence of form $((1,y_k))_{y_k \in Y}$ which is uniformly supported by $supp(f)$. Since 0 and 1 are constants, this means there should exist at least an infinite uniformly supported sequence of elements from X, or an infinite uniformly supported sequence of elements from Y. This contradicts the hypothesis that neither X nor Y is FSM Dedekind infinite.

Remark 9.2 Similarly, it can be proved that $X+Y$ does not contain an infinite uniformly supported subset if neither X nor Y contain an infinite uniformly supported subset. □

8. Let us suppose that $\wp_{fs}(X)$ is FSM Dedekind infinite. Assume that $(X_n)_{n \in \mathbb{N}}$ is an infinite countable family of different subsets of X such that the mapping $n \mapsto X_n$ is finitely supported. Thus, each X_n is supported by the same set $S = supp(n \mapsto X_n)$. We define a countable family $(Y_n)_{n \in \mathbb{N}}$ of subsets of X that are non-empty and pairwise disjoint. A ZF construction of such a family belongs to Kuratowski, and can also be found in Lemma 4.11 from [34]. This approach works also in FSM in the view of the S-finite support principle because every Y_k is defined only involving elements in the family $(X_n)_{n \in \mathbb{N}}$, and so whenever $(X_n)_{n \in \mathbb{N}}$ is uniformly supported (meaning that all X_n are supported by the same set of atoms), we get that $(Y_n)_{n \in \mathbb{N}}$ is uniformly supported. Formally the sequence $(Y_n)_{n \in \mathbb{N}}$ is recursively constructed as below. For $n \in \mathbb{N}$, assume that Y_m is defined for any $m < n$ such that the set $\{X_k \setminus \underset{m<n}{\cup} Y_m \,|\, k \geq n\}$ is infinite. Define $n' = min\{k \,|\, k \geq n \text{ and } X_k \setminus \underset{m<n}{\cup} Y_m \neq \emptyset \text{ and } (X \setminus X_k) \setminus \underset{m<n}{\cup} Y_m \neq \emptyset\}$. With n' selected, we define $Y_n = \begin{cases} X_{n'} \setminus \underset{m<n}{\cup} Y_m, \text{ if } \{X_k \setminus (X_{n'} \cup \underset{m<n}{\cup} Y_m) \,|\, k > n'\} \text{ is infinite;} \\ (X \setminus X_{n'}) \setminus \underset{m<n}{\cup} Y_m, \text{ otherwise.} \end{cases}$

Obviously, Y_1 is supported by $S \cup supp(X)$. By induction, assume that Y_m is supported by $S \cup supp(X)$ for each $m < n$. Since Y_n is defined as a set combination of X_i's (which are all S-supported) and Y_m's with $m < n$, we get that Y_n is supported by $S \cup supp(X)$ according to the S-finite support principle. Therefore the family $(Y_i)_{i \in \mathbb{N}}$ is uniformly supported by $S \cup supp(X)$. Let $U_i = Y_0 \cup \ldots \cup Y_i$ for all $i \in \mathbb{N}$. Clearly, all U_i are supported by $S \cup supp(X)$, and $U_0 \subsetneq U_1 \subsetneq U_2 \subsetneq \ldots \subsetneq X$. Let $V_n = (X \setminus \underset{i \in \mathbb{N}}{\cup} U_i) \cup U_n$. Clearly, $X = \underset{n \in \mathbb{N}}{\cup} V_n$. Moreover, V_n is supported by $S \cup supp(X)$ for all $n \in \mathbb{N}$. Therefore, the mapping $n \mapsto V_n$ is finitely supported. Obviously, $V_0 \subsetneq V_1 \subsetneq V_2 \subsetneq \ldots \subsetneq X$. However, there does not exist $n \in \mathbb{N}$ such that $X = V_n$, and so X is FSM ascending infinite.
The converse holds since if X is FSM ascending infinite, there is a finitely supported increasing countable chain of finitely supported sets $X_0 \subseteq X_1 \subseteq \ldots \subseteq X_n \subseteq \ldots$ with $X \subseteq \cup X_n$, but there does not exist $n \in \mathbb{N}$ such that $X \subseteq X_n$. In this sequence there should exist infinitely many different elements of form X_i (otherwise their union is a term of the sequence), and the result follows from Proposition 9.13.

9. Suppose X is FSM Dedekind infinite. Therefore, $\wp_{fs}(X)$ is FSM Dedekind infinite. According to item 8, we have that X is FSM ascending infinite. The reverse implication is not valid because, as we prove in Proposition 9.8, $\wp_{fin}(A)$ is FSM ascending infinite while it is not FSM Dedekind infinite. □

Corollary 9.1 *The following sets and all of their FSM classical infinite subsets are FSM classical infinite, but they are not FSM Dedekind infinite.*

1. *The invariant set A of atoms.*
2. *The powerset $\wp_{fs}(A)$ of the set of atoms.*
3. *The set $T_{fin}(A)$ of all finite injective tuples of atoms.*
4. *The invariant set A^A_{fs} of all finitely supported functions from A to A.*
5. *The invariant set of all finitely supported functions $f : A \to A^n$, where $n \in \mathbb{N}$.*
6. *The invariant set of all finitely supported functions $f : A \to T_{fin}(A)$.*
7. *The invariant set of all finitely supported functions $f : A \to \wp_{fs}(A)$.*
8. *The powerset $\wp_{fs}(A \times A)$ of the Cartesian product of the set of atoms with itself.*
9. *The invariant sets $FM_0(A)$, $FM_1(A)$ and $FM_2(A)$ used in the construction of $FM(A)$ (when A is the set of atoms in ZFA), and those sets similarly defined when A is the fixed ZF set adjoined to ZF in order to define FSM.*
10. *The sets $\wp_{fin}(A)$, $\wp_{fin}(\wp_{fs}(A))$, $\wp_{fin}(\wp_{fin}(A))$, $\wp_{fin}(A^A_{fs})$, $\wp_{fin}(\wp_{fs}(A)^A_{fs})$.*
11. *Constructions of form $\wp_{fin}(\ldots \wp_{fin}(\wp_{fs}(A)))$, $\wp_{fin}(\ldots \wp_{fin}(\wp_{fs}(A)^A_{fs}))$.*
12. *Every finite Cartesian combination between the set A, $\wp_{fin}(A)$, $\wp_{cofin}(A)$, $\wp_{fs}(A)$ and A^A_{fs}.*
13. *The disjoint unions $A + A^A_{fs}$, $A + \wp_{fs}(A)$, $\wp_{fs}(A) + A^A_{fs}$ and $A + \wp_{fs}(A) + A^A_{fs}$.*

Proof 1. A does not contain an infinite uniformly supported subset, and so it is not FSM Dedekind infinite (according to Theorem 9.2(4)).

2. $\wp_{fs}(A)$ does not contain an infinite uniformly supported subset because for any finite set S of atoms there exist only finitely many elements of $\wp_{fs}(A)$ supported

by S, namely the subsets of S and the supersets of $A \setminus S$. Thus, $\wp_{fs}(A)$ is not FSM Dedekind infinite (Theorem 9.2(4)).

3. $T_{fin}(A)$ does not contain an infinite uniformly supported subset because the finite injective tuples of atoms supported by a finite set S are only those injective tuples formed by elements of S, being at most $1+A^1_{|S|}+A^2_{|S|}+\ldots+A^{|S|}_{|S|}$ such tuples, where $A^k_n = n(n-1)\ldots(n-k+1)$.
4. We prove the following lemmas.

Lemma 9.2 *Let $S=\{s_1,\ldots,s_n\}$ be a finite subset of an invariant set $(U,\cdot)$ and X a finitely supported subset of an invariant set $(V,\diamond)$. Then if X does not contain an infinite uniformly supported subset, we have that X^S_{fs} does not contain an infinite uniformly supported subset.* □

Proof of Lemma 9.2 Firstly, we prove that there is an FSM injection g from X^S_{fs} into $X^{|S|}$. For $f \in X^S_{fs}$ define $g(f)=(f(s_1),\ldots,f(s_n))$. Clearly, g is injective (and it is also surjective). Let $\pi \in Fix(supp(s_1)\cup\ldots\cup supp(s_n)\cup supp(X))$. Thus, $g(\pi\widetilde{\star}f)=(\pi\diamond f(\pi^{-1}\cdot s_1),\ldots,\pi\diamond f(\pi^{-1}\cdot s_n))=(\pi\diamond f(s_1),\ldots,\pi\diamond f(s_n))$ $=\pi\otimes g(f)$ for all $f\in X^S_{fs}$, where $\otimes$ is the S_A-action on $X^{|S|}$ defined as in Proposition 2.2 and $\widetilde{\star}$ is the S_A-action on X^S_{fs} defined as in Proposition 2.7. Hence g is finitely supported, and the conclusion follows from Theorem 9.2(1) and by repeatedly applying similar arguments as in Theorem 9.2(6) (if we slightly modify the proof of the theorem, using the fact that $supp(x)\cup supp(y)=supp((x,y))$ for all $x,y\in X$, we show that the $|S|$-time Cartesian product of X, i.e. $X^{|S|}$ does not contain an infinite uniformly supported subset; otherwise X should contain itself an infinite uniformly supported subset, which contradicts the hypothesis - see also Remark 9.1).

Lemma 9.3 *Let $S=\{s_1,\ldots,s_n\}$ be a finite subset of an invariant set $(U,\cdot)$, and X a finitely supported subset of an invariant set $(V,\diamond)$. Then if X is not FSM Dedekind infinite, we have that X^S_{fs} is not FSM Dedekind-infinite.* □

Proof of Lemma 9.3 We proved that there is an FSM injection g from X^S_{fs} into $X^{|S|}$. The conclusion follows from Theorem 9.2(1) and by repeatedly applying Theorem 9.2(6) (from which we know that the $|S|$-time Cartesian product of X, i.e. $X^{|S|}$, is not FSM Dedekind-infinite).

Lemma 9.4 *Let $f: A\to A$ be a function that is finitely supported by a certain finite set of atoms S. Then either $f|_{A\setminus S}=Id$ or $f|_{A\setminus S}$ is a one-element subset of S.* □

Proof of Lemma 9.4 Let $f: A\to A$ be a function that is finitely supported by the finite set of atoms S. We distinguish two cases:

I. There is $a\notin S$ with $f(a)=a$. Then for each $b\notin S$ we have that $(a\,b)\in Fix(S)$, and so $f(b)=f((a\,b)(a))=(a\,b)(f(a))=(a\,b)(a)=b$. Thus, $f|_{A\setminus S}=Id$.

II. For all $a\notin S$ we have $f(a)\neq a$. We claim that $f(a)\in S$ for all $a\notin S$. Suppose by contradiction that $f(a)=b\in A\setminus S$ for a certain $a\notin S$. Thus, $(a\,b)\in Fix(S)$, and so $f(b) = f((ab)(a)) = (ab)(f(a)) = (ab)(b) = a$. Let us consider $c \in A \setminus S$, $c \neq a,b$. Thus, $(ac) \in Fix(S)$, and so $f(c) = f((ac)(a)) = (ac)(f(a)) = (ac)(b) = b$. Furthermore, $(bc) \in Fix(S)$, and so we obtain $f(b) = f((bc)(c)) = (bc)(f(c)) = (bc)(b) = c$. However, $f(b) = a$ which contradicts the functionality of f. Thus, $f(a) \in S$ for any $a \notin S$. If $x,y \notin S$, then we should have $f(x), f(y) \in S$, and so, because $(xy) \in Fix(S)$, we get $f(x) = f((xy)(y)) = (xy)(f(y)) = f(y)$ since both x and y belong to $A \setminus S$ which means they are different from $f(y)$ belonging to S. Therefore there is $x_0 \in S$ such that $f|_{A \setminus S} = \{x_0\}$.

Proof of this item. Assume by contradiction that A^A_{fs} contains an infinite, uniformly supported subset, meaning that there are infinitely many functions from A to A supported by the same finite set S. According to Lemma 9.4, any S-supported function $f : A \to A$ should have the property that either $f|_{A \setminus S} = Id$ or $f|_{A \setminus S}$ is a one-element subset of S. A function from A to A is precisely characterized by the set of values it takes on the elements of S and on the elements of $A \setminus S$, respectively. For each possible definition of such an f on S we have at most $|S| + 1$ possible ways to define f on $A \setminus S$. Since we assumed that there exist infinitely many finitely supported functions from A to A supported by the same set S, there should exist infinitely many finitely supported functions from S to A supported by the set S. But this is a contradiction according to Lemma 9.2, which states that A^S_{fs} does not contain an infinite uniformly supported subset (because A does not contain an infinite uniformly supported subset).

5. There is an equivariant bijective mapping between $(A^n)^A_{fs}$ and $(A^A_{fs})^n$ defined as follows: if $f : A \to A^n$ is a finitely supported function with $f(a) = (a_1, \dots, a_n)$, we associate to f the Cartesian pair $(f_1, \dots, f_n)$, where for each $i \in \mathbb{N}$, $f_i : A \to A$ is defined by $f_i(a) = a_i$ for all $a \in A$. We omit the technical details because they are based only on the application of Proposition 2.9. We proved above that A^A_{fs} does not contain an infinite uniformly supported subset, and so neither $(A^A_{fs})^n$ contains an infinite uniformly supported subset by involving a similar proof as of Theorem 9.2(6) (see the proof of Lemma 9.2).
6. Assume by contradiction that $T_{fin}(A)^A$ contains an infinite S-uniformly supported subset. If $f : A \to T_{fin}(A)$ is a function supported by S, then consider $f(a) = x$ for some $a \notin S$. For $b \notin S$ we have $(ab) \in Fix(S)$, and so $f(b) = f((ab)(a)) = (ab) \otimes f(a) = (ab) \otimes x$ which means $|f(a)| = |f(b)|$ for all $a,b \notin S$. Each S-supported function $f : A \to T_{fin}(A)$ is fully described the values it takes on the elements of S and on the elements of $A \setminus S$, respectively, i.e., by the elements of $f(S)$ and of $f(A \setminus S)$, that is it can be decomposed into two S-supported functions $f|_S$ and $f|_{A \setminus S}$ (because both S and $A \setminus S$ are supported by S). However $f(A \setminus S) \subseteq A'^n$ for some $n \in \mathbb{N}$, where A'^n is the set of all injective n-tuples of A. According to Lemma 9.2 we have at most finitely many S-supported functions from S to $T_{fin}(A)$. According to item 5 we have at most finitely many S-supported functions from $A \setminus S$ to A'^n for each fixed $n \in \mathbb{N}$. This is because A'^n is a subset of A^n and $A \setminus S$ is a subset of A, and so by involving Proposition 5.3(3) and (4) we find a finitely supported injection φ from $(A'^n)^{A \setminus S}$ and $(A^n)^A$; if $\mathscr{X}$ was an infinite subset in $(A'^n)^{A \setminus S}$ uniformly supported by T, then $\varphi(\mathscr{X})$ would be an infinite subset of $(A^n)^A$ uniformly supported by $T \cup supp(\varphi)$. Therefore there should

exist an infinite subset $M \subseteq \mathbb{N}$ such that we have at least one S-supported function $g : A \setminus S \to A'^k$ for any $k \in M$. Fix $a \in A \setminus S$. For each of the above g's (that form an S-supported family $\mathscr{F}$) we have that $g(a)$'s form a uniformly supported family (by $S \cup \{a\}$) of $T_{fin}(A)$, which is also infinite because tuples having different cardinalities are different and M is infinite. However, we contradict the proof of item 3 stating that $T_{fin}(A)$ does not contain an infinite uniformly supported subset. Alternatively, one can remark that if $|S \cup \{a\}| = l$, then there is $m \in M$ with $m > l$. Moreover, $g(a)$ for some $g : A \setminus S \to A'^m$ in $\mathscr{F}$, which is an injective m-tuple of atoms, cannot be supported by $S \cup \{a\}$; thus, the set of all $g(a)$'s cannot be infinite and uniformly supported.

7. We can use a similar approach as in item 6, to prove that there exist at most finitely many S-supported functions from A to $\wp_{fin}(A)$. For this we just replace A'^n with the set of all n-sized subsets of A, $\wp_n(A)$. All it remains is to prove that for each $n \in \mathbb{N}$ there cannot exist infinitely many functions $g : A \to \wp_n(A)$ supported by the same set S'. Fix $n \in \mathbb{N}$. Assume by contradiction that there exist infinitely many functions $g : A \to \wp_n(A)$ supported by the same set S'. According to Lemma 9.2 there are only finitely many functions from S' to $\wp_n(A)$ supported by the same set of atoms, and so there should exist infinitely many functions $g : (A \setminus S') \to \wp_n(A)$ supported by S'. For such a g, let us fix an element $a \in A$ with $a \notin S'$. There exist $x_1, \dots, x_n \in A$ fixed (depending only on the fixed a) and different such that $g(a) = \{x_1, \dots, x_n\}$. Let b be an arbitrary element from $A \setminus S'$, and so $(a b) \in Fix(S')$ which means $g(b) = g((a b)(a)) = (a b) \star g(a) = (a b) \star \{x_1, \dots, x_n\} = \{(a b)(x_1), \dots, (a b)(x_n)\}$.
We analyze the two possibilities:
Case 1: One of $x_1, \dots, x_n$ coincides to a; assume $x_1 = a$. We claim that $x_2, \dots, x_n \in S'$. Assume the contrary, that there exists $i \in \{2, \dots, n\}$ such that $x_i \notin S'$. Without losing the generality suppose $x_2 \notin S'$, which means $(a x_2) \in Fix(S')$, and so $g(x_2) = g((a x_2)(a)) = (a x_2) \star g(a) = (a x_2) \star \{a, x_2, \dots, x_n\} = \{a, x_2, \dots, x_n\}$. Let $c \in A \setminus S'$ with c different from $a, x_2, \dots, x_n$. We have $g(c) = g((a c)(a)) = (a c) \star g(a) = (a c) \star \{a, x_2, \dots, x_n\} = \{c, x_2, \dots, x_n\}$, and hence $g(x_2) = g((c x_2)(c)) = (c x_2) \star g(c) = (c x_2) \star \{c, x_2, \dots, x_n\} = \{c, x_2, \dots, x_n\}$ which contradicts the functionality of g. Therefore, $g(b) = \{b, x_2, \dots, x_n\}$ for all $b \in A \setminus S'$, and so only the selection of $x_2, \dots, x_n$ provides the distinction between g's. Since S' is finite, $\{x_2, \dots, x_n\}$ can be selected in $C_{|S'|}^{n-1}$ ways if $|S'| \geq n-1$ or in 0 ways otherwise.
Case 2: Consider now that all $x_1, \dots, x_n$ are different from a.

$$\text{Then } g(b) = \begin{cases} \{x_1, \dots, x_n\}, & \text{if } b \neq x_1, \dots, x_n\ ; \\ \{a, x_2, \dots, x_n\}, & \text{if } x_1 \notin S' \text{ and } b = x_1\ ; \\ \dots \\ \{x_1, \dots, x_{n-1}, a\}, & \text{if } x_n \notin S' \text{ and } b = x_n\ . \end{cases}$$

Since $x_1, \dots, x_n, a$ are fixed atoms, then $g(A \setminus S')$ is finite. However $Im(g)$ should be supported by S'. According to Proposition 2.5, since $Im(g)$ is finite, it should be uniformly supported by S'. We obtain that $x_1, \dots, x_n \in S'$, and so $g(A \setminus S') = \{x_1, \dots, x_n\}$. Otherwise, if some $x_i \notin S'$, we would get $\{x_1, \dots, a, \dots, x_n\} \in Im(g)$ (where a replaces x_i) and so $\{x_1, \dots, a, \dots, x_n\}$ is supported by S'. Again by Proposition 2.5 we would have that a is supported by S' which means $\{a\} =$

$supp(a) \subseteq S'$ contradicting the choice of a. Alternatively, for proving that all $x_1,\ldots,x_n \in S'$, assume by contradiction that one of them (say x_1) does not belong to S'. Let c be an atom from $A \setminus S'$ with c different from $a,x_1,x_2,\ldots,x_n$. We have $g(c) = g((ac)(a)) = (ac) \star g(a) = (ac) \star \{x_1,\ldots,x_n\} = \{x_1,\ldots,x_n\}$, and hence $g(x_1) = g((cx_1)(c)) = (cx_1) \star g(c) = (cx_1) \star \{x_1,x_2,\ldots,x_n\} = \{c,x_2,\ldots,x_n\}$. However $g(x_1) = \{a,x_2,\ldots,x_n\}$ which contradicts the functionality of g. Since S' is finite, $\{x_1,\ldots,x_n\}$ can be selected in $C^n_{|S'|}$ ways $|S'| \geq n$ or in 0 ways otherwise. In either case, it could not exist infinitely many g's supported by S'.
Similarly, there exist at most finitely many S-supported functions from A to $\wp_{cofin}(A)$ (using eventually the fact that there is an equivariant bijection $X \mapsto A \setminus X$ between $\wp_{fin}(A)$ and $\wp_{cofin}(A)$). Assume by contradiction that $\wp_{fs}(A)^A$ contains an infinite S-uniformly supported subset. If $f : A \to \wp_{fs}(A)$ is a function supported by S, then consider $f(a) = X$ for some $a \notin S$. For $b \notin S$ we have $f(b) = (ab) \star X$ which means $f(A \setminus S)$ is formed only by finite subsets of atoms if X is finite, and $f(A \setminus S)$ is formed only by cofinite subsets of atoms if X is cofinite. Thus, whenever $f : A \to \wp_{fs}(A)$ is a function supported by S, we have either $f(A \setminus S) \subseteq \wp_{fin}(A)$ or $f(A \setminus S) \subseteq \wp_{cofin}(A)$. Each S-supported function $f : A \to \wp_{fs}(A)$ is fully described by $f(S)$ and $f(A \setminus S)$. According to Lemma 9.2 we have at most finitely many S-supported functions from S to $\wp_{fs}(A)$. Furthermore, we have at most finitely many S-supported functions from $A \setminus S$ to $\wp_{fin}(A)$, and at most finitely many S-supported functions from $A \setminus S$ to $\wp_{cofin}(A)$. Thus, $\wp_{fs}(A)^A$ does not contain an infinite uniformly supported subset.

8. According to Theorem 2.3 and Theorem 2.4, we have $|\wp_{fs}(A)^A_{fs}| = |\wp_{fs}(A)|^{|A|} = |\{0,1\}^A_{fs}|^{|A|} = (|\{0,1\}|^{|A|})^{|A|} = |\{0,1\}^{A \times A}_{fs}| = |\wp_{fs}(A \times A)|$, and so there exists a finitely supported bijection between $\wp_{fs}(A \times A)$ and $\wp_{fs}(A)^A_{fs}$. The result follows from item 7.
9. Let A be the set of atoms in ZFA. The set $FM_0(A)$ coincide with the empty set, and so it is not FSM Dedekind infinite. $FM_1(A) = A + \{\emptyset\}$. If there was an infinite, finitely supported injective (countable) sequence in $FM_1(A)$, then there would be an infinite injective sequence of atoms all supported by the same finite set, which is a contradiction. Now, suppose that there is a countable injective sequence of elements from $FM_2(A) = A + \wp_{fs}(A + \{\emptyset\})$, all whose terms are supported by the same set S. Thus, at least one of the sets A and $\wp_{fs}(A + \{\emptyset\})$ contains an infinite uniformly supported family of elements. However, this is a contradiction (direct calculation).
10. The related sets do not contain infinite uniformly supported subsets, and so they are not FSM Dedekind infinite (Theorem 9.2(5)).
11. Directly from Theorem 9.2(5).
12. According to Theorem 9.2(6).
13. According to Theorem 9.2(7). □

Corollary 9.2 *There exist two FSM sets whose cardinalities are incomparable via the relation $\leq$ (over cardinalities), and none of them is FSM Dedekind infinite.*

Proof According to Corollary 9.1, none of the sets $A \times A$ and $\wp_{fs}(A)$ is FSM Dedekind infinite. According to Theorem 5.5(2), there does not exist a finitely supported injective mapping $f : A \times A \to \wp_{fs}(A)$. According to Lemma 11.10 from [37] that is preserved in FSM (proof omitted) there does not exist a finitely supported injective mapping $f : \wp_{fs}(A) \to A \times A$. □

Corollary 9.3 *The following sets and all of their supersets, their powersets and the families of their finite subsets, are both FSM classical infinite and FSM Dedekind infinite.*

1. *The invariant sets $\wp_{fs}(\wp_{fs}(A))$, $\wp_{fs}(\wp_{fin}(A))$ and $\mathbb{N}$.*
2. *The set of all finitely supported mappings from X to Y, and the set of all finitely supported mappings from Y to X, where X is a finitely supported subset of an invariant set with at least two elements, and Y is an FSM Dedekind infinite set.*
3. *The set of all finitely supported functions $f : \wp_{fin}(Y) \to X$ and the set of all finitely supported functions $f : \wp_{fs}(Y) \to X$, where Y is an infinite, finitely supported subset of an invariant set, and X is a finitely supported subset of an invariant set with at least two elements.*
4. *The set $T^{\delta}_{fin}(A) = \bigcup_{n\in\mathbb{N}} A^n$ of all finite tuples of atoms (not necessarily injective).*
5. *The invariant sets $FM_{\alpha}(A)$ used in the construction of $FM(A)$ (when A is the set of atoms in ZFA), whenever $\alpha > 2$ and α is not as limit ordinal, and those sets similarly defined when A is the fixed ZF set adjoined to ZF in order to define FSM.*
6. *The invariant set $FM_{\lambda}(A)$ used in the construction of $FM(A)$ (when A is the set of atoms in ZFA), whenever λ is a limit ordinal, and those sets similarly defined when A is the fixed ZF set adjoined to ZF in order to define FSM.*
7. *The invariant FM universe $FM(A)$.*

Proof 1. This follows from Theorem 9.2(3) and Theorem 9.2(2).

2. Let $(y_n)_{n\in\mathbb{N}}$ be an injective, uniformly supported, countable sequence in Y (that exists from Theorem 9.2(1)). Thus, each y_n is supported by the same set S of atoms. In Y^X we consider the injective family $(f_n)_{n\in\mathbb{N}}$ of functions from X to Y where for each $i \in \mathbb{N}$ we define $f_i(x) = y_i$ for all $x \in X$. According to Proposition 2.9, each f_i is supported by S, and so is the infinite family $(f_n)_{n\in\mathbb{N}}$, meaning that there is an S-supported injective mapping from $\mathbb{N}$ to Y^X. In this case it is necessary to require only that X is non-empty.
   Fix two different elements $x_1, x_2 \in X$. Take $\mathscr{F} = (y_n)_{n\in\mathbb{N}}$ an injective, uniformly supported, countable sequence in Y. In X^Y we consider the injective family $(g_n)_{n\in\mathbb{N}}$ of functions from Y to X where for each $i \in \mathbb{N}$ we define $g_i(y) = \begin{cases} x_1 \text{ if } y = y_i \\ x_2 \text{ if } y = y_j \text{ with } j \neq i, \text{ or } y \notin \mathscr{F} \end{cases}$. According to Proposition 2.9, each g_i is supported by the finite set $supp(x_1) \cup supp(x_2) \cup supp(\mathscr{F})$, and so the infinite family $(g_n)_{n\in\mathbb{N}}$ is uniformly supported meaning that there is an injective mapping from $\mathbb{N}$ to X^Y supported by $supp(x_1) \cup supp(x_2) \cup supp(\mathscr{F})$.

3. From Theorem 2.3, there exists a one-to-one mapping from $\wp_{fs}(U)$ onto $\{0,1\}^U_{fs}$ for an arbitrary finitely supported subset of an invariant set U. Fix two distinct elements $x_1, x_2 \in X$. There exists a finitely supported (by $supp(x_1) \cup supp(x_2)$) bijective mapping from $\{0,1\}^U_{fs}$ to $\{x_1,x_2\}^U_{fs}$ which associates to each $f \in \{0,1\}^U_{fs}$ an element $g \in \{x_1,x_2\}^U_{fs}$ defined by $g(x) = \begin{cases} x_1 \text{ for } f(x) = 0 \\ x_2 \text{ for } f(x) = 1 \end{cases}$ for all $x \in U$ and supported by $supp(x_1) \cup supp(x_2) \cup supp(f)$. Obviously, there is a finitely supported injection between $\{x_1,x_2\}^U_{fs}$ and X^U_{fs}. Thus, there is a finitely supported injection from $\wp_{fs}(U)$ into X^U_{fs}. If we take $U = \wp_{fin}(Y)$ or $U = \wp_{fs}(Y)$, the result follows from Theorem 9.2(1), Theorem 9.2(2) and Theorem 9.2(3).
4. Fix $a \in A$ and $i \in \mathbb{N}$. We consider the tuple $x_i = (a, \ldots, a) \in A^i$. Clearly, x_i is supported by $\{a\}$ for each $i \in \mathbb{N}$, and so $(x_n)_{n\in\mathbb{N}}$ is a uniformly supported subset of $T^{\delta}_{fin}(A)$.
5. Denote by $\alpha - 1$ the predecessor of α. We have $FM_\alpha(A) = A + \wp_{fs}(A + \wp_{fs}(FM_{\alpha-2}(A)))$. Since $FM_{\alpha-2}(A)$ is an infinite invariant set for $\alpha > 2$, from Theorem 9.2(3), we obtain that $\wp_{fs}(\wp_{fs}(FM_{\alpha-2}(A)))$ is FSM Dedekind infinite. Since we have $\wp_{fs}(\wp_{fs}(FM_{\alpha-2}(A))) \subseteq A + \wp_{fs}(A + \wp_{fs}(FM_{\alpha-2}(A)))$, it follows that $FM_\alpha(A)$ is FSM Dedekind infinite as a superset of an FSM Dedekind infinite set.
6. This follows because $FM_\lambda(A)$ is a superset of some FSM Dedekind infinite set $FM_\alpha(A)$ with $\alpha > 2$.
7. $FM(A)$ is a superset of an FSM Dedekind infinite set. □

We provided a full proof of Corollary 9.1 in order to help the non-expert readers to become familiarized with FSM techniques, and in order to emphasize how particular atomic mappings look like. Actually, from the proof of Corollary 9.1 (particularly from Lemma 9.4) we are able to provide a full characterization of finitely supported mappings from A to A, and of finitely supported mappings from A to $\wp_{fs}(A)$. However, below we present a general result from which many items in Corollary 9.1 could follow directly.

Theorem 9.3 *Let X be a finitely supported subset of an invariant set $(Y,\cdot)$ such that X does not contain an infinite uniformly supported subset. The set X^A_{fs} does not contain an infinite uniformly supported subset, and so it is not FSM Dedekind infinite.*

Proof Assume, by contradiction, that for a certain finite set $S \subseteq A$ there exist infinitely many functions $f : A \to X$ that are supported by S. Each S-supported function $f : A \to X$ can be uniquely decomposed into two S-supported functions $f|_S$ and $f|_{A\setminus S}$ (this follows from Proposition 2.9 and because both S and $A \setminus S$ are supported by S). Since there exist only finitely many functions from S to X supported by S (see Lemma 9.2), there should exist an infinite family $\mathcal{F}$ of functions $g : (A \setminus S) \to X$ which are supported by S (the functions g are the restrictions of the functions f to $A \setminus S$). Let us fix an element $a \in A \setminus S$. Consider an arbitrary S-supported function $g : (A \setminus S) \to X$. For each $\pi \in Fix(S \cup \{a\})$, according to Proposition 2.9, we have $\pi \cdot g(a) = g(\pi(a)) = g(a)$ which means that $g(a)$ is supported by $S \cup \{a\}$. However, in X there are at most finitely many elements supported by $S \cup \{a\}$, and because the family $\{h(a) \,|\, h \in \mathscr{F}\} \subseteq X$ is uniformly supported by $S \cup \{a\}$, this family should be finite. Therefore, there is $n \in \mathbb{N}$ such that $h_1(a), \dots, h_n(a)$ are distinct elements in X with $h_1, \dots, h_n \in \mathscr{F}$, and $h(a) \in \{h_1(a), \dots, h_n(a)\}$ for all $h \in \mathscr{F}$. Fix some $h \in \mathscr{F}$ and an arbitrary $b \in A \setminus S$ (which means that the transposition $(a\,b)$ fixes S pointwise). We have that there is $i \in \{1, \dots, n\}$ such that $h(a) = h_i(a)$. Since h, h_i are supported by S and $(a\,b) \in Fix(S)$, from Proposition 2.9 we have $h(b) = h((a\,b)(a)) = (a\,b) \cdot h(a) = (a\,b) \cdot h_i(a) = h_i((a\,b)(a)) = h_i(b)$, which finally leads to $h = h_i$ since b was arbitrarily chosen from their domain of definition. Thus, the family $\mathscr{F} = \{h_1, \dots, h_n\}$, which means that $\mathscr{F}$ is finite, and so we get a contradiction. □

Corollary 9.4 *The set $\wp_{fs}(A^n)$, where A^n is the n-times Cartesian product of A, does not contain an infinite uniformly supported subset, and so it is not FSM Dedekind infinite for any $n \in \mathbb{N}$.*

Proof We prove the result by induction on n. For $n = 1$, the result is obvious. Assume that $\wp_{fs}(A^{n-1})$ does not contain an infinite uniformly supported subset. According to Theorem 2.3 and Theorem 2.4, we have $|\wp_{fs}(A^n)| = |\{0,1\}^{A^n}_{fs}| = |\{0,1\}^{A^{n-1} \times A}_{fs}| = |\{0,1\}|^{|A^{n-1}| \cdot |A|} = |(\{0,1\}^{A^{n-1}}_{fs})^A_{fs}|$, which means that there is a finitely supported bijection between $\wp_{fs}(A^n)$ and $(\{0,1\}^{A^{n-1}}_{fs})^A_{fs}$. Furthermore, there exists a finitely supported bijection between $(\{0,1\}^{A^{n-1}}_{fs})^A_{fs}$ and $(\wp_{fs}(A^{n-1}))^A_{fs}$, and the last set does not contain an infinite uniformly supported subset according to the inductive hypothesis and to Theorem 9.3. The result follows now. □

We can provide a more general result by using the proving method in Corollary 9.4.

Theorem 9.4 *Let X be a finitely supported subset of an invariant set $(Y, \cdot)$ such that X does not contain an infinite uniformly supported subset. The set $X^{A^n}_{fs} = \{f : A^n \to X \,|\, f$ is finitely supported$\}$ does not contain an infinite uniformly supported subset, and so it is not FSM Dedekind infinite for any $n \in \mathbb{N}$.*

Proof We prove the result by induction on n. For $n = 1$, the result follows from Theorem 9.3. Assume that $X^{A^{n-1}}_{fs}$ does not contain an infinite uniformly supported subset. According to Theorem 2.4, we have $|X^{A^n}_{fs}| = |X^{(A^{n-1} \times A)}_{fs}| = |X|^{|A^{n-1} \times A|} = |X|^{|A^{n-1}| \cdot |A|} = (|X|^{|A^{n-1}|})^{|A|} = |(X^{A^{n-1}}_{fs})^A_{fs}|$, which means that there is a finitely supported bijection between $X^{A^n}_{fs}$ and $(X^{A^{n-1}}_{fs})^A_{fs}$. However, by Theorem 9.3, $(X^{A^{n-1}}_{fs})^A_{fs}$ does not contain an infinite uniformly supported subset (since the set $Z = X^{A^{n-1}}_{fs}$ does not contain an infinite uniformly supported subset according to the inductive hypothesis). The result follows now. □

Proposition 9.1 *1. Let X be a finitely supported subset of an invariant set $(Y, \cdot)$ such that $\wp_{fs}(X)$ does not contain an infinite uniformly supported subset. Let*

$Z=\{z_1,\dots,z_n\}$ be a finite subset of an invariant set $(U,\diamond)$. Then the set Z^X_{fs} does not contain an infinite uniformly supported subset, and so it is not FSM Dedekind infinite.

2. *Let S be a finite subset of A and X a finitely supported subset of an invariant set $(Y,\cdot)$ such that $\wp_{fs}(X)$ does not contain an infinite S-uniformly supported subset. The set S^X_{fs} does not contain an infinite S-uniformly supported subset.*

Proof 1. Let us consider a function $f:X\to Z$ supported by a certain finite set S. For each $i\in\{1,\dots,n\}$ we claim that $f^{-1}(\{z_i\})$ is supported by $S\cup supp(z_i)$. Indeed, let $\pi\in Fix(S\cup supp(z_i))$. We have $x\in f^{-1}(\{z_i\})\Leftrightarrow f(x)=z_i\Leftrightarrow \pi\diamond f(x)=\pi\diamond z_i=z_i\Leftrightarrow f(\pi\cdot x)=z_i\Leftrightarrow \pi\cdot x\in f^{-1}(\{z_i\})$. Therefore, $f^{-1}(\{z_j\})$ is supported by $T=S\cup supp(z_1)\cup supp(z_2)\cup\ldots\cup supp(z_n)$ for all $j\in\{1,\dots,n\}$.

Each S-supported function $f:X\to Z$ is fully described by the T-supported tuple $\mathscr{P}_f=(f^{-1}(\{z_1\}),\dots,f^{-1}(\{z_n\}))$. Let us consider an S-uniformly supported family $\mathscr{F}$ of mappings from X to Z. The T-uniformly supported family $\mathscr{F}_i=\{f^{-1}(\{z_i\})\,|\,f\in\mathscr{F}\}\subseteq\wp_{fs}(X)$ should be finite for each $i\in\{1,\dots,n\}$. Therefore, the T-uniformly supported set $\{\mathscr{P}_f\,|\,f\in\mathscr{F}\}$ is finite (by using the classical product rule, its cardinality is equal with at most the product between $|\mathscr{F}_i|$ for $i\in\{1,\dots,n\}$). Since the mapping $f\mapsto\mathscr{P}_f$ is T-supported and injective, we have that $\mathscr{F}$ should be finite.

2. Consider $S=\{a_1,\dots,a_n\}$. As above $f^{-1}(\{a_j\})$ is supported by $S\cup supp(a_1)\cup supp(a_2)\cup\ldots\cup supp(a_n)=S$ for all $j\in\{1,\dots,n\}$. Each S-supported function $f:X\to S$ is fully described by the S-supported tuple $\mathscr{P}_f=(f^{-1}(\{a_1\}),\dots,f^{-1}(\{a_n\}))$. Let us consider an S-uniformly supported family $\mathscr{F}$ of mappings from X to S. The S-uniformly supported family $\mathscr{F}_i=\{f^{-1}(\{a_i\})\,|\,f\in\mathscr{F}\}\subseteq\wp_{fs}(X)$ should be finite for each $i\in\{1,\dots,n\}$. Therefore, the S-uniformly supported set $\{\mathscr{P}_f\,|\,f\in\mathscr{F}\}$ is finite. Since the mapping $f\mapsto\mathscr{P}_f$ is S-supported and injective, we have that $\mathscr{F}$ should be finite. □

Proposition 9.2 *1. Let S be a finite subset of A and X a finitely supported subset of an invariant set $(Y,\cdot)$. Let $\mathscr{F}$ be an S-uniformly supported family of mappings from X to $A\setminus S$. If there is $a\in A\setminus S$ such that the family $\mathscr{F}'=\{f^{-1}(\{a\})\,|\,f\in\mathscr{F}\}$ is finite, then $\mathscr{F}$ is finite having the same cardinality as $\mathscr{F}'$.*

2. *Let S be a finite subset of A and X a finitely supported subset of an invariant set $(Y,\cdot)$ having the property that there exists $a\in A\setminus S$ such that $\wp_{fs}(X)$ does not contain an infinite $(S\cup\{a\})$-uniformly supported subset. Then the set A^X_{fs} does not contain an infinite S-uniformly supported subset.*

Proof 1. There exists $n\in\mathbb{N}$ such that the subsets $h_1^{-1}(\{a\}),\dots,h_n^{-1}(\{a\})$ of X are distinct with $h_1,\dots,h_n\in\mathscr{F}$, and $h^{-1}(\{a\})\in\{h_1^{-1}(\{a\}),\dots,h_n^{-1}(\{a\})\}$ for all $h\in\mathscr{F}$. Fix some $h\in\mathscr{F}$ and an arbitrary $b\in A\setminus S$ (which means that the transposition $(a b)$ fixes S pointwise). We have $h^{-1}(\{a\})=h_i^{-1}(\{a\})$ for some $i\in\{1,\dots,n\}$. From Proposition 2.9, because $(a b)\in Fix(S)$ and S supports both h,h_i, we have $h^{-1}(\{b\})=\{x\in X\,|\,h(x)=b\}=\{x\in X\,|\,h(x)=(a b)(a)\}=\{x\in X\,|\,(a b)(h(x))=(a b)((a b)(a))=a\}=\{x\in X\,|\,h((a b)\cdot x)=a\}\overset{(a b)\cdot x:=y}{=}\{(a b)\cdot y\in X\,|\,h(y)=$

$a\} = (ab) \star \{y \in X \,|\, h(y) = a\} = (ab) \star h^{-1}(\{a\}) = (ab) \star h_i^{-1}(\{a\}) = (ab) \star \{y \in X \,|\, h_i(y) = a\} = \{(ab) \cdot y \in X \,|\, h_i(y) = a\}$ $=\{x \in X \,|\, h_i((ab) \cdot x) = a\}$=$\{x \in X \,|\, (ab)(h_i(x)) = a\}$= $\{x \in X \,|\, h_i(x) = (ab)(a) = b\}$=$h_i^{-1}(\{b\})$. Thus, $h^{-1}(\{c\}) = h_i^{-1}(\{c\})$ for all $c \in A \setminus S$. Now, let us choose an arbitrary $v \in X$. If $h(v) = d$ for some $d \in A \setminus S$, then $v \in h^{-1}(\{d\}) = h_i^{-1}(\{d\})$, which means $h_i(v) = d$. Thus, $h = h_i$, and so $\mathscr{F}$ is finite.

2. Firstly, we remark that $\wp_{fs}(X)$ does not contain an infinite S-uniformly supported subset. Let us consider a function $f : X \to A$ supported by S, and define U_f=$\{x \in X \,|\, f(x) \in S\}$. We have that U_f is supported by S. Indeed for $\pi \in Fix(S)$ and $x \in U_f$ (i.e. $f(x) \in S$), according to Proposition 2.9, we have $\pi \cdot x \in X$ and $f(\pi \cdot x) = \pi(f(x)) = f(x) \in S$, which means $\pi \cdot x \in U_f$. Similarly, the set $V_f = \{x \in X \,|\, f(x) \notin S\}$ which is the complementary of U_f is supported by S. Indeed, for $\pi \in Fix(S)$ and $x \in V_f$ we have $\pi \cdot x \in X$ and $f(x) \notin S$, and so $f(\pi \cdot x) = \pi(f(x)) \notin \pi \star S = S$, which means $\pi \cdot x \in V_f$. Assume, by contradiction, that there are infinitely many S-supported functions f from X to A forming an S-uniformly supported family $\mathscr{H}$. Each S-supported function $f : X \to A$ can be uniquely decomposed into two S-supported functions $f|_{U_f} : U_f \to S$ and $f|_{V_f} : V_f \to A \setminus S$ (this follows from Proposition 2.9 and because both U_f and V_f are supported by S). Since U_f is an S-supported element in $\wp_{fs}(X)$, we have that $U_f \in \{X_1, \dots, X_m\}$ where $X_1, \dots, X_m$ are the only S-supported subsets of X. By applying m times Proposition 9.1(2), there exist only finitely many S-supported functions from U_f to S, and so there should exist an infinite family of functions $f|_{V_f}$ from V_f to $(A \setminus S)$ which are supported by S. However, $V_f \in \{X_1, \dots, X_m\}$ for all $f \in \mathscr{H}$ since V_f is an S-supported subset of X. Therefore, there should exist $k \in \{1, \dots, m\}$ such that there exists an infinite family $\mathscr{F}$ of functions $g : X_k \to (A \setminus S)$ which are supported by S.

Let $g \in \mathscr{F}$. We claim that $g^{-1}(\{a\})$ is supported by $S \cup \{a\}$. Indeed, let $\pi \in Fix(S \cup \{a\})$. According to Proposition 2.9 and since $\pi \cdot x \in X_k$ whenever $x \in X_k$ (this is because X_k is S-supported), we have $x \in g^{-1}(\{a\}) \Leftrightarrow g(x) = a \Leftrightarrow \pi(g(x)) = \pi(a) = a \Leftrightarrow g(\pi \cdot x) = a \Leftrightarrow \pi \cdot x \in g^{-1}(\{a\})$. Therefore the family $\mathscr{F}' = \{h^{-1}(\{a\}) \,|\, h \in \mathscr{F}\} \subseteq \wp_{fs}(X_k) \subseteq \wp_{fs}(X)$ is uniformly supported by $S \cup \{a\}$, and so this family should be finite. By item 1, we have that $\mathscr{F}$ is finite, a contradiction. □

Theorem 9.5 *1. Let X be a finitely supported subset of an invariant set $(Y, \cdot)$ such that $\wp_{fs}(X)$ does not contain an infinite uniformly supported subset. The set $(\wp_{fs}(A^n))^X_{fs}$ does not contain an infinite uniformly supported subset, and so it is not FSM Dedekind infinite for any $n \in \mathbb{N}$.*

2. Let X be a finitely supported subset of an invariant set $(Y, \cdot)$ such that $\wp_{fs}(X)$ does not contain an infinite uniformly supported subset. The set $(\wp_{fs}(A))^X_{fs}$ does not contain an infinite uniformly supported subset, and so it is not FSM Dedekind infinite.

3. Let X be a finitely supported subset of an invariant set $(Y, \cdot)$ such that $\wp_{fs}(X)$ does not contain an infinite uniformly supported subset. Then the set A^X_{fs} does not contain an infinite uniformly supported subset, and so it is not FSM Dedekind infinite.

Proof We prove only the first item since the second and the third ones follow obviously. According to Theorem 9.4, the set $(\wp_{fs}(X))^{A^n}_{fs}$ does not contain an infinite uniformly supported subset. However, according to Theorem 2.3 and Theorem 2.4, we have $|(\wp_{fs}(X))^{A^n}_{fs}| = |(\{0,1\}^X_{fs})^{A^n}_{fs}| = (|\{0,1\}|^{|X|})^{|A^n|} = |\{0,1\}|^{|X|\cdot|A^n|} = |\{0,1\}|^{|A^n|\cdot|X|} = (|\{0,1\}|^{|A^n|})^{|X|} = |(\{0,1\}^{A^n}_{fs})^X_{fs}| = |(\wp_{fs}(A^n))^X_{fs}|$. Thus, there is a finitely supported bijection between $(\wp_{fs}(X))^{A^n}_{fs}$ and $(\wp_{fs}(A^n))^X_{fs}$, and so the result follows. □

Corollary 9.5 *The set $A^{A^n}_{fs} = \{f : A^n \to A \,|\, f$ is finitely supported$\}$ does not contain an infinite uniformly supported subset, and so it is not FSM Dedekind infinite for any $n \in \mathbb{N}$.*

Proof According to Corollary 9.4, we have that $\wp_{fs}(A^n)$ does not contain an infinite uniformly supported subset. The result follows by applying Theorem 9.5(3). □

According to Proposition 9.3, there exist two atomic FSM sets that are incomparable via the relation $\leq$ (over cardinalities), and one of them is FSM Dedekind infinite, while the other one is not FSM Dedekind infinite.

Proposition 9.3 *There do not exist neither a finitely supported injective mapping $f : \wp_{fin}(A) \to T^{\delta}_{fin}(A)$, nor a finitely supported injective mapping $f : T^{\delta}_{fin}(A) \to \wp_{fin}(A)$.*

Proof For the first part, let us assume by contradiction that $f : \wp_{fin}(A) \to T^{\delta}_{fin}(A)$ is finitely supported and injective. Let $X \in \wp_{fin}(A)$. Since the support of a finite subset of atoms coincides with the related subset, and the support of a finite tuple of atoms is represented by the set of atoms forming the related tuple (see Proposition 2.6), according to Proposition 2.9, for any permutation of atoms $\pi \in Fix(supp(f) \cup supp(X)) = Fix(supp(f) \cup X)$ we have $\pi \otimes f(X) = f(\pi \star X) = f(X)$, where $\otimes$ is the canonical action on $T^{\delta}_{fin}(A)$ constructed as in Proposition 2.2(5). This means $supp(f) \cup X$ supports $f(X)$, that is $supp(f(X)) \subseteq supp(f) \cup X$, and so the atoms forming $f(X)$ are contained in $supp(f) \cup X$ (claim 1). Let us take two distinct atoms $b_1, b_2 \in A \setminus supp(f)$. We consider the cases:

Case 1. The tuple $f(\{b_1,b_2\})$ contains only elements from $supp(f)$. Let $c_1 \in A \setminus supp(f)$ distinct from b_1, b_2. Then c_1 does not appear in the tuple $f(\{b_1,b_2\})$, and so the transposition $(b_2\, c_1)$ fixes the tuple $f(\{b_1,b_2\})$. Since $(b_2\, c_1) \in Fix(supp(f))$ according to Proposition 2.9, we get $f(\{b_1,c_1\}) = f((b_2\, c_1) \star \{b_1,b_2\}) = (b_2\, c_1) \otimes f(\{b_1,b_2\}) = f(\{b_1,b_2\})$, contradicting the injectivity of f.

Case 2. The tuple $f(\{b_1,b_2\})$ contains an element outside $supp(f)$. Connecting this assertion with (claim 1), we have that at least b_1 or b_2 appear (possibly multiple times) in the tuple $f(\{b_1,b_2\})$. Say b_1 is in the tuple $f(\{b_1,b_2\})$. Since $(b_1\, b_2) \in Fix(supp(f))$, from Proposition 2.9 we get $f(\{b_1,b_2\}) = f((b_1\, b_2) \star \{b_1,b_2\}) = (b_1\, b_2) \otimes f(\{b_1,b_2\}) = (b_1\, b_2)(f(\{b_1,b_2\}))$, which is a contradiction because b_2 replaces b_1 in the (ordered) tuple $f(\{b_1,b_2\})$ under the effect of the transposition $(b_1\, b_2)$.

The second part of the result can be proved by contradiction using the second part of Proposition 2.14 (stating that there is no finitely supported injection from $T_{fin}(A)$ to $\wp_{fin}(A)$) and the obvious remark that there exists the equivariant trivial/identity injection from $T_{fin}(A)$ into $T^{\delta}_{fin}(A)$. □

Proposition 9.4 *Let X be a finitely supported subset of an invariant set. If $\wp_{fs}(X)$ is not FSM Dedekind infinite, then each finitely supported surjective mapping $f : X \to X$ should be injective. The reverse implication does not hold.*

Proof Let $f : X \to X$ be a finitely supported surjection. Since f is surjective, we can define the function $g : \wp_{fs}(X) \to \wp_{fs}(X)$ by $g(Y) = f^{-1}(Y)$ for all $Y \in \wp_{fs}(X)$ which is finitely supported and injective according to Lemma 5.5. Since $\wp_{fs}(X)$ is not FSM Dedekind infinite, it follows that g is surjective.

Now let us consider two elements $a, b \in X$ such that $f(a) = f(b)$. We prove by contradiction that $a = b$. Suppose that $a \neq b$. Let us consider $Y = \{a\}$ and $Z = \{b\}$. Obviously, $Y, Z \in \wp_{fs}(X)$. Since g is surjective, for Y and Z there exist $Y_1, Z_1 \in \wp_{fs}(X)$ such that $f^{-1}(Y_1) = g(Y_1) = Y$ and $f^{-1}(Z_1) = g(Z_1) = Z$. We know that $f(Y) \cap f(Z) = \{f(a)\}$. Thus, $f(a) \in f(Y) = f(f^{-1}(Y_1)) \subseteq Y_1$. Similarly, $f(a) = f(b) \in f(Z) = f(f^{-1}(Z_1)) \subseteq Z_1$, and so $f(a) \in Y_1 \cap Z_1$. Thus, $a \in f^{-1}(Y_1 \cap Z_1) = f^{-1}(Y_1) \cap f^{-1}(Z_1) = Y \cap Z$. However, since we assumed that $a \neq b$, we have that $Y \cap Z = \emptyset$, which represents a contradiction. It follows that $a = b$, and so f is injective.

In order to prove the invalidity of the reverse implication, we prove that any finitely supported surjective mapping $f : \wp_{fin}(A) \to \wp_{fin}(A)$ is also injective, while $\wp_{fs}(\wp_{fin}(A))$ is FSM Dedekind infinite. Let us consider a finitely supported surjection $f : \wp_{fin}(A) \to \wp_{fin}(A)$. Let $X \in \wp_{fin}(A)$. Then $supp(X) = X$. By Proposition 2.9, for any $\pi \in Fix(supp(f) \cup supp(X)) = Fix(supp(f) \cup X)$ we have $\pi \star f(X) = f(\pi \star X) = f(X)$ which means $supp(f) \cup X$ supports $f(X)$, that is $f(X) = supp(f(X)) \subseteq supp(f) \cup X$ (1). If $X \subseteq supp(f)$, we get $f(X) \subseteq supp(f)$ (2).

For a fixed $m \geq 1$, let us fix m (arbitrarily chosen) atoms $b_1, \dots, b_m \in A \setminus supp(f)$. Let us consider the family $U = \{\{a_1, \dots, a_n, b_1, \dots, b_m\} \,|\, a_1, \dots, a_n \in supp(f), n \geq 1\} \cup \{\{b_1, \dots, b_m\}\}$. The set U is finite since $supp(f)$ is finite and $b_1, \dots, b_m \in A \setminus supp(f)$ are fixed. Let us consider $Y \in U$, that is $Y \setminus supp(f) = \{b_1, \dots, b_m\}$. There exists $Z \in \wp_{fin}(A)$ such that $f(Z) = Y$. According to relations (1) and (2), Z should be of form $Z = \{c_1, \dots, c_k, b_{i_1}, \dots, b_{i_l}\}$ with $c_1, \dots, c_k \in supp(f)$ and $b_{i_1}, \dots, b_{i_l} \in A \setminus supp(f)$ or of form $Z = \{b_{i_1}, \dots, b_{i_l}\}$ with $b_{i_1}, \dots, b_{i_l} \in A \setminus supp(f)$. Furthermore, by (1), in either case, $\{b_1, \dots, b_m\} \subseteq \{b_{i_1}, \dots, b_{i_l}\}$. We prove that $l = m$. Assume by contradiction that there exists b_{i_j} with $j \in \{1, \dots, l\}$ such that $b_{i_j} \notin \{b_1, \dots, b_m\}$. Then $(b_{i_j}\ b_1) \star Z = Z$ since both $b_{i_j}, b_1 \in Z$ and Z is a finite subset of atoms (b_{i_j} and b_1 are interchanged in Z under the effect of the transposition $(b_{i_j}\ b_1)$, but the whole Z is left invariant). Moreover, since $b_{i_j}, b_1 \notin supp(f)$ we have $(b_{i_j}\ b_1) \in Fix(supp(f))$, and by Proposition 2.9 we get $f(Z) = f((b_{i_j}\ b_1) \star Z) = (b_{i_j}\ b_1) \star f(Z)$ which is a contradiction because $b_1 \in f(Z)$ while $b_{i_j} \notin f(Z)$. Thus, $\{b_{i_1}, \dots, b_{i_l}\} = \{b_1, \dots, b_m\}$, and so $Z \in U$. Thus, $U \subseteq f(U)$, which means $|U| \leq |f(U)|$. However, since f is a function and U is finite, we get $|f(U)| \leq |U|$. We obtain $|U| = |f(U)|$, and because U is finite with $U \subseteq f(U)$, we get $U = f(U)$, which means $f|_U : U \to U$ is surjective (3). Since U is finite, $f|_U$ should be injective, which means $f(U_1) \neq f(U_2)$ whenever $U_1, U_2 \in U$ with $U_1 \neq U_2$ (4).

If $d_1, \ldots, d_\nu \in A \setminus supp(f)$ with$\{d_1, \ldots, d_\nu\} \neq \{b_1, \ldots, b_m\}$, $\nu \geq 1$, and we consider $V = \{\{a_1, \ldots, a_n, d_1, \ldots, d_\nu\} \,|\, a_1, \ldots, a_n \in supp(f), n \geq 1\} \cup \{\{d_1, \ldots, d_\nu\}\}$, then U and V are disjoint. Whenever $U_1 \in U$ and $V_1 \in V$ we have $f(U_1) \in U$ and $f(V_1) \in V$ (using the same arguments we involved to prove relation (3)), and so $f(U_1) \neq f(V_1)$ (5).

If $T = \{\{a_1, \ldots, a_n\} \,|\, a_1, \ldots, a_n \in supp(f)\}$ and $Y \in T$, then there is $T' \in \wp_{fin}(A)$ such that $Y = f(T')$. According to (3) we should have $T' \in T$ (otherwise, if T' belongs to some V considered in the above paragraph, i.e. if T' contains an element outside $supp(f)$, we would get the contradiction $Y = f(T') \in V$, i.e. we would get that Y contains an element outside $supp(f)$), and so $T \subseteq f(T)$ from which $T = f(T)$ since T is finite (using similar arguments as those we involved to prove relation (3) from $U \subseteq f(U)$). Thus, $f|_T : T \to T$ is surjective. Since T is finite, $f|_T$ should also be injective which means $f(T_1) \neq f(T_2)$ whenever $T_1, T_2 \in T$ with $T_1 \neq T_2$ (6). The case $supp(f) = \emptyset$ is included in the above analysis and leads to $f(\emptyset) = \emptyset$ and $f(X) = X$ for all $X \in \wp_{fin}(A)$.

Clearly, we have $f(T_1) \neq f(V_1)$ whenever $T_1 \in T$ and $V_1 \in V$ since $f(T_1) \in T$, $f(V_1) \in V$ and T and V are disjoint (7). Since $b_1, \ldots, b_m$ and $d_1, \ldots, d_\nu$ were arbitrarily chosen from $A \setminus supp(f)$, the injectivity of f follows from (4), (5), (6) and (7) that describe all the possible cases for considering two different finite subsets of atoms for which we compare the values of f on them. □

Proposition 9.5 *1. Let X be a finitely supported subset of an invariant set. If $\wp_{fin}(X)$ is FSM Dedekind infinite, then X should be FSM non-uniformly amorphous, meaning that X should contain two disjoint, infinite, uniformly supported subsets.*

2. Let X be a finitely supported subset of an invariant set. If $\wp_{fs}(X)$ is FSM Dedekind infinite, then X should be FSM non-amorphous, meaning that X should contain two disjoint, infinite, finitely supported supported subsets. The reverse implication is not valid.

Proof 1. Assume that $(X_n)_{n\in\mathbb{N}}$ is a countable family of different finite subsets of X such that the mapping $n \mapsto X_n$ is finitely supported. Thus, each X_n is supported by the same set $S = supp(n \mapsto X_n)$. Since each X_n is finite (and the support of a finite set coincides with the union of the supports of its elements), as in the proof of Lemma 9.1, we have that $\underset{n\in\mathbb{N}}{\cup} X_n$ is uniformly supported by S. Furthermore, $\underset{n\in\mathbb{N}}{\cup} X_n$ is infinite since all X_i are pairwise different. Moreover, the countable sequence $(Y_n)_{n\in\mathbb{N}}$ defined by $Y_n = X_n \setminus \underset{m<n}{\cup} X_m$ is a uniformly supported (by S) sequence of pairwise disjoint uniformly supported sets with $\underset{n\in\mathbb{N}}{\cup} X_n = \underset{n\in\mathbb{N}}{\cup} Y_n$. Again since each Y_n is finite (and the support of a finite set coincides with the union of the supports of its elements), any element belonging to a set from the sequence $(Y_n)_{n\in\mathbb{N}}$ is S-supported. Since the union of all Y_n is infinite, and each Y_n is finite, there should exist infinitely many terms from the sequence $(Y_n)_{n\in\mathbb{N}}$ that are non-empty. Assume that $(Y_n)_{n\in M\subseteq\mathbb{N}}$ with M infinite is a subset of $(Y_n)_{n\in\mathbb{N}}$ formed by non-empty terms. Let

$U_1 = \{\cup Y_k \,|\, k \in M, k \text{ is odd}\}$ and $U_2 = \{\cup Y_k \,|\, k \in M, k \text{ is even}\}$. Then U_1 and U_2 are disjoint, uniformly S-supported and infinite subsets of X.

2. Assume that $\wp_{fs}(X)$ is FSM Dedekind infinite. As in the proof of Theorem 9.2(8), we can define a uniformly supported, countable family $(Y_n)_{n\in\mathbb{N}}$ of subsets of X that are non-empty and pairwise disjoint. Let $V_1 = \{\cup Y_k \,|\, k \text{ is odd}\}$ and $V_2 = \{\cup Y_k \,|\, k \text{ is even}\}$. Then V_1 and V_2 are disjoint, infinite subsets of X. Since each Y_i is supported by $S' = supp(n \mapsto Y_n)$ we have $\pi \star Y_i = Y_i$ for all $i \in \mathbb{N}$ and $\pi \in Fix(S')$. Fix $\pi \in Fix(S')$ and $x \in V_1$. Thus, there is $l \in \mathbb{N}$ such that $x \in Y_{2l+1}$. We obtain $\pi \cdot x \in \pi \star Y_{2l+1} = Y_{2l+1}$, and so $\pi \cdot x \in V_1$. Thus, V_1 is supported by S'. Similarly, V_2 is supported by S', and so X is FSM non-amorphous.

Conversely, the set $A + A = \{0,1\} \times A$ (disjoint union of A and A) is obviously non-amorphous because $\{(0,a) \,|\, a \in A\}$ is equivariant, infinite and coinfinite. One can define the equivariant bijection $f : \wp_{fs}(A) \times \wp_{fs}(A) \to \wp_{fs}(\{0,1\} \times A)$ by $f(U,V) = \{(0,x) \,|\, x \in U\} \cup \{(1,y) \,|\, y \in V\}$ for all $U,V \in \wp_{fs}(A)$. Clearly, f is equivariant because for each $\pi \in S_A$ we have $f(\pi \star U, \pi \star V) = \pi \star f(U,V)$. However, $\wp_{fs}(A) \times \wp_{fs}(A)$ is not FSM Dedekind infinite according to Corollary 9.1(2) and Theorem 9.2(6). □

It is worth noting that non-uniformly amorphous FSM sets are non-amorphous FSM sets, since uniformly supported sets are obviously finitely supported. The converse however is not valid since $\wp_{fin}(A)$ is non-amorphous but it has no infinite uniformly supported subset (the only finite subsets of atoms supported by a finite set S of atoms being the subsets of S), and so it cannot be non-uniformly amorphous.

Corollary 9.6 *Let X be a finitely supported amorphous subset of an invariant set (i.e. any finitely supported subset of X is either finite or cofinite). Then each finitely supported surjective mapping $f : X \to X$ should be injective.*

Proof Since any finitely supported subset of X is either finite or cofinite, then any uniformly supported subset of X is either finite or cofinite. From Proposition 9.5, $\wp_{fin}(X)$ is not FSM Dedekind infinite. For the rest of the proof we follow step-by-step the proof of Proposition 9.4 (and of Lemma 5.5). If X is finite, we are done, so assume X is infinite. If $Y \in \wp_{fin}(X)$, then $f^{-1}(Y) \in \wp_{fs}(X)$ (supported by $supp(f) \cup supp(X) \cup supp(Y)$). Since X is amorphous, it follows that $f^{-1}(Y)$ is either finite or cofinite. If $f^{-1}(Y)$ is cofinite, then its complement $\{x \in X \mid f(x) \notin Y\}$ is finite. This means that all but finitely many elements in X would have their image under f belonging to the finite set Y. Therefore, $Im(f)$ would be a finite subset of X, which contradicts the surjectivity of f. Thus, $f^{-1}(Y)$ is a finite subset of X. In this sense we can well-define the function $g : \wp_{fin}(X) \to \wp_{fin}(X)$ by $g(Y) = f^{-1}(Y)$ which is supported by $supp(f) \cup supp(X)$ and injective. Since $\wp_{fin}(X)$ is not FSM Dedekind infinite, it follows that g is surjective, and so f is injective exactly as in the last paragraph of the proof of Proposition 9.4. □

Proposition 9.6 *1. Let X be an FSM Dedekind infinite set. Then there exists a finitely supported surjection $j : X \to \mathbb{N}$. The reverse implication is not valid.*

2. *If X is a finitely supported subset of an invariant set such that there exists a finitely supported surjection $j : X \to \mathbb{N}$, then $\wp_{fs}(X)$ is FSM Dedekind infinite. The reverse implication is also valid.*

Proof 1. Let X be an FSM Dedekind infinite set. According to Theorem 9.2(1), there is a finitely supported injection $i : \mathbb{N} \to X$. Let us fix $n_0 \in \mathbb{N}$. We define the function $j : X \to \mathbb{N}$ by

$$j(x) = \begin{cases} i^{-1}(x), & \text{if}\, x \in Im(i) \,; \\ n_0, & \text{if}\, x \notin Im(i) \,. \end{cases}$$

Since $Im(i)$ is supported by $supp(i)$ and n_0 is empty supported, by verifying the condition in Proposition 2.9 we have that j is supported by $supp(i) \cup supp(X)$. Indeed, when $\pi \in Fix(supp(i) \cup supp(X))$, then $x \in Im(i) \Leftrightarrow \pi \cdot x \in Im(i)$, and $n = i^{-1}(\pi \cdot x) \Leftrightarrow i(n) = \pi \cdot x \Leftrightarrow \pi^{-1} \cdot i(n) = x \Leftrightarrow i(\pi^{-1} \diamond n) = x \Leftrightarrow i(n) = x \Leftrightarrow n = i^{-1}(x)$, where $\diamond$ is the trivial action on $\mathbb{N}$; similarly, $y \notin Im(i) \Leftrightarrow \pi \cdot y \notin Im(i)$ and $j(\pi \cdot y) = n_0 = \pi \diamond n_0 = \pi \diamond j(y)$. Clearly, j is surjective. However, the reverse implication is not valid because the mapping $f : \wp_{fin}(A) \to \mathbb{N}$ defined by $f(X) = |X|$ for all $X \in \wp_{fin}(A)$ is equivariant and surjective, but $\wp_{fin}(A)$ is not FSM Dedekind infinite.

2. Suppose now there exists a finitely supported surjection $j : X \to \mathbb{N}$. Clearly, for any $n \in \mathbb{N}$, the set $j^{-1}(\{n\})$ is non-empty and supported by $supp(j)$. Define $f : \mathbb{N} \to \wp_{fs}(X)$ by $f(n) = j^{-1}(\{n\})$. For $\pi \in Fix(supp(j))$ and an arbitrary $n \in \mathbb{N}$ we have $j(x) = n \Leftrightarrow j(\pi^{-1} \cdot x) = n$, and so $x \in j^{-1}(\{n\}) \Leftrightarrow \pi^{-1} \cdot x \in j^{-1}(\{n\})$, which means $f(n) = \pi \star f(n)$ for all $n \in \mathbb{N}$, and so f is supported by $supp(j)$. Since f is also injective, by Theorem 9.2(1) we have that $\wp_{fs}(X)$ is FSM Dedekind infinite.

Conversely, assume that $\wp_{fs}(X)$ is FSM Dedekind infinite. As in the proof of Theorem 9.2(8), we can define a uniformly supported, countable family $(Y_n)_{n \in \mathbb{N}}$ of subsets of X that are non-empty and pairwise disjoint. The mapping f can be defined by $f(x) = \begin{cases} n, & \text{if}\ \exists n. x \in Y_n; \\ 0, & \text{otherwise} \end{cases}$, and obviously, f is supported by $supp(n \mapsto Y_n)$. □

Proposition 9.7 *Let X be an infinite, finitely supported subset of an invariant set. Then there exists a finitely supported surjection $f : \wp_{fs}(X) \to \mathbb{N}$.*

Proof Let X_i be the set of all i-sized subsets from X, i.e. $X_i = \{Z \subseteq X \,|\, |Z| = i\}$. The family $(X_i)_{i \in \mathbb{N}}$ is uniformly supported by $supp(X)$ and all X_i are non-empty and pairwise disjoint. Define the mapping f by $f(Y) = \begin{cases} n, & \text{if}\, Y \in X_n; \\ 0, & \text{if}\, Y \text{ is infinite.} \end{cases}$ According to Proposition 2.9, f is supported by $supp(X)$ (since any X_n is supported by $supp(X)$) and it is surjective. We actually proved the existence of a finitely supported surjection from $\wp_{fin}(X)$ onto $\mathbb{N}$. □

The sets A and $\wp_{fin}(A)$ are both FSM classical infinite and none of them is FSM Dedekind infinite. We prove below that A is not FSM ascending infinite, while $\wp_{fin}(A)$ is FSM ascending infinite.

Proposition 9.8

- *The set A is not FSM ascending infinite.*
- *Let X be a finitely supported subset of an invariant set U. If X is FSM classical infinite, then the set $\wp_{fin}(X)$ is FSM ascending infinite.*

Proof In order to prove that A is not FSM ascending infinite, we firstly prove that each finitely supported increasing countable chain of finitely supported subsets of A must be stationary. Indeed, if there exists an increasing countable chain $X_0 \subseteq X_1 \subseteq \ldots \subseteq A$ such that $n \mapsto X_n$ is finitely supported, then, according to Proposition 2.8 and because $\mathbb{N}$ is a trivial invariant set, each element X_i of the chain must be supported by the same $S = supp(n \mapsto X_n)$. However, there are only finitely many such subsets of A, namely the subsets of S and the supersets of $A \setminus S$. Therefore the chain is finite, and because it is ascending, there exists $n_0 \in \mathbb{N}$ such that $X_n = X_{n_0}$ for all $n \geq n_0$. Now, let $Y_0 \subseteq Y_1 \subseteq \ldots \subseteq Y_n \subseteq \ldots$ be a finitely supported countable chain with $A \subseteq \bigcup_{n \in \mathbb{N}} Y_n$. Then $A \cap Y_0 \subseteq A \cap Y_1 \subseteq \ldots \subseteq A \cap Y_n \subseteq \ldots \subseteq A$ is a finitely supported countable chain of subsets of A (supported by $supp(n \mapsto Y_n)$) which should be stationary (finite). Furthermore, since $\bigcup_{i \in \mathbb{N}} (A \cap Y_i) = A \cap (\bigcup_{i \in \mathbb{N}} Y_i) = A$, there is some k_0 such that $A \cap Y_{k_0} = A$, and so $A \subseteq Y_{k_0}$. Thus, A is not FSM ascending infinite.

We know that $\wp_{fin}(X)$ is a subset of the invariant set $\wp_{fin}(U)$ supported by $supp(X)$. Let us consider $X_n = \{Z \in \wp_{fin}(X) \,|\, |Z| \leq n\}$. Clearly, $X_0 \subseteq X_1 \subseteq \ldots \subseteq X_n \subseteq \ldots$. Furthermore, because permutations of atoms are bijective, we have that for an arbitrary $k \in \mathbb{N}$, $|\pi \star Y| = |Y|$ for all $\pi \in S_A$ and all $Y \in X_k$, and so $\pi \star Y \in X_k$ for all $\pi \in Fix(supp(X))$ and all $Y \in X_k$. Thus, each X_k is a subset of $\wp_{fin}(X)$ finitely supported by $supp(X)$, and so $(X_n)_{n \in \mathbb{N}}$ is finitely (uniformly) supported by $supp(X)$. Obviously, $\wp_{fin}(X) = \bigcup_{n \in \mathbb{N}} X_n$. However, there exists no $n \in \mathbb{N}$ such that $\wp_{fin}(X) = X_n$. Thus, $(\wp_{fin}(X), \star)$ is FSM ascending infinite. □

Theorem 9.6 *Let X be a finitely supported subset of an invariant set $(Z, \cdot)$.*

1. *If X is FSM Dedekind infinite, then X is FSM Mostowski infinite.*
2. *If X is FSM Mostowski infinite, then X is FSM Tarski II infinite.*
   The reverse implication is not valid.

Proof 1. Suppose X is FSM Dedekind infinite. According to Theorem 9.2(1) there exists a uniformly supported infinite injective sequence $T = (x_n)_{n \in \mathbb{N}}$ of elements from X. Thus, each element of T is supported by $supp(T)$ and there is a bijective correspondence between $\mathbb{N}$ and T defined as $n \mapsto x_n$ which is supported by $supp(T)$. If we define the relation $\sqsubset$ on T by: $x_i \sqsubset x_j$ if and only if $i < j$, we have that $\sqsubset$ is a (strict) total order relation supported by $supp(T)$. Thus, T is an infinite, finitely supported (strictly) totally ordered subset of X, and so X is FSM Mostowski infinite because any strict total order can be extended to a total order.

2. Suppose that X is not FSM Tarski II infinite. Then every non-empty, finitely supported family of finitely supported subsets of X which is totally ordered by inclusion has a maximal element under inclusion. Let $(U, <)$ be a finitely supported

strictly totally ordered subset of X (any total order relation induces a strict total order relation). We prove that U is finite, and so X is not FSM Mostowski infinite. In this sense it is sufficient to prove that $<$ and $>$ are well-orderings. Since both of them are (strict) total orderings, we need to prove that any finitely supported subset of U has a least and a greatest element with respect to $<$, i.e. a minimal and a maximal element (because $<$ is total). Let Y be a finitely supported subset of U. The set $\downarrow z = \{y \in Y \,|\, y < z\}$ is supported by $supp(z) \cup supp(Y) \cup supp(<)$ for all $z \in Y$. The family $T = \{\downarrow z \,|\, z \in Y\}$ is itself finitely supported by $supp(Y) \cup supp(<)$ because for all $\pi \in Fix(supp(Y) \cup supp(<))$ we have $\pi \cdot \downarrow z = \downarrow \pi \cdot z$. Since $<$ is transitive, we have that T is (strictly) totally ordered by inclusion, and so it has a maximal element, which means Y has a maximal element. Similarly, the set $\uparrow z = \{y \in Y \,|\, z < y\}$ is supported by $supp(z) \cup supp(Y) \cup supp(<)$ for all $z \in Y$ and the family $T' = \{\uparrow z \,|\, z \in Y\}$ is itself finitely supported by $supp(Y) \cup supp(<)$ because for all $\pi \in Fix(supp(Y) \cup supp(<))$ we have $\pi \cdot \uparrow z = \uparrow \pi \cdot z$. The family T' is (strictly) totally ordered by inclusion, and so it has a maximal element, from which Y has a minimal element. We used the obvious properties $z < t$ if and only if $\downarrow z \subset \downarrow t$, and $z < t$ if and only if $\uparrow t \subset \uparrow z$.

Conversely, according to Proposition 9.8, $\wp_{fin}(A)$ is FSM ascending infinite, and so it is FSM Tarski II infinite. However $\wp_{fin}(A)$ is not FSM Mostowski infinite according to Corollary 9.8 □

Proposition 9.9 *Let X be a finitely supported subset of an invariant set $(Z, \cdot)$. If X is FSM Mostowski infinite, then X is non-amorphous meaning that X can be expressed as a disjoint union of two infinite, finitely supported subsets. The reverse implication is not valid.*

Proof Suppose that there is an infinite, finitely supported totally ordered subset $(Y, \leq)$ of X. Assume by contradiction that Y is amorphous, meaning that any finitely supported subset of Y is either finite or cofinite. As in the proof of Theorem 9.6 (without making the requirement that $\leq$ is strict, which anyway would not essentially change the proof), for $z \in Y$ we define the finitely supported subsets $\downarrow z = \{y \in Y \,|\, y \leq z\}$ and $\uparrow z = \{y \in Y \,|\, z \leq y\}$ for all $z \in Y$. We have that the mapping $z \mapsto \downarrow z$ from Y to $T = \{\downarrow z \,|\, z \in Y\}$ is itself finitely supported by $supp(Y) \cup supp(\leq)$. Furthermore, it is bijective, and so T is amorphous. Thus, any subset Z of T is either finite or cofinite, and obviously any subset Z of T is finitely supported. Similarly, the mapping $z \mapsto \uparrow z$ from Y to $T' = \{\uparrow z \,|\, z \in Y\}$ is finitely supported and bijective, which means that any subset of T' is either finite or cofinite, and clearly, any subset of T' is finitely supported.

We distinguish the following two cases:

1. There are only finitely many elements $x_1, \dots, x_n \in Y$ such that $\downarrow x_1, \dots, \downarrow x_n$ are finite. Thus, for $y \in U = Y \setminus \{x_1, \dots, x_n\}$ we have $\downarrow y$ infinite. Since $\downarrow y$ is a subset of Y, it should be cofinite, and so $\uparrow y$ is finite (because $\leq$ is a total order relation). Let $M = \{\uparrow y \,|\, y \in U\}$. As in Theorem 9.6 we have that M is totally ordered with respect to sets inclusion. Furthermore, for an arbitrary $y \in U$ we cannot have $y \leq x_k$ for some $k \in \{1, \dots, n\}$ because $\downarrow y$ is infinite, while $\downarrow x_k$ is finite, and so $\uparrow y$ is a subset of U. Thus, M is an infinite, finitely supported (by $supp(U) \cup supp(\leq)$),

totally ordered family formed by finite subsets of U. Since M is finitely supported, for each $y \in U$ and each $\pi \in Fix(supp(M))$ we have $\pi \cdot \uparrow y \in M$. Since $\uparrow y$ is finite, we have that $\pi \cdot \uparrow y$ is finite having the same number of elements as $\uparrow y$. Since $\pi \cdot \uparrow y$ and $\uparrow y$ are comparable via inclusion, they should be equal. Thus, M is uniformly supported. Since $\leq$ is a total order, for $\pi \in Fix(supp(\uparrow y))$ we have $\uparrow \pi \cdot y = \pi \cdot \uparrow y = \uparrow y$, and so $\pi \cdot y = y$, from which $supp(y) \subseteq supp(\uparrow y)$. Thus, U is uniformly supported. Since any element of U has only a finite number of successors (leading to the conclusion that $\geq$ is an well-ordering on U uniformly supported by $supp(U)$) and U is *uniformly supported*, we can define an order monomorphism between $\mathbb{N}$ and U which is supported by $supp(U)$. For example, choose $u_0 \neq u_1 \in U$, then let u_2 be *the greatest element* (with respect to $\leq$) in $U \setminus \{u_0, u_1\}$, u_3 be *the greatest element* in $U \setminus \{u_0, u_1, u_2\}$ (no choice principle is used since $\geq$ is an well-ordering, and so such a *greatest* element is precisely defined), and so on, and find an infinite, uniformly supported countable sequence $u_0, u_1, u_2, \ldots$. Since $\mathbb{N}$ is non-amorphous (being expressed as the union between the even elements and the odd elements), we conclude that U is non-(uniformly) amorphous containing two infinite uniformly supported disjoint subsets.

2. We have cofinitely many elements z such that $\downarrow z$ is finite. Thus, there are only finitely many elements $y_1, \ldots, y_m \in Y$ such that $\downarrow y_1, \ldots, \downarrow y_m$ are infinite. Since every infinite subset of Y is cofinite, only $\uparrow y_1, \ldots, \uparrow y_m$ are finite. Let $z \in Y \setminus \{y_1, \ldots, y_m\}$ which means $\uparrow z$ infinite. Since $\uparrow z$ is a subset of Y it should be cofinite, and so $\downarrow z$ is finite. As in the above item, the set $M' = \{\downarrow z \,|\, z \in Y \setminus \{y_1, \ldots, y_m\}\}$ is an infinite, finitely supported, totally ordered (by inclusion) family of finite sets, and so it has to be uniformly supported, from which $Y \setminus \{y_1, \ldots, y_m\}$ is uniformly supported, and so $\leq$ is an FSM well-ordering on $Y \setminus \{y_1, \ldots, y_m\}$. Therefore, $Y \setminus \{y_1, \ldots, y_m\}$ has an infinite, uniformly supported, countable subset, and so $Y \setminus \{y_1, \ldots, y_m\}$ is non-(uniformly) amorphous containing two infinite uniformly supported disjoint subsets.

Thus, Y is non-amorphous, and so X is non-amorphous.

Conversely, the set $A + A$ (disjoint union of A and A) is obviously non-amorphous because because $\{(0,a) \,|\, a \in A\}$ is equivariant, infinite and coinfinite. However, if we assume there exists a finitely supported total order relation on an infinite subset of $A + A$, then there should exist an infinite, finitely supported, total order on at least one of the sets $\{(0,a) \,|\, a \in A\}$ or $\{(1,a) \,|\, a \in A\}$, which leads to an infinite, finitely supported total order relation on A. However A is not FSM Mostowski infinite by Corollary 9.8. □

Theorem 9.7 *Let X be a finitely supported subset of an invariant set $(Z, \cdot)$. If X contains no infinite uniformly supported subset, then X is not FSM Mostowski infinite.*

Proof Assume by contradiction that X is FSM Mostowski infinite, meaning that X contains an infinite, finitely supported, totally ordered subset $(Y, \leq)$. We claim that Y is uniformly supported by $supp(\leq) \cup supp(Y)$. Let $\pi \in Fix(supp(\leq) \cup supp(Y))$ and let $y \in Y$ an arbitrary element. Since π fixes $supp(Y)$ pointwise and $supp(Y)$ supports Y, we obtain that $\pi \cdot y \in Y$, and so we should have either $y < \pi \cdot y$, or $y = \pi \cdot y$, or $\pi \cdot y < y$. If $y < \pi \cdot y$, then, because π fixes $supp(\leq)$ pointwise and because the mapping $z \mapsto \pi \cdot z$ is bijective from Y to $\pi \star Y$, we get $y < \pi \cdot y < \pi^2 \cdot y < \ldots < \pi^n \cdot y$

for all $n \in \mathbb{N}$. However, since any permutation of atoms interchanges only finitely many atoms, it has a finite order in the group S_A, and so there is $m \in \mathbb{N}$ such that $\pi^m = Id$. This means $\pi^m \cdot y = y$, and so we get $y < y$ which is a contradiction. Similarly, the assumption $\pi \cdot y < y$, leads to the relation $\pi^n \cdot y < \ldots < \pi \cdot y < y$ for all $n \in \mathbb{N}$ which is also a contradiction since π has finite order. Therefore, $\pi \cdot y = y$, and because y was arbitrary chosen form Y, Y should be a uniformly supported infinite subset of X. □

Looking to the proof of Proposition 9.9, the next result follows directly.

Corollary 9.7 *Let X be a finitely supported subset of an invariant set $(Z, \cdot)$. If X is FSM Mostowski infinite, then X is non-uniformly amorphous meaning that X has two disjoint, infinite, uniformly supported subsets.*

Remark 9.3 In a permutation model of set theory with atoms, a set can be well-ordered if and only if there is a one-to-one mapping of the related set into the kernel of the model. Furthermore, in a permutation model defined by using supports, a set U is well-orderable in the model if and only if there exists a set supporting every element in U. Moreover, we note that the axiom of choice is valid in the kernel of the model [37]. Therefore, totally ordered, finitely supported sets in Basic Fraenkel Model (that are proved to be uniformly supported similarly as in Theorem 9.7, and so they are well-orderable) can be embedded into the kernel of the model, and they should contain countable (uniformly supported) subsets; this provides an equivalence between Dedekind infinity and Mostowski infinity in the related permutation model. Although FSM (or the theory of nominal sets) is somehow related to (has connections with) permutation models of set theory with atoms, it is independently developed over ZF without being necessary to relax the axioms of extensionality or foundation. FSM sets are ZF sets together with group actions satisfying a certain finite support requirement, and such a theory makes sense over ZF *without being necessary to require the validity of the axiom of choice on (non-atomic) ZF sets.* Thus, FSM is the entire ZF together with atomic sets with finite supports (where the set of atoms is a fixed ZF set formed by elements whose internal structure is ignored and which are basic in the higher-order construction). There may exist infinite ZF sets that do not contain infinite countable subsets, and there may also exist infinite uniformly supported FSM sets (particularly such ZF sets) that do not contain infinite countable, uniformly supported subsets.

Corollary 9.8

1. *The sets A, $A+A$ and $A \times A$ are FSM classical infinite, but they are not FSM Mostowski infinite, nor FSM Tarski II infinite.*
2. *None of the sets $\wp_{fin}(A)$, $\wp_{cofin}(A)$, $\wp_{fs}(A)$ and $\wp_{fin}(\wp_{fs}(A))$ is Mostowski infinite in FSM.*
3. *None of the sets A^A_{fs}, $T_{fin}(A)^A_{fs}$ and $\wp_{fs}(A)^A_{fs}$ is FSM Mostowski infinite.*

Proof In the view of Theorem 9.7 it is sufficient to prove that none of the sets A, $\wp_{fin}(A)$, $\wp_{cofin}(A)$, $\wp_{fs}(A)$, $A+A$, $A \times A$, A^A_{fs},$T_{fin}(A)^A_{fs}$ and $\wp_{fs}(A)^A_{fs}$ contain infinite uniformly supported subsets. For A, $\wp_{fin}(A)$, $\wp_{cofin}(A)$ and $\wp_{fs}(A)$, this is obvious because for any finite set S of atoms there are at most finitely many subsets of A supported by S, namely the subsets of S and the supersets of $A \setminus S$. Moreover, $\wp_{fin}(\wp_{fs}(A))$ does not contain an infinite uniformly supported subset according to Lemma 9.1 since $\wp_{fs}(A)$ does not contain an infinite uniformly supported subset.

Regarding A^A_{fs}, the things are also similar with Corollary 9.1(4). According to Lemma 9.4, any S-supported function $f : A \to A$ should have the property that either $f|_{A \setminus S} = Id$ or $f|_{A \setminus S}$ is a one-element subset of S. For each possible definition of such an f on S we have at most $|S|+1$ possible ways to define f on $A \setminus S$, and so at most $|S|+1$ possible ways to completely define f on A. If there was an infinite uniformly S-supported sequence of finitely supported functions from A to A, there should exist infinitely many finitely supported functions from S to A supported by the same finite set S. However, this contradicts the fact that $A^{|S|}$ does not contain an infinite uniformly supported subset (this follows by applying finitely many times the result that $X \times X$ does not contain an infinite uniformly supported subset whenever X does not contain an infinite uniformly supported subset). Analyzing the proofs of Corollary 9.1(6) and (7), we also conclude that $T_{fin}(A)^A_{fs}$ and $\wp_{fs}(A)^A_{fs}$ do not contain infinite uniformly supported subsets.

We also have that A is not FSM Tarski II infinite because $\wp_{fs}(A)$ contains no infinite uniformly supported subsets, and so every totally ordered subset (particularly via inclusion) of $\wp_{fs}(A)$ should be finite meaning that it should have a maximal element. Furthermore, we have that there is an equivariant bijection between $\wp_{fs}(A+A)$ and $\wp_{fs}(A) \times \wp_{fs}(A)$. Since $\wp_{fs}(A)$ does not contain an infinite uniformly supported subset, we have that $\wp_{fs}(A) \times \wp_{fs}(A)$ does not contain an infinite uniformly supported subset (the proof is quasi-identical to the one of Theorem 9.2(6) without considering the countability of the related infinite uniformly supported family). Therefore, any infinite totally ordered (via inclusion) uniformly supported family of $\wp_{fs}(A+A)$ should be finite containing a maximal element. There is an equivariant bijection between $\wp_{fs}(A)^A_{fs}$ and $\wp_{fs}(A \times A)$ (see Theorem 2.3 and Theorem 2.4). Therefore any uniformly supported totally ordered subset of $\wp_{fs}(A \times A)$ should be finite containing a maximal element. □

Corollary 9.9 *Let X be a finitely supported subset of an invariant set Y such that X does not contain an infinite uniformly supported subset. Then the set $\wp_{fin}(X)$ is not FSM Mostowski infinite.*

Proof According to Lemma 9.1, $\wp_{fin}(X)$ does not contain an infinite uniformly supported subset. Thus, by Theorem 9.7, $\wp_{fin}(X)$ is not FSM Mostowski infinite.□

Theorem 9.8 *Let X be a finitely supported subset of an invariant set $(Y, \cdot)$.*

1. *If X is FSM Tarski I infinite, then X is FSM Tarski III infinite. The converse does not hold. However if X is FSM Tarski III infinite, then $\wp_{fs}(X)$ is FSM Tarski I infinite.*
2. *If X is FSM Tarski III infinite, then X is FSM Dedekind infinite. The converse does not hold. However if X is FSM Dedekind infinite, then $\wp_{fs}(X)$ is FSM Tarski III infinite.*

Proof 1. We consider the case when X has at least two elements (otherwise the theorem is trivial). Let X be FSM Tarski I infinite. Then $|X \times X| = |X|$. Fix two elements $x_1, x_2 \in X$ with $x_1 \neq x_2$. We can define an injection $f : X \times \{0,1\} \to X \times X$ by $f(u) = \begin{cases} (x, x_1) \text{ for } u = (x,0) \\ (x, x_2) \text{ for } u = (x,1) \end{cases}$. Clearly, by checking the condition in Proposition 2.9 and using Proposition 2.2, we have that f is supported by $supp(X) \cup supp(x_1) \cup supp(x_2)$ (since $\{0,1\}$ is necessarily a trivial invariant set), and so $|X \times \{0,1\}| \leq |X \times X|$. Thus, $|X \times \{0,1\}| \leq |X|$. Obviously, there is an injection $i : X \to X \times \{0,1\}$ defined by $i(x) = (x,0)$ for all $x \in X$ which is supported by $supp(X)$. According to Lemma 5.2, we get $2|X| = |X \times \{0,1\}| = |X|$.

Let us consider $X = \mathbb{N} \times A$. We remark that $|\mathbb{N} \times \mathbb{N}| = |\mathbb{N}|$ by considering the equivariant injection $h : \mathbb{N} \times \mathbb{N} \to \mathbb{N}$ defined by $h(m,n) = 2^m 3^n$ and using Lemma 5.2. Similarly, $|\{0,1\} \times \mathbb{N}| = |\mathbb{N}|$ by considering the equivariant injection $h' : \mathbb{N} \times \{0,1\} \to \mathbb{N}$ defined by $h'(n,0) = 2^n$ and $h'(n,1) = 3^n$ and using Lemma 5.2. We have $2|X| = 2|\mathbb{N}||A| = |\mathbb{N}||A| = |X|$. However, we prove that $|X \times X| \neq |X|$. Assume the contrary, and so we have $|\mathbb{N} \times (A \times A)| = |\mathbb{N} \times A \times \mathbb{N} \times A| = |\mathbb{N} \times A|$. Thus, there is a finitely supported injection $g : A \times A \to \mathbb{N} \times A$, and by Proposition 5.1 there is a finitely supported surjection $f : \mathbb{N} \times A \to A \times A$. Let us consider three different atoms $a, b, c \notin supp(f)$. There exists $(i,x) \in \mathbb{N} \times A$ such that $f(i,x) = (a,b)$. Since $(ab) \in Fix(supp(f))$ and $\mathbb{N}$ is trivial invariant set, we have $f(i,(ab)(x)) = (ab)f(i,x) = (ab)(a,b) = ((ab)(a),(ab)(b)) = (b,a)$. We should have $x = a$ or $x = b$, otherwise f is not a function. Assume without losing the generality that $x = a$, which means $f(i,a) = (a,b)$. Therefore $f(i,b) = f(i,(ab)(a)) = (ab)f(i,a) = (ab)(a,b) = (b,a)$. Similarly, since $(ac),(bc) \in Fix(supp(f))$, we have $f(i,c) = f(i,(ac)(a)) = (ac)f(i,a) = (ac)(a,b) = (c,b)$ and so we obtain $f(i,b) = f(i,(bc)(c)) = (bc)f(i,c) = (bc)(c,b) = (b,c)$. But $f(i,b) = (b,a)$ contradicting the functionality of f. Therefore, X is FSM Tarski III infinite, but it is not FSM Tarski I infinite.

Now, suppose that X is FSM Tarski III infinite, which means $|\{0,1\} \times X| = |X|$. We define the mapping $\psi : \wp_{fs}(X) \times \wp_{fs}(X) \to \wp_{fs}(\{0,1\} \times X)$ by $f(U,V) = \{(0,x) \,|\, x \in U\} \cup \{(1,y) \,|\, y \in V\}$ for all $U, V \in \wp_{fs}(X)$. Clearly, ψ is well-defined and bijective, and for each $\pi \in Fix(supp(X))$ we have $\psi(\pi \star U, \pi \star V) = \pi \star \psi(U,V)$ which means ψ is finitely supported. Thus, $|\wp_{fs}(X) \times \wp_{fs}(X)| = |\wp_{fs}(\{0,1\} \times X)|$ $= |\wp_{fs}(X)|$. The last equality follows by applying twice Lemma 5.5 (using the fact that there is a finitely supported surjection from X onto $X \times \{0,1\}$ and a finitely supported surjection from $X \times \{0,1\}$ onto X, we obtain there is a finitely supported injection from $\wp_{fs}(X \times \{0,1\})$ into $\wp_{fs}(X)$, and a finitely supported injection from $\wp_{fs}(X)$ into $\wp_{fs}(X \times \{0,1\})$) and Lemma 5.2.

2. Now suppose X is FSM Tarski III infinite. Let us consider an element y_1 belonging to an invariant set (whose action is also denoted by $\cdot$) with $y_1 \notin X$ (such an element can be, for example, a non-empty element in $\wp_{fs}(X) \setminus X$). Fix $y_2 \in X$. One can define a mapping $f : X \cup \{y_1\} \to X \times \{0,1\}$ by $f(x) = \begin{cases} (x,0) & \text{for } x \in X \\ (y_2,1) & \text{for } x = y_1 \end{cases}$. Clearly, f is injective and it is supported by $S = supp(X) \cup supp(y_1) \cup supp(y_2)$ because for all π fixing S pointwise we have $f(\pi \cdot x) = \pi \cdot f(x)$ for all $x \in X \cup \{y_1\}$.

Therefore, $|X \cup \{y_1\}| \leq |X \times \{0,1\}| = |X|$, and so there is a finitely supported injection $g : X \cup \{y_1\} \to X$. The mapping $h : X \to X$ defined by $h(x) = g(x)$ is injective, supported by $supp(g) \cup supp(X)$, and $g(y_1) \in X \setminus h(X)$, which means h is not surjective. It follows that X is FSM Dedekind infinite.

Let us consider $X = A \cup \mathbb{N}$. Since A and $\mathbb{N}$ are disjoint, we have that X is an invariant set (similarly as in Proposition 2.2). Clearly, X is FSM Dedekind infinite. Assume by contradiction that $|X| = 2|X|$, that is $|A \cup \mathbb{N}| = |A + A + \mathbb{N}| = |(\{0,1\} \times A) \cup \mathbb{N}|$. Thus, there is a finitely supported injection $f : (\{0,1\} \times A) \cup \mathbb{N} \to A \cup \mathbb{N}$, and so there exists a finitely supported injection $f : (\{0,1\} \times A) \to A \cup \mathbb{N}$. We prove that whenever $\varphi : A \to A \cup \mathbb{N}$ is finitely supported and injective, for $a \notin supp(\varphi)$ we have $\varphi(a) \in A$. Assume by contradiction that there is $a \notin supp(\varphi)$ such that $\varphi(a) \in \mathbb{N}$. Since $supp(\varphi)$ is finite, there exists $b \notin supp(\varphi)$, $b \neq a$. Thus, $(a\,b) \in Fix(supp(\varphi))$, and so $\varphi(b) = \varphi((a\,b)(a)) = (a\,b) \diamond \varphi(a) = \varphi(a)$ since $(\mathbb{N}, \diamond)$ is a trivial invariant set. This contradicts the injectivity of φ. We can consider the mappings $\varphi_1, \varphi_2 : A \to A \cup \mathbb{N}$ defined by $\varphi_1(a) = f(0,a)$ for all $a \in A$ and $\varphi_2(a) = f(1,a)$ for all $a \in A$, that are injective and supported by $supp(f)$. Therefore, $f(\{0\} \times A) = \varphi_1(A)$ contains at most finitely many element from $\mathbb{N}$, and $f(\{1\} \times A) = \varphi_2(A)$ also contains at most finitely many element from $\mathbb{N}$. Thus, f is an injection from $(\{0,1\} \times A)$ to $A \cup Z$ where Z is a finite subset of $\mathbb{N}$. It follows that $f(\{0\} \times A)$ contains an infinite finitely supported subset of atoms U, and $f(\{1\} \times A)$ contains an infinite finitely supported subset of atoms V. Since f is injective, it follows that U and V are infinite, disjoint, finitely supported subsets of A, which contradicts the fact that A is amorphous.

Now, if X is FSM Dedekind infinite, we have that there is a finitely supported injection h from X onto a finitely supported proper subset Z of X. Consider an element y_1 belonging to an invariant set with $y_1 \notin X$. We can define an injection $h' : X \cup \{y_1\} \to X$ by taking $h'(x) = h(x)$ for all $x \in X$ and $h'(y_1) = b$ with $b \in X \setminus Z$. Clearly, h' is supported by $supp(h) \cup supp(y_1) \cup supp(b)$. Since there also exists an $supp(X)$-supported injection from X to $X \cup \{y_1\}$, according to Lemma 5.2, one can define a finitely supported bijection ψ from X to $X \cup \{y_1\}$. According to Lemma 5.5, the mapping $g : \wp_{fs}(X \cup \{y_1\}) \to \wp_{fs}(X)$ defined by $g(V) = f^{-1}(V)$ for all $V \in \wp_{fs}(X \cup \{y_1\})$ is finitely supported and injective. Therefore, $2^{|X|} \geq 2^{|X|+1} = 2 \cdot 2^{|X|}$ which in the view of Lemma 5.2 (since we also have $2^{|X|} \leq 2 \cdot 2^{|X|}$) leads to the conclusion that $\wp_{fs}(X)$ is FSM Tarski III infinite. □

Corollary 9.10 *The following sets are FSM classical infinite, but they are not FSM Tarski I infinite, nor FSM Tarski III finite.*

1. *The invariant set A.*
2. *The invariant set $\wp_{fs}(A)$.*
3. *The invariant sets $\wp_{fin}(A)$ and $\wp_{cofin}(A)$.*
4. *The set $\wp_{fin}(X)$ where X is a finitely supported subset of an invariant set containing no infinite uniformly supported subset.*

Proof The result follows directly because, according to Theorem 9.2 and Corollary 9.1, the related sets are not FSM Dedekind infinite. □

Corollary 9.11 *The sets $A^{\mathbb{N}}_{fs}$ and $\mathbb{N}^{A}_{fs}$ are FSM Tarski I infinite, and so they are also Tarski III infinite.*

Proof There is an equivariant bijection ψ between $(A^{\mathbb{N}})^2_{fs}$ and $A^{\mathbb{N}\times\{0,1\}}_{fs}$ that associates to each Cartesian pair (f,g) of mappings from $\mathbb{N}$ to A a mapping $h: \mathbb{N}\times\{0,1\}\to A$ defined as follows:

$$h(u)=\begin{cases} f(n) \text{ if } u=(n,0) \\ g(n) \text{ if } u=(n,1) \end{cases}.$$

The equivariance of ψ follows from Proposition 2.9 because if $\pi\in S_A$ we have $\psi(\pi\widetilde{\star}f,\pi\widetilde{\star}g)=h'$ where $h'(n,0)=(\pi\widetilde{\star}f)(n)=\pi(f(n))$ and $h'(n,1)=(\pi\widetilde{\star}g)(n)=\pi(g(n))$. Thus, $h'(u)=\pi(h(u))$ for all $u\in\mathbb{N}\times\{0,1\}$ which means $h'=\pi\widetilde{\star}h=\pi\widetilde{\star}\psi(f,g)$.

There also exists an equivariant bijection φ between $(\mathbb{N}^A)^2_{fs}$ and $(\mathbb{N}\times\mathbb{N})^A_{fs}$ that associates to each Cartesian pair (f,g) of mappings from A to $\mathbb{N}$ a mapping $h: A\to\mathbb{N}\times\mathbb{N}$ defined by $h(a)=(f(a),g(a))$ for all $a\in A$. The equivariance of φ follows because if $\pi\in S_A$ then $\varphi(\pi\widetilde{\star}f,\pi\widetilde{\star}g)=h'$ with $h'(a)=((\pi\widetilde{\star}f)(a),(\pi\widetilde{\star}g)(a))=(f(\pi^{-1}(a)),g(\pi^{-1}(a)))=h(\pi^{-1}(a))=(\pi\widetilde{\star}h)(a)$ for all $a\in A$, and so $h'=\pi\widetilde{\star}h=\pi\widetilde{\star}\varphi(f,g)$. Therefore $|(A^{\mathbb{N}})^2_{fs}|=|A^{\mathbb{N}\times\{0,1\}}_{fs}|=|A^{\mathbb{N}}_{fs}|$. Thus, $|(\mathbb{N}^A)^2_{fs}|=|(\mathbb{N}\times\mathbb{N})^A_{fs}|=|\mathbb{N}^A_{fs}|$ according to Proposition 5.3(3) and Lemma 5.2 (we used $|\mathbb{N}\times\mathbb{N}|=|\mathbb{N}|$). □

Corollary 9.12 *Let X be a finitely supported subset of an invariant set $(Y,\cdot)$ such that X does not contain an infinite uniformly supported subset. The set $X^{nA}_{fs}=\{f: nA\to X\,|\,f$ is finitely supported$\}$ does not contain an infinite uniformly supported subset, and so it is not FSM Dedekind infinite, whenever $n\in\mathbb{N}^*$, where nA is the n-times disjoint union between A and itself (i.e. $nA=A+\ldots+A$).*

Proof Using the proof of Theorem 9.8(1) and an induction after n, we have that $|nA|\leq|A^n|$ for all integers $n\geq 1$. According to Proposition 5.3(4) we have $|X^{nA}_{fs}|\leq|X^{A^n}_{fs}|$. Therefore, there exists a finitely supported injection from X^{nA}_{fs} to $X^{A^n}_{fs}$. However, $X^{A^n}_{fs}$ does not contain an infinite uniformly supported subset according to Theorem 9.4, and so the result follows. □

Theorem 9.9 *Let X be a finitely supported subset of an invariant set $(Y,\cdot)$. If $\wp_{fs}(X)$ is FSM Tarski I infinite, then $\wp_{fs}(X)$ is FSM Tarski III infinite. The converse does not hold.*

Proof The direct implication is an immediate consequence of Theorem 9.8(1). Thus, we focus on the proof of the invalidity of the reverse implication.

Firstly, we remark that whenever U,V are finitely supported subsets of an invariant set with $U\cap V=\emptyset$, we have that there is a finitely supported (by $supp(U)\cup supp(V)$) bijection from $\wp_{fs}(U\cup V)$ into $\wp_{fs}(U)\times\wp_{fs}(V)$ that maps each $X\in\wp_{fs}(U\cup V)$ into the pair $(X\cap U,X\cap V)$. Similarly, whenever B,C are invariant sets there is an equivariant bijection from $\wp_{fs}(B)\times\wp_{fs}(C)$ into $\wp_{fs}(B+C)$ that maps each pair $(B_1,C_1) \in \wp_{fs}(B) \times \wp_{fs}(C)$ into the set $\{(0,b)\,|\,b \in B_1\} \cup \{(1,c)\,|\,c \in C_1\}$. This follows directly by verifying the conditions in Proposition 2.9.

Let us consider the set $A \cup \mathbb{N}$ which is FSM Dedekind infinite. According to Theorem 9.8(2), we have that $\wp_{fs}(A \cup \mathbb{N})$ is FSM Tarski III infinite. We prove that it is not FSM Tarski I infinite. Assume by contradiction that $|\wp_{fs}(A \cup \mathbb{N}) \times \wp_{fs}(A \cup \mathbb{N})| = |\wp_{fs}(A \cup \mathbb{N})|$ which means $|\wp_{fs}(A+\mathbb{N}+A+\mathbb{N})| = |\wp_{fs}(A \cup \mathbb{N})|$, and so $|\wp_{fs}(A+A+\mathbb{N})| = |\wp_{fs}(A \cup \mathbb{N})|$. Thus, according to Proposition 5.3(4), there is a finitely supported injection from $\wp_{fs}(A+A)$ to $\wp_{fs}(A \cup \mathbb{N})$, which means there is a finitely supported injection from $\wp_{fs}(A) \times \wp_{fs}(A)$ to $\wp_{fs}(A) \times \wp_{fs}(\mathbb{N})$, and so there is a finitely supported injection from $A \times A$ to $\wp_{fs}(A) \times \wp_{fs}(\mathbb{N})$. According to Proposition 5.1, there should exist a finitely supported surjection $f : \wp_{fs}(A) \times \wp_{fs}(\mathbb{N}) \to A \times A$. Let us consider two atoms $a, b \notin supp(f)$ with $a \neq b$. It follows that $(a b) \in Fix(supp(f))$. Since f is surjective, there exists $(X,M) \in \wp_{fs}(A) \times \wp_{fs}(\mathbb{N})$ such that $f(X,M) = (a,b)$. According to Proposition 2.9 and because $\mathbb{N}$ is a trivial invariant set meaning that $(a b) \star M = M$, we have $f((a b) \star X, M) = f((a b) \otimes (X,M)) = (a b) \otimes f(X,M) = (a b) \otimes (a,b) = ((a b)(a),(a b)(b)) = (b,a)$. Due to the functionality of f we should have $((a b) \star X, M) \neq (X,M)$, which means $(a b) \star X \neq X$.

We prove that if both $a,b \in supp(X)$, then $(a b) \star X = X$. Indeed, suppose $a,b \in supp(X)$. Since $X \in \wp_{fs}(A)$, we have that X is either finite or cofinite. If X is finite, then $supp(X) = X$, and so $a,b \in X$. Therefore, $(a b) \star X = \{(a b)(x)\,|\,x \in X\} = \{(a b)(a)\} \cup \{(a b)(b)\} \cup \{(a b)(c)\,|\,c \in X \setminus \{a,b\}\} = \{b\} \cup \{a\} \cup (X \setminus \{a,b\}) = X$. Now, if X is cofinite, then $supp(X) = A \setminus X$, and so $a,b \in A \setminus X$. Since $a,b \notin X$, we have $a,b \neq x$ for all $x \in X$, which means $(a b)(x) = x$ for all $x \in X$, and again $(a b) \star X = X$.

Thus, one of a or b does not belong to $supp(X)$. Assume $b \notin supp(X)$. Let us consider $c \neq a,b$, $c \notin supp(f)$, $c \notin supp(X)$. Then $(b c) \in Fix(supp(X))$, and so $(b c) \star X = X$. Moreover, $(b c) \in Fix(supp(f))$, and by Proposition 2.9 we have $(a,b) = f(X,M) = f((b c) \star X, M) = f((b c) \otimes (X,M)) = (b c) \otimes f(X,M) = (b c) \otimes (a,b) = (a,c)$ which is a contradiction because $b \neq c$. □

Proposition 9.10 *Let X be a finitely supported subset of an invariant set $(Y,\cdot)$. If X is FSM Tarski III infinite, then there exists a finitely supported bijection $g : \mathbb{N} \times X \to X$. The reverse implication is also valid.*

Proof By hypothesis, there is a finitely supported bijection $\varphi : \{0,1\} \times X \to X$. Let us consider $f_1, f_2 : X \to X$ defined by $f_1(x) = \varphi(0,x)$ for all $x \in X$ and $f_2(x) = \varphi(1,x)$ for all $x \in X$; these mappings are injective and supported by $supp(\varphi)$ (according to Proposition 2.9). Since φ is injective we also have $Im(f_1) \cap Im(f_2) = \emptyset$, and because φ is surjective we get $Im(f_1) \cup Im(f_2) = X$. We prove by induction that the n-times auto-composition of f_2, denoted by f_2^n, is supported by $supp(f_2)$ for all $n \in \mathbb{N}$. For $n = 1$ this is obvious. So assume that f_2^{n-1} is supported by $supp(f_2)$. By Proposition 2.9 we must have $f_2^{n-1}(\sigma \cdot x) = \sigma \cdot f_2^{n-1}(x)$ for all $\sigma \in Fix(supp(f_2))$ and $x \in X$. Let us fix $\pi \in Fix(supp(f_2))$. According to Proposition 2.9, we have $f_2^n(\pi \cdot x) = f_2(f_2^{n-1}(\pi \cdot x)) = f_2(\pi \cdot f_2^{n-1}(x)) = \pi \cdot f_2(f_2^{n-1}(x)) = \pi \cdot f_2^n(x)$ for all $x \in X$, and so f_2^n is finitely supported from Proposition 2.9. Define $f : \mathbb{N} \times X \to X$

by $f((n,x)) = f_2^n(f_1(x))$. Let $\pi \in Fix(supp(f_1) \cup supp(f_2))$. According to Proposition 2.9 and because $(\mathbb{N}, \diamond)$ is a trivial invariant set we get $f(\pi \otimes (n,x)) = f((n, \pi \cdot x)) = f_2^n(f_1(\pi \cdot x)) = f_2^n(\pi \cdot f_1(x)) = \pi \cdot f_2^n(f_1(x)) = \pi \cdot f((n,x))$ for all $(n,x) \in \mathbb{N} \times X$, which means f is supported by $supp(f_1) \cup supp(f_2)$. We prove the injectivity of f. Assume $f((n,x)) = f((m,y))$ which means $f_2^n(f_1(x)) = f_2^m(f_1(y))$. If $n > m$ this leads to $f_2^{n-m}(f_1(x)) = f_1(y)$ (since f_2 is injective) which is in contradiction with the relation $Im(f_1) \cap Im(f_2) = \emptyset$. Similarly, we cannot have $n < m$. Thus, $n = m$ which leads to $f_1(x) = f_1(y)$, and so $x = y$ due to the injectivity of f_1. Therefore, f is injective. Since we obviously have a finitely supported injection from X into $\mathbb{N} \times X$ (e.g $x \mapsto (0,x)$ which is supported by $supp(X)$), in the view of Lemma 5.2 we can find a finitely supported bijection between X and $\mathbb{N} \times X$.

The reverse implication is almost trivial. There is a finitely supported injection from $\{0,1\} \times X$ into $\mathbb{N} \times X$. If there is a finitely supported injection from $\mathbb{N} \times X$ into X, then there is a finitely supported injection from $\{0,1\} \times X$ into X. The desired result follows from Lemma 5.2. □

In the table below we present the forms of infinity satisfied by classical infinite sets.

Set	TI inf	TIII inf	D inf	M inf	A inf	TII inf	N-am
A	No	No	No	No	No	No	No
$A+A$	No	No	No	No	No	No	Yes
$A \times A$	No	No	No	No	No	No	Yes
$\wp_{fin}(A)$	No	No	No	No	Yes	Yes	Yes
$T_{fin}(A)$	No	No	No	No	Yes	Yes	Yes
$\wp_{fs}(A)$	No	No	No	No	Yes	Yes	Yes
$\wp_{fin}(\wp_{fs}(A))$	No	No	No	No	Yes	Yes	Yes
A_{fs}^A	No	No	No	No	Yes	Yes	Yes
$T_{fin}(A)_{fs}^A$	No	No	No	No	Yes	Yes	Yes
$\wp_{fs}(A)_{fs}^A$	No	No	No	No	Yes	Yes	Yes
$\wp_{fs}(A^n), n \in \mathbb{N}$	No	No	No	No	Yes	Yes	Yes
$A_{fs}^{A^n}, n \in \mathbb{N}$	No	No	No	No	Yes	Yes	Yes
$A \cup \mathbb{N}$	No	No	Yes	Yes	Yes	Yes	Yes
$A \times \mathbb{N}$	No	Yes	Yes	Yes	Yes	Yes	Yes
$\wp_{fs}(A \cup \mathbb{N})$	No	Yes	Yes	Yes	Yes	Yes	Yes
$\wp_{fs}(\wp_{fs}(A))$	?	Yes	Yes	Yes	Yes	Yes	Yes
$\wp_{fs}(\wp_{fs}(\wp_{fs}(A)))$	Yes	Yes	Yes	Yes	Yes	Yes	Yes
$A_{fs}^{\mathbb{N}}$	Yes	Yes	Yes	Yes	Yes	Yes	Yes
$\mathbb{N}_{fs}^A$	Yes	Yes	Yes	Yes	Yes	Yes	Yes

9.2 Countability in Finitely Supported Mathematics

Definition 9.2 Let Y be a finitely supported subset of an invariant set X. Then Y is *countable in FSM (or FSM countable)* if there exists a finitely supported onto mapping $f : \mathbb{N} \to Y$.

Proposition 9.11 *Let Y be a finitely supported countable subset of an invariant set $(X, \cdot)$. Then Y is uniformly supported.*

Proof There exists a finitely supported onto mapping $f : \mathbb{N} \to Y$. Thus, for each arbitrary $y \in Y$, there exists $n \in \mathbb{N}$ such that $f(n) = y$. According to Proposition 2.9, for each $\pi \in Fix(supp(f))$ we have $\pi \cdot y = \pi \cdot f(n) = f(\pi \diamond n) = f(n) = y$, where $\diamond$ is the necessarily trivial action on $\mathbb{N}$. Thus, Y is uniformly supported by $supp(f)$.□

Proposition 9.12 *Let Y be a finitely supported subset of an invariant set X. Then Y is countable in FSM if and only if there exists a finitely supported one-to-one mapping $g : Y \to \mathbb{N}$.*

Proof Suppose that Y is countable in FSM. Then there exists a finitely supported onto mapping $f : \mathbb{N} \to Y$. We define $g : Y \to \mathbb{N}$ by $g(y) = min[f^{-1}(\{y\})]$, for all $y \in Y$. According to Proposition 2.9, g is supported by $supp(f) \cup supp(Y)$. Obviously, g is one-to-one. Conversely, if there exists a finitely supported one-to-one mapping $g : Y \to \mathbb{N}$, then $g(Y)$ is supported is equivariant as a subset of the trivial invariant set $\mathbb{N}$. Thus, there exists a finitely supported bijection $g : Y \to g(Y)$, where $g(Y) \subseteq \mathbb{N}$. We define $f : \mathbb{N} \to Y$ by

$$f(n) = \begin{cases} g^{-1}(n) & \text{if} n \in g(Y) \\ t & \text{if} n \in \mathbb{N} \setminus g(Y) \end{cases},$$

where t is a fixed element of Y. According to Proposition 2.9, we have that f is supported by $supp(g) \cup supp(Y) \cup supp(t)$. Moreover, f is onto. □

Proposition 9.13 *Let Y be an infinite, finitely supported, countable subset of an invariant set X. Then there exists a finitely supported bijective mapping $g : Y \to \mathbb{N}$.*

Proof Firstly, we prove that for any infinite subset B of $\mathbb{N}$, there is an injection from $\mathbb{N}$ into B. Fix such a B. It follows that B is well-ordered. Define $f :$ $\mathbb{N} \to B$ by: $f(1) = min(B)$, $f(2) = min(B \setminus f(1))$, and recursively $f(m) = min(B \setminus \{f(1), f(2), ..., f(m-1)\})$ for all $m \in \mathbb{N}$ (since B is infinite). Since $\mathbb{N}$ is well-ordered, choice is not involved. Obviously, since both B and $\mathbb{N}$ are trivial invariant sets, we have that f is equivariant. Since B is a subset of $\mathbb{N}$ we also have an equivariant injective mapping $h : B \to \mathbb{N}$. According to Lemma 5.1, there is an equivariant bijection between B and $\mathbb{N}$ (we can even prove that f is bijective).

Since Y is countable, according to Proposition 9.12 there exists a finitely supported one-to-one mapping $u : Y \to \mathbb{N}$. Thus, the mapping $u : Y \to u(Y)$ is finitely supported and bijective. Since $u(Y) \subseteq N$, we have that there is an equivariant bijection v between $u(Y)$ and $\mathbb{N}$, and so there exists a finitely supported bijective mapping $g : Y \to \mathbb{N}$ defined by $g = v \circ u$. □

From Chapter 3 we know that the (in)consistency of the choice principle **CC(fin)** in FSM is an open problem, meaning that we do not know whether this principle is consistent or not with respect to the FSM axioms. A relationship between countable union principles and countable choice principles is presented in ZF in [34]. Below we prove that such a relationship is preserved in FSM.

Definition 9.3 1. The Countable Union Theorem for finite sets **CUT(fin)** has in FSM the form "Given any invariant set X and any countable family $\mathscr{F} = (X_n)_n$ of finite subsets of X such that the mapping $n \mapsto X_n$ is finitely supported, there exists a finitely supported onto mapping $f : \mathbb{N} \to \underset{n}{\cup} X_n$".

2. The Countable Union Theorem for k-element sets **CUT(k)** has in FSM the form "Given any invariant set X and any countable family $\mathscr{F} = (X_n)_n$ of k-element subsets of X such that the mapping $n \mapsto X_n$ is finitely supported, there exists a finitely supported onto mapping $f : \mathbb{N} \to \underset{n}{\cup} X_n$".
3. The Countable Choice principle for k-element sets **CC(k)** has in FSM the form "Given any invariant set X and any countable family $\mathscr{F} = (X_n)_n$ of k-element subsets of X such that the mapping $n \mapsto X_n$ is finitely supported, there exists a finitely supported choice function on $\mathscr{F}$".

Proposition 9.14 *In FSM, the following equivalences hold:*

1. ***CUT(fin)*** $\Leftrightarrow$ ***CC(fin)***;
2. ***CUT(2)*** $\Leftrightarrow$ ***CC(2)***;
3. ***CUT(n)*** $\Leftrightarrow$ ***CC(i)*** *for all* $i \leq n$.

Proof 1. Let us assume that **CUT(fin)** is valid in FSM. We consider the finitely supported countable family $\mathscr{F} = (X_n)_n$ in FSM, where each X_n is a non-empty finite subset of an invariant set X in FSM.

From **CUT(fin)**, there exists a finitely supported onto mapping $f : \mathbb{N} \to \underset{n}{\cup} X_n$. Since f is onto and each X_n is non-empty, we have that $f^{-1}(X_n)$ is a non-empty subset of $\mathbb{N}$ for each $n \in \mathbb{N}$. Consider the function $g : \mathscr{F} \to \cup\mathscr{F}$, defined by $g(X_n) = f(min[f^{-1}(X_n)])$. We claim that $supp(f) \cup supp(n \mapsto X_n)$ supports g. Let $\pi \in Fix(supp(f) \cup supp(n \mapsto X_n))$. According to Proposition 2.9, and because $\mathbb{N}$ is a trivial invariant set and each element X_n is supported by $supp(n \mapsto X_n)$, we have $\pi \cdot g(X_n) = \pi \cdot f(min[f^{-1}(X_n)]) = f(\pi \diamond min[f^{-1}(X_n)]) = f(min[f^{-1}(X_n)]) = g(X_n) = g(\pi \star X_n)$, where by $\star$ we denoted the S_A-action on $\mathscr{F}$, by $\cdot$ we denoted the S_A-action on $\cup\mathscr{F}$ and by $\diamond$ we denoted the trivial action on $\mathbb{N}$. Therefore, g is finitely supported. Moreover, $g(X_n) \in X_n$, and so g is a choice function on $\mathscr{F}$.

Conversely, let $\mathscr{F} = (X_n)_n$ be a countable family of finite subsets of X such that the mapping $n \mapsto X_n$ is finitely supported. Thus, each X_n is supported by the same set $S = supp(n \mapsto X_n)$. Since each X_n is finite (and the support of a finite set coincides with the union of the supports of its elements), as in the proof of Lemma 9.1, we have that $Y = \underset{n \in \mathbb{N}}{\cup} X_n$ is uniformly supported by S. Moreover, the countable sequence $(Y_n)_{n \in \mathbb{N}}$ defined by $Y_n = X_n \setminus \underset{m<n}{\cup} X_m$ is a uniformly supported (by S) sequence of

pairwise disjoint uniformly supported sets with $Y = \underset{n\in\mathbb{N}}{\cup} Y_n$. Consider the infinite family $M \subseteq \mathbb{N}$ such that all the terms of $(Y_n)_{n\in M}$ are non-empty.

For each $n \in M$, the set T_n of total orders on Y_n is finite, non-empty, and uniformly supported by S. Thus, by applying **CC(fin)** to $(T_n)_{n\in M}$, there is a choice function f on $(T_n)_{n\in M}$ which is also supported by S. Furthermore, $f(T_n)$ is supported by $supp(f) \cup supp(T_n) = S$ for all $n \in M$. One can define a uniformly supported (by S) total order relation on Y (which is also a well-order relation on Y) as follows: $x \leq y$ if and only if $\begin{cases} x \in Y_n \text{ and } y \in Y_m \text{ with } n < m \\ \quad\text{or} \\ x, y \in Y_n \text{ and } x f(T_n) y \end{cases}$.

Clearly, if Y is infinite, then there is an S-supported order isomorphism between $(Y, \leq)$ and M with the natural order, which means (in the view of Proposition 9.13) that Y is countable.

2. As in the above item **CUT(2)** $\Rightarrow$ **CC(2)**.

For proving **CC(2)** $\Rightarrow$ **CUT(2)**, let $\mathscr{F} = (X_n)_n$ be a countable family of 2-element subsets of X such that the mapping $n \mapsto X_n$ is finitely supported. According to **CC(2)** we have that there exists a finitely supported choice function g on $(X_n)_n$. Let $x_n = g(X_n) \in X_n$. As in the above item, we have that $supp(n \mapsto X_n)$ supports x_n for all $n \in \mathbb{N}$.

For each n, let y_n be the unique element of $X_n \setminus \{x_n\}$. Since for any n both x_n and X_n are supported by the same set $supp(n \mapsto X_n)$, it follows that y_n is also supported by $supp(n \mapsto X_n)$ for all $n \in \mathbb{N}$.

Define $f : \mathbb{N} \to \underset{n}{\cup} X_n$ by $f(n) = \begin{cases} x_{\frac{n}{2}} & \text{if } n \text{ is even} \\ y_{\frac{n-1}{2}} & \text{if } n \text{ is odd} \end{cases}$. We can equivalently describe f as being defined by $f(2k) = x_k$ and $f(2k+1) = y_k$. Clearly, f is onto. Furthermore, because all x_n and all y_n are uniformly supported by $supp(n \mapsto X_n)$, we have that $f(n) = \pi \cdot f(n)$, for all $\pi \in Fix(supp(n \mapsto X_n))$ and all $n \in \mathbb{N}$. Thus, according to Proposition 2.8, we obtain that f is also supported by $supp(n \mapsto X_n)$, and so $\underset{n}{\cup} X_n$ is FSM countable.

3. As in the proof of item 1. □

We can easily remark that under **CC(fin)** a finitely supported subset X of an invariant set is FSM Dedekind infinite if and only if $\wp_{fin}(X)$ is FSM Dedekind infinite.

Proposition 9.15 *Let Y be a finitely supported countable subset of an invariant set X. Then the set $\underset{n\in\mathbb{N}}{\cup} Y^n$ is countable, where Y^n is defined as the n-time Cartesian product of Y.*

Proof Since Y is countable, we can order it as a sequence $Y = \{x_1, \ldots, x_n, \ldots\}$. The other sets of form Y^k are *uniquely* represented with respect to the previous enumeration of the elements of Y. Since Y is finitely supported and countable, all the elements of Y are supported by the same set S of atoms. Thus, in the view of Proposition 2.2, for each $k \in \mathbb{N}$, all the elements of Y^k are supported by S. Fix $n \in \mathbb{N}$.

On Y^n define the S-supported strict well-order relation $\sqsubset$ by: $(x_{i_1},x_{i_2},\dots,x_{i_n}) \sqsubset$

$$(x_{j_1},x_{j_2},\dots,x_{j_n}) \text{ if and only if } \begin{cases} i_1 < j_1 \\ \text{or} \\ i_1 = j_1 \text{ and } i_2 < j_2 \\ \text{or} \\ \dots \\ \text{or} \\ i_1 = j_1,\dots,i_{n-1} = j_{n-1} \text{ and } i_n < j_n \end{cases}.$$

Now, define an S-supported strict well-order relation $\prec$ on $\underset{n\in\mathbb{N}}{\cup} Y^n$ by

$$u \prec v \text{ if and only if } \begin{cases} u \in Y^n \text{ and } v \in Y^m \text{ with } n < m \\ \text{or} \\ u,v \in Y_n \text{ and } u \sqsubset v \end{cases}.$$

Thus, there is an S-supported order isomorphism between $(\underset{n\in\mathbb{N}}{\cup} Y^n, \prec)$ and $(\mathbb{N},<)$.□

9.3 FSM Sets Containing Infinite Uniformly Supported Subsets

The results presented in this section follow the approach in [13].

Definition 9.4 Let X be a finitely supported subset of an invariant set Y. X is called *FSM uniformly infinite* if there exists an infinite, uniformly supported subset of X. Otherwise, we call X *FSM non-uniformly infinite.*

FSM uniformly infinite sets are related FSM Dedekind infinite sets. Obviously, an FSM Dedekind infinite set is FSM uniformly infinite, while the converse is not necessarily valid in the absence of the axiom of choice over non-atomic sets (see Remark 9.3).

Theorem 9.10

1. *Let X be a finitely supported subset of an invariant set $(Y,\cdot)$ such that X is not FSM uniformly infinite. Then the set $\wp_{us}(X)$ is not FSM uniformly infinite.*
2. *Let X be a finitely supported subset of an invariant set $(Y,\cdot)$ such that X is not FSM uniformly infinite. Then the set $\wp_{fin}(X)$ is not FSM uniformly infinite.*

Proof 1. Suppose by contradiction that the set $\wp_{us}(X)$ contains an infinite subset $\mathscr{F}$ such that all the elements of $\mathscr{F}$ are different and supported by the same finite set S. By convention, without assuming that $i \mapsto X_i$ is finitely supported, we understand $\mathscr{F}$ as $\mathscr{F} = (X_i)_{i\in I}$ with the properties that $X_i \neq X_j$ whenever $i \neq j$ and $supp(X_i) \subseteq S$ for all $i \in I$. Let us fix an arbitrary $j \in I$. According to Proposition 2.5, $supp(X_j) = \underset{x\in X_j}{\cup} supp(x)$. Therefore, because $supp(X_j) \subseteq S$, X_j has the property that $supp(x) \subseteq S$ for all $x \in X_j$. Since j has been arbitrarily chosen from I, it follows that $\underset{i\in I}{\cup} X_i$ is a uniformly supported subset of X (all its elements being supported by S).

Furthermore, $\bigcup_{i\in I} X_i$ is infinite since the family $(X_i)_{i\in I}$ is infinite and $X_i \neq X_j$ whenever $i \neq j$. This contradicts the hypothesis.

2. This item was presented as Lemma 9.1 and it was independently proved (similarly as item 1). However, it is easy to derive it from item 1. We always have that $\wp_{fin}(X) \subseteq \wp_{us}(X)$ because any finite subset of X of form $\{x_1,\dots,x_n\}$ is uniformly supported by $supp(x_1) \cup \ldots \cup supp(x_n)$. Since $\wp_{us}(X)$ does not contain an infinite uniformly supported subset, it follows that neither $\wp_{fin}(X)$ contains an infinite uniformly supported subset. □

Theorem 9.11 *Let X be a finitely supported subset of an invariant set $(Y,\cdot)$.*

1. *If X is not FSM uniformly infinite, then any finitely supported monotone (with respect to the inclusion relation) function $f : \wp_{us}(X) \to \wp_{us}(X)$ has a least fixed point supported by $supp(f) \cup supp(X)$.*
2. *If X is not FSM uniformly infinite, then any finitely supported monotone (with respect to the inclusion relation) function $f : \wp_{fin}(X) \to \wp_{fin}(X)$ has a least fixed point supported by $supp(f) \cup supp(X)$.*

Proof 1. Let $f : \wp_{us}(X) \to \wp_{us}(X)$ be a finitely supported order preserving function. Firstly, since $\wp_{us}(X)$ is a subset of $\wp_{fs}(Y)$ supported by $supp(X)$, we have $\pi \star \emptyset, \pi^{-1} \star \emptyset \in \wp_{us}(X)$ for any permutation of atoms $\pi \in Fix(supp(X))$. Thus, $\emptyset \subseteq \pi \star \emptyset$ and $\emptyset \subseteq \pi^{-1} \star \emptyset$. Since the relation $\subseteq$ on $\wp_{us}(X)$ is supported by $supp(X)$, we get $\pi \star \emptyset \subseteq \pi \star (\pi^{-1} \star \emptyset) = (\pi \circ \pi^{-1}) \star \emptyset = \emptyset$, and so $\emptyset = \pi \cdot \emptyset$ which means that $\emptyset$ is an element in $\wp_{us}(X)$ supported by $supp(X)$. Actually, $\emptyset$ belongs to $\wp_{fin}(X)$ that is a subset of $\wp_{us}(X)$.

Since $\emptyset \subseteq f(\emptyset)$ and f is monotone, we can define the ascending sequence $\emptyset \subseteq f(\emptyset) \subseteq f^2(\emptyset) \subseteq \ldots \subseteq f^n(\emptyset) \subseteq \ldots$, where $f^n(\emptyset) = f(f^{n-1}(\emptyset))$ and $f^0(\emptyset) = \emptyset$. We prove by induction that $(f^n(\emptyset))_{n\in\mathbb{N}}$ is uniformly supported by $supp(f) \cup supp(X)$, namely $supp(f^n(\emptyset)) \subseteq supp(f) \cup supp(X)$ for each $n \in \mathbb{N}$. We have $supp(f^0(\emptyset)) = supp(\emptyset) \subseteq supp(X) \subseteq supp(f) \cup supp(X)$. Let us assume that $supp(f^n(\emptyset)) \subseteq supp(f) \cup supp(X)$ for some $n \in \mathbb{N}$. We have to prove that $supp(f^{n+1}(\emptyset)) \subseteq supp(f) \cup supp(X)$. Let $\pi \in Fix(supp(f) \cup supp(X))$. From the inductive hypothesis, we have $\pi \in Fix(supp(f^n(\emptyset)))$ and so $\pi \star f^n(\emptyset) = f^n(\emptyset)$. Since π fixes $supp(f)$ pointwise, according to Proposition 2.9, we have $\pi \star f^{n+1}(\emptyset) = \pi \star f(f^n(\emptyset)) = f(\pi \star f^n(\emptyset)) = f(f^n(\emptyset)) = f^{n+1}(\emptyset)$. Therefore, $(f^n(\emptyset))_{n\in\mathbb{N}} \subseteq \wp_{us}(X)$ is uniformly supported by $supp(f) \cup supp(X)$. Thus, according to Theorem 9.10(1), $(f^n(\emptyset))_{n\in\mathbb{N}}$ should be finite, and so there exists $n_0 \in \mathbb{N}$ such that $f^n(\emptyset) = f^{n_0}(\emptyset)$ for all $n \geq n_0$. Thus, $f(f^{n_0}(\emptyset)) = f^{n_0+1}(\emptyset) = f^{n_0}(\emptyset)$, and so $f^{n_0}(\emptyset)$ is a fixed point of f. It is supported by $supp(f) \cup supp(X)$, and obviously it is the least one.

2. A similar argument allows us to prove the second item of the proposition. This time we use Theorem 9.10(2) to prove that the uniformly supported ascending family $(f^n(\emptyset))_{n\in\mathbb{N}} \subseteq \wp_{fin}(X)$ is finite, and so it is stationary. □

The following results represents the formal proof of Remark 9.1 and Remark 9.2. The proving technique is quite similar with the techniques involved in the proof of Theorem 9.2.

Theorem 9.12 *The following properties of FSM uniformly infinite sets hold.*

1. *Let X be an infinite, finitely supported subset of an invariant set Y. Then the sets $\wp_{fs}(\wp_{fin}(X))$ and $\wp_{fs}(T_{fin}(X))$ are FSM uniformly infinite.*
2. *Let X be an infinite, finitely supported subset of an invariant set Y. Then the set $\wp_{fs}(\wp_{fs}(X))$ is FSM uniformly infinite.*
3. *Let X and Y be two finitely supported subsets of an invariant set Z. If neither X nor Y is FSM uniformly infinite, then $X \times Y$ is not FSM uniformly infinite.*
4. *Let X and Y be two finitely supported subsets of an invariant set Z. If neither X nor Y is FSM uniformly infinite, then $X + Y$ is not FSM uniformly infinite.*

Proof Items 1 and 2 follow from Theorem 9.2(2) and Theorem 9.2(3), respectively. In order to prove that $\wp_{fs}(T_{fin}(X))$ is FSM uniformly infinite, we consider Y_i the set of all i-sized injective tuples formed by elements of X, and we have that each Y_i is a subset of $T_{fin}(X)$ supported by $supp(X)$; furthermore, the family $(Y_i)_{i\in\mathbb{N}}$ is an infinite, uniformly supported, subset of $\wp_{fs}(T_{fin}(X))$.

3. Suppose by contradiction that $X \times Y$ is FSM uniformly infinite. Thus, there exists an infinite injective family $((x_i, y_i))_{i\in I} \subseteq X \times Y$ and a finite $S \subseteq A$ with the property that $supp((x_i, y_i)) \subseteq S$ for all $i \in I$ (1). Fix some $j \in I$. We have that $supp((x_j, y_j)) = supp(x_j) \cup supp(y_j)$. According to relation (1) we obtain, $supp(x_i) \cup supp(y_i) \subseteq S$ for all $i \in I$. Thus, $supp(x_i) \subseteq S$ for all $i \in I$ and $supp(y_i) \subseteq S$ for all $i \in I$ (2). Since the family $((x_i, y_i))_{i\in I}$ is infinite and injective, then at least one of the uniformly supported families $(x_i)_{i\in I}$ and $(y_i)_{i\in I}$ is infinite, a contradiction.

4. Suppose by contradiction that $X + Y$ is FSM uniformly infinite. Thus, there exists an infinite injective family $(z_i)_{i\in I} \subseteq X \times Y$ and a finite $S \subseteq A$ such that $supp(z_i) \subseteq S$ for all $i \in I$. According to the construction of the disjoint union of two S_A-sets (see Proposition 2.2), there should exist an infinite family of $(z_i)_i$ of form $((0, x_j))_{x_j\in X}$ which is uniformly supported by S, or an infinite family of form $((1, y_k))_{y_k\in Y}$ which is uniformly supported by S. Since 0 and 1 are constants, there should exist at least an infinite uniformly supported family of elements from X, or an infinite uniformly supported family of elements from Y, a contradiction. □

From the proofs of Corollaries 9.1 and 9.4, and Theorem 9.2 we get the next results.

Theorem 9.13 *All the sets presented below are FSM non-uniformly infinite (i.e. none of them contains infinite uniformly supported subsets).*

1. *The invariant set A of atoms.*
2. *The powerset $\wp_{fs}(A)$ of the set of atoms.*
3. *The set $T_{fin}(A)$ of all finite injective tuples of atoms.*
4. *The invariant set of all finitely supported functions $f : A \to \wp_{fs}(A)$.*
5. *The invariant set A^A_{fs} of all finitely supported functions from A to A.*
6. *The invariant set of all finitely supported functions $f : A \to A^n$, where $n \in \mathbb{N}$ and A^n is the n-times Cartesian product of A.*
7. *The invariant set of all finitely supported functions $f : A \to T_{fin}(A)$.*
8. *The invariant set $\wp_{fs}(A^n)$, whenever $n \in \mathbb{N}$.*
9. *The invariant set A^{A^n}, whenever $n \in \mathbb{N}$.*

From the proof of Theorem 9.2(1) and Proposition 9.4 we get the following result.

Theorem 9.14

1. *Let X be a finitely supported subset of an invariant set. If X is not FSM uniformly infinite, then each finitely supported injective mapping $f : X \to X$ should be surjective.*
2. *Let X be a finitely supported subset of an invariant set. If $\wp_{fs}(X)$ is not FSM uniformly infinite, then each finitely supported surjective mapping $f : X \to X$ should be injective. The converse does not hold since every finitely supported surjective mapping $f : \wp_{fin}(A) \to \wp_{fin}(A)$ is also injective, while $\wp_{fs}(\wp_{fin}(A))$ is FSM uniformly infinite.*

From Theorem 9.7, the next result follows directly.

Theorem 9.15 *1. Let X be a finitely supported subset of an invariant set $(Z,\cdot)$. If X is totally ordered, then it is uniformly supported.*
2. Let X be a finitely supported subset of an invariant set $(Z,\cdot)$. If X contains an infinite, finitely supported, totally ordered subset, then it is FSM uniformly infinite.

Theorem 9.16 *Let $(X,\sqsubseteq,\cdot)$ be a finitely supported partially ordered set with the property that every uniformly supported subset of X has a least upper bound. Let $f : X \to X$ be a finitely supported function with the property that $x \sqsubseteq f(x)$ for all $x \in X$. Then there exists $x \in X$ such that $f(x) = x$.*

Proof According to Theorem 9.15(1) every finitely supported, totally ordered subset of X is uniformly supported. Thus, from theorem's hypothesis, every finitely supported, totally ordered subset of X has a least upper bound. According to Theorem 6.8, f should have a fixed point. □

From the proof of Theorem 9.8(2), by involving Theorem 9.14(1), we get:

Theorem 9.17 *Let X be a finitely supported subset of an invariant set $(Y,\cdot)$. If there exists a finitely supported bijection between X and $X+X$, then X is FSM uniformly infinite. The converse does not hold.*

9.4 Pictorial Summary

In Figure 9.1 we present some of the relationships between the FSM definitions for infinity. The 'ultra thick arrows' symbolize *strict* implications (p implies q, but q does not imply p), while 'thin dashed arrows' symbolize implications for which we have not proved yet if they are strict or not (the validity of the reverse implications follows when assuming choice principles over non-atomic ZF sets – analyze this with respect to Remark 9.3). 'Thick arrows' match equivalences.

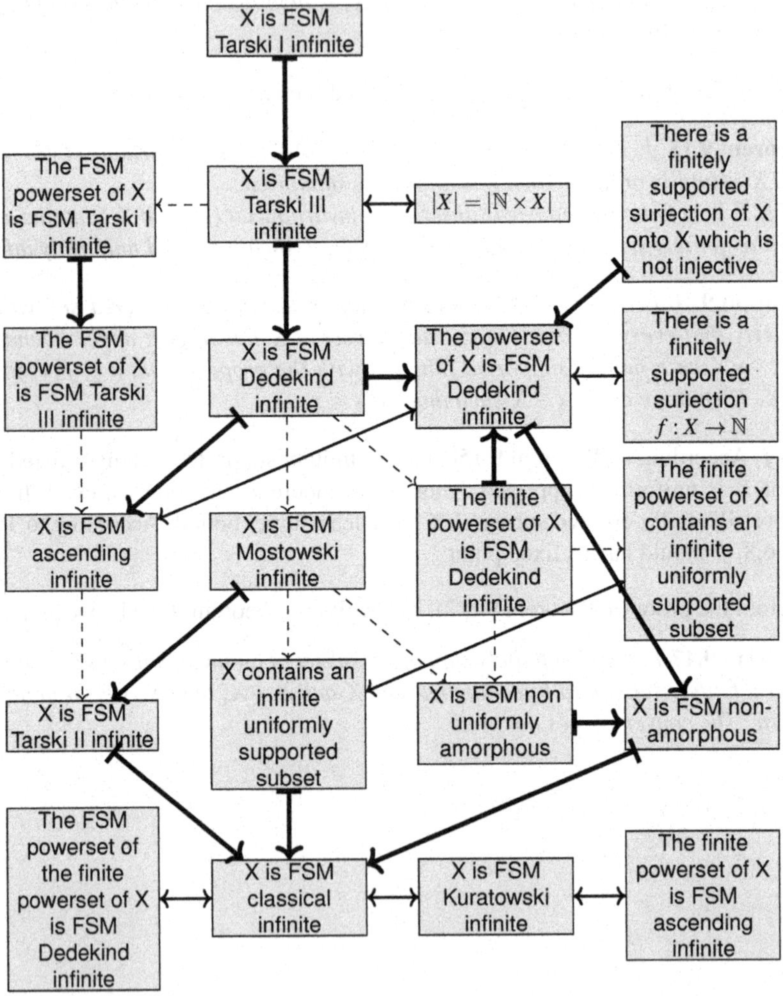

Fig. 9.1 Relationships between various forms of infinity

Chapter 10
Properties of Atoms in Finitely Supported Mathematics

Abstract We present a large collection of properties of the set of atoms, of its (finite or cofinite) powerset and of its (finite) higher-order powerset in the world of finitely supported algebraic structures. Firstly, we prove that atomic sets have many specific FSM properties (that are not translated from ZF). We can structure these specific properties into five main groups, presenting the relationship between atomic and non-atomic sets, specific finiteness properties of atomic sets, specific (order) properties of cardinalities in FSM, surprising fixed point properties of self-mappings on the (finite) powerset of atoms, and the inconsistency of various choice principles for specific atomic sets. Other properties of atoms are obtained by translating classical (non-atomic) ZF results into FSM, by replacing 'non-atomic object' with 'atomic finitely supported object'. Furthermore, we also proved that the powerset of atoms satisfies some choice principles such as prime ideal theorem and ultrafilter theorem, although these principles are generally not valid in FSM. Ramsey theorem for the set of atoms and Kurepa antichain principle for the powerset of atoms also hold, and admit constructive proofs. The properties presented in this chapter are also valid in the frameworks of nominal sets and in the framework of Fraenkel-Mostowski sets. This chapter is based on [11, 15].

10.1 Several Specific Properties of Atoms

Theorem 10.1 *Even if A is infinite, we have the following specific FSM properties:*

1. *A finitely supported function $f : A \to A$ is surjective if and only if it is injective.*
2. *A finitely supported function $f : \wp_{fin}(A) \to \wp_{fin}(A)$ is surjective if and only if it is injective.*
3. *Any finitely supported injection $f : \wp_{fs}(A) \to \wp_{fs}(A)$ is also surjective.*
4. *Any finitely supported injective mapping $f : \wp_{fin}(\wp_{fs}(A)) \to \wp_{fin}(\wp_{fs}(A))$ is also surjective.*
5. *There is no finitely supported surjection $f : A \to A \times A$.*

A. Alexandru, G. Ciobanu, *Foundations of Finitely Supported Structures*,
https://doi.org/10.1007/978-3-030-52962-8_10

6. *There is no finitely supported injection* $f : A \times A \to A$.
7. *There is no finitely supported surjective mapping* $f : \wp_{fs}(A) \to \wp_{fs}(A) \times \wp_{fs}(A)$.
8. *There is no finitely supported injection* $f : \wp_{fs}(A) \times \wp_{fs}(A) \to \wp_{fs}(A)$.
9. *For a finitely supported bijection* $f : A \to A$*, we have that* $\{a \in A \,|\, f(a) \neq a\}$ *is finite and* $supp(f) = \{a \in A \,|\, f(a) \neq a\}$.
10. *Every finitely supported injection* $f : A^A_{fs} \to A^A_{fs}$ *is also surjective.*
11. *There are no finitely supported injections* $f : A \to X$ *and* $f : X \to A$ *between A and an arbitrary infinite ZF set X constructed without involving atoms.*
12. *There are no finitely supported injections* $f : \wp_{fs}(A) \to X$ *and* $f : X \to \wp_{fs}(A)$ *between* $\wp_{fs}(A)$ *and an arbitrary infinite ZF set X constructed without involving atoms.*
13. *There are no finitely supported surjective mappings* $f : A \to X$ *and* $f : X \to A$ *between A and an arbitrary infinite ZF set X constructed without involving atoms.*
14. *There does not exist a finitely supported surjective mapping* $f : X \to \wp_{fs}(A)$ *from an arbitrary infinite ZF set X constructed without involving atoms onto* $\wp_{fs}(A)$*, but there exists a finitely supported (equivariant) surjective mapping* $f : \wp_{fs}(A) \to \mathbb{N}$ *from* $\wp_{fs}(A)$ *onto the set of all positive integers* $\mathbb{N}$*, which associates to each* $X \in \wp_{fs}(A)$ *the cardinality of its support.*
15. *There does not exist a finitely supported surjective mapping* $f : \wp_{fs}(A) \to A \times A$.
16. *There does not exist a finitely supported injection* $f : A \times A \to \wp_{fs}(A)$.
17. *A function* $f : A \to X$ *between A and an arbitrary infinite ZF set X constructed without involving atoms is finitely supported if and only if there exists* $x \in X$ *such that* $\{y \in A \mid f(y) \neq x\}$ *is finite. In this case,* $supp(f) = \{y \in A \mid f(y) \neq x\}$.
18. *The set A of atoms has the property that for all increasing countable chain* $X_0 \subseteq X_1 \subseteq \ldots \subseteq X_n \subseteq \ldots$ *of finitely supported subsets such that* $n \mapsto X_n$ *is finitely supported there exists* $n \in \mathbb{N}$ *such that* $X \subseteq X_n$.
19. *There does not exist a finitely supported choice function defined on* $\wp_{fin}(A)$*, i.e. a finitely supported function* $f : \wp_{fin}(A) \to A$ *with the property that* $f(X) \in X$ *for all* $X \in \wp_{fin}(A)$.
20. *There does not exist a finitely supported choice function defined on* $\wp_{cofin}(A)$*, i.e. a finitely supported function* $f : \wp_{cofin}(A) \to A$ *with the property that* $f(X) \in X$ *for all* $X \in \wp_{cofin}(A)$.
21. *There do not exist finitely supported total order relations (and so neither finitely supported well-order relations) defined on infinite, finitely supported subsets of the sets A,* $\wp_{fin}(A)$*,* $\wp_{cofin}(A)$*,* $\wp_{fs}(A)$*, or* $\wp_{fin}(\wp_{fs}(A))$.
22. $|\wp_{fs}(A)| = 2|\wp_{fin}(A)|$.
23. $\wp_{fin}(A)$ *is infinite, but* $|\wp_{fin}(A)| \neq 2|\wp_{fin}(A)|$.
24. $|A| < |\wp_n(A)| < |2^A_{fs}|$ *for all* $n \in \mathbb{N}$*,* $n \geq 2$*, where* $\wp_n(A) \stackrel{def}{=} \{X \subseteq A \,|\, |X| = n\}$.
25. $|A| < |\wp_{fin}(A)| < |2^A_{fs}|$.
26. *Weak Tarski fixed point theorem for* $\wp_{fin}(A)$*:*
    Let $f : \wp_{fin}(A) \to \wp_{fin}(A)$ *be a finitely supported monotone function. Then there exists a least* $X_0 \in \wp_{fin}(A)$ *supported by* $supp(f)$ *such that* $f(X_0) = X_0$.
27. *Weak Tarski point theorem for* $\wp_{fin}(\wp_{fs}(A))$*:*

Let $f : \wp_{fin}(\wp_{fs}(A)) \to \wp_{fin}(\wp_{fs}(A))$ *be a finitely supported monotone function. Then there exists a least* $X_0 \in \wp_{fin}(\wp_{fs}(A))$ *supported by* $supp(f)$ *such that* $f(X_0) = X_0$.

28. *Bourbaki-Witt fixed point theorem for* $\wp_{fin}(A)$*:*
Let $f : \wp_{fin}(A) \to \wp_{fin}(A)$ *be a finitely supported function with the property that* $X \subseteq f(X)$ *for all* $X \in \wp_{fin}(A)$*. Then there is* $X_0 \in \wp_{fin}(A)$ *such that* $f(X_0) = X_0$.
29. *Bourbaki-Witt fixed point theorem for* $\wp_{fin}(\wp_{fs}(A))$*:*
Let $f : \wp_{fin}(\wp_{fs}(A)) \to \wp_{fin}(\wp_{fs}(A))$ *be a finitely supported function with the property that* $X \subseteq f(X)$ *for all* $X \in \wp_{fin}(\wp_{fs}(A))$*. Then there exists* $X_0 \in \wp_{fin}(\wp_{fs}(A))$ *such that* $f(X_0) = X_0$.
30. $(\wp_{fin}(A), \star, \subseteq)$ *is an invariant set having the property that any finitely supported totally ordered subset of* $\wp_{fin}(A)$ *has an upper bound in* $\wp_{fin}(A)$*, but* $\wp_{fin}(A)$ *does not have a maximal element.*
31. $(\wp_{fin}(\wp_{fs}(A)), \star, \subseteq)$ *is an invariant set having the property that any finitely supported totally ordered subset of* $\wp_{fin}(\wp_{fs}(A))$ *has an upper bound, but* $\wp_{fin}(\wp_{fs}(A))$ *does not have a maximal element.*
32. *There is a finitely supported surjection from the family of all finite one-to-one tuples of atoms onto the family of all finite one-to-one non-empty tuples of atoms, and a finitely supported surjection from the family of all finite one-to-one non-empty tuples of atoms onto the family of all finite one-to-one tuples of atoms, but there is no finitely supported bijection between these two families.*
33. *There is a finitely supported bijection between the powerset of the family of all finite one-to-one tuples of atoms and the powerset of the family of all finite one-to-one non-empty tuples of atoms, but there is no finitely supported bijection between the family of all finite one-to-one tuples of atoms and the family of all finite one-to-one non-empty tuples of atoms.*

Proof 1. Suppose $g : A \to A$ is a finitely supported injection. Since A is not FSM Dedekind infinite (Corollary 9.1(1)), it follows that g is surjective. Since $\wp_{fs}(A)$ is not FSM Dedekind infinite, from Proposition 9.4 we should have that any finitely supported surjection $f : A \to A$ is also injective; for this fact we provide below an alternative proof.

Let $f : A \to A$ be a finitely supported surjective function. Firstly, it is easy to verify that $f[supp(f)] = supp(f)$. Indeed, let $c \in supp(f)$. Since f is onto (by hypothesis), there is $x^* \in A$ such that $f(x^*) = c$. If $x^* \in supp(f)$, then we are done; so assume that $x^* \notin supp(f)$. But then it is fairly obvious that $Im(f)$ is finite. Indeed, for any $y \in A \setminus (supp(f) \cup \{x^*\})$, consider the transposition $\pi = (x^*\, y)$ and note that $\pi \in Fix(supp(f) \cup \{c\})$. Thus, $f(y) = f(\pi \cdot x^*) = \pi \cdot f(x^*) = \pi \cdot c = c$. This contradicts the hypothesis that $f : A \to A$ is onto, and so $supp(f) \subseteq f[supp(f)]$. However, it is obvious that $f(X)$ is supported by $supp(f) \cup supp(X)$ whenever $X \in \wp_{fs}(A)$, and so $f[supp(f)]$ is supported by $supp(f) \cup supp(supp(f)) = supp(f)$. This means $supp(f[supp(f)]) \subseteq supp(f)$. Since $f[supp(f)]$ is a finite subset of A, we have $f[supp(f)] = supp(f[supp(f)]) \subseteq supp(f)$, from which we get $supp(f) = f[supp(f)]$. It follows that the restriction $f|_{supp(f)} : supp(f) \to supp(f)$ is a bijection (recall that $supp(f)$ is a finite set). Now we show that $f|_{A \setminus supp(f)}$ is the identity mapping. Fix $a \in A \setminus supp(f)$ and by way of contradiction assume $f(a) \neq a$, say $f(a) = b$. Then arguing as above, it can be easily shown that $f[A \setminus (supp(f) \cup \{b\})] = \{b\}$. Indeed, take any $c \in A \setminus (supp(f) \cup \{a,b\})$ and consider the transposition $\pi = (a\,c) \in Fix(supp(f))$, then $f(c) = f((a\,c) \cdot a) = (a\,c) \cdot f(a) = (a\,c) \cdot b = b$. Thus, $Im(f)$ is finite, a contradiction. Hence $f|_{A \setminus supp(f)}$ is the identity mapping, and consequently in view of the first observation (that the restriction $f|_{supp(f)} : supp(f) \to supp(f)$ is bijective), f is injective.

2. If $f : \wp_{fin}(A) \to \wp_{fin}(A)$ is surjective, then it is also injective according to the proof of Proposition 9.4. Conversely, if $f : \wp_{fin}(A) \to \wp_{fin}(A)$ is injective, then it should also be surjective because $\wp_{fin}(A)$ is not FSM Dedekind infinite (more exactly, $\wp_{fin}(A)$ does not contain an infinite uniformly supported subset).

3. This follows from Corollary 9.1(2).

4. This follows from Corollary 9.1(7).

5. This follows from Theorem 5.5(3).

6. This follows from Theorem 5.5(4).

7. This follows from Theorem 5.5(9).

8. This follows from Theorem 5.5(10).

9. This follows from Proposition 2.11.

10. This follows from Corollary 9.1(4).

11/13. This follows from the proof of Lemma 5.4.

12. Suppose by contradiction that there exists a finitely supported injection $f : \wp_{fs}(A) \to X$. Since there also is an equivariant injection $a \mapsto \{a\}$ from A into $\wp_{fs}(A)$, this contradicts the statement in item 10 that there is no finitely supported injection from A to X.

Furthermore, we prove by contradiction that there does not exist a finitely supported injective mapping $g : X \to \wp_{fs}(A)$. Suppose that there is a finitely supported injection $g : X \to \wp_{fs}(A)$. Then $g(X)$ is an infinite family of subsets of A with the property that all the elements of $g(X)$ are supported by the same finite set $supp(g)$. This is because $\pi \star g(x) = g(\pi \diamond x) = g(x)$ for all $x \in X$ and all $\pi \in Fix(supp(g))$. However, there are only finitely many subsets of A supported by $supp(g)$, namely the subsets of $supp(g)$ and the supersets of $A \setminus supp(g)$.

14. Suppose by contradiction that there exists a finitely supported surjective mapping $f : X \to \wp_{fs}(A)$. Obviously, one can define a finitely supported surjection $s : \wp_{fs}(A) \to A$ as below. Let us fix an atom $x \in A$. We define s by $s(X) = \begin{cases} a, \text{ if} X \text{ is a one-element set } \{a\}\,; \\ x, \text{ if} X \text{ has more than one element}. \end{cases}$ Clearly, s is supported by $supp(x) = \{x\}$. Then, $s \circ f$ is a surjection from X onto A which contradicts item 12.

According to the proof of the inconsistency of the choice principle **SIP** with FSM (see Theorem 3.1), there exists an equivariant surjection $h : \wp_{fs}(A) \to \mathbb{N}$ defined by $h(B) = |supp(B)|$ for any finitely supported subset B of A.

15. This follows from Theorem 5.5(1).

16.This follows from Theorem 5.5(2).

17. This follows from Proposition 2.10.

18. This follows from Proposition 9.8.

19. By contradiction, assume that f is a finitely supported choice function function on $\wp_{fin}(A)$. Let S be a finite set (of atoms) supporting f. We may select a pair $Y := \{a,b\}$ from $\wp_{fin}(A)$ such that a and b do not belong to S. Let π be a permutation of atoms which fixes S pointwise, and interchanges a and b. Since f satisfies the property that $f(X) \in X$ for all $X \in \wp_{fin}(A)$, we have $f(Y) = a$ or $f(Y) = b$. Since π interchanges a and b, we have $\pi \cdot f(Y) = \pi(f(Y)) \neq f(Y)$. However, $\pi \star Y = \{\pi(a), \pi(b)\} = \{b,a\} = Y$. Since π fixes S pointwise and S supports f, we have $\pi(f(Y)) = \pi \cdot f(Y) = f(\pi \star Y) = f(Y)$, a contradiction.

20. Suppose by contradiction that there is a finitely supported function $g : \wp_{cofin}(A) \to A$ with the property that $g(Y) \in Y$ for all $Y \in \wp_{cofin}(A)$. Thus, $g(A \setminus X) \in A \setminus X$ for all $X \in \wp_{fin}(A)$. We prove that $A \setminus (\pi \star X) = \pi \star (A \setminus X)$ for all $\pi \in S_A$ and all $X \in \wp_{fin}(A)$. Indeed, let $y \in A \setminus (\pi \star X)$. We can express y as $y = \pi \cdot (\pi^{-1} \cdot y)$. If $\pi^{-1} \cdot y \in X$, then $y \in \pi \star X$, which is a contradiction. Thus, $\pi^{-1} \cdot y \in (A \setminus X)$, and so $y \in \pi \star (A \setminus X)$. Conversely, if $y \in \pi \star (A \setminus X)$, then $y = \pi \cdot x$ with $x \in A \setminus X$. Suppose $y \in \pi \star X$. Then $y = \pi \cdot z$ with $z \in X$. Thus, $x = z$ which is a contradiction, and so $y \in A \setminus (\pi \star X)$. Now, since g is finitely supported, according to Proposition 2.8, for all $\pi \in Fix(supp(g))$ we have $g(A \setminus (\pi \star X)) = g(\pi \star (A \setminus X)) = \pi \cdot g(A \setminus X)$ for all $X \in \wp_{fin}(A)$. Let us define $f : \wp_{fin}(A) \to \wp_{fin}(A)$ by $f(X) = X \cup \{g(A \setminus X)\}$. Let us fix $\pi \in Fix(supp(g))$. We have $\pi \star f(X) = \pi \star X \cup \{\pi \cdot g(A \setminus X)\} = \pi \star X \cup \{g(A \setminus (\pi \star X))\} = f(\pi \star X)$ for all $X \in \wp_{fin}(X)$. Therefore, f is finitely supported by $supp(g)$. Furthermore, because $g(A \setminus X) \notin X$ for all $X \in \wp_{fin}(A)$, we have $X \subsetneq f(X)$ for all $X \in \wp_{fin}(A)$, which contradicts item 27 that will be independently proved below.

21. This follows from proof of the inconsistency of the choice principle **OP** with FSM (see Theorem 3.1) and from Corollary 9.8.

22. Let us consider the function $f : \wp_{fin}(A) \to \wp_{cofin}(A)$ defined by $f(X) = A \setminus X$ for all $X \in \wp_{fin}(A)$. From the proof of Theorem 5.4, f is equivariant and bijective, and so $|\wp_{fin}(A)| = |\wp_{cofin}(A)|$. However, every finitely supported subset of A is either finite or cofinite, and so $\wp_{fs}(A)$ is the union of the disjoint subsets $\wp_{fin}(A)$ and $\wp_{cofin}(A)$. Thus, $|\wp_{fs}(A)| = 2|\wp_{fin}(A)|$.

23. Obviously, $\wp_{fin}(A)$ is infinite because there exist an equivariant injection $i : A \to \wp_{fin}(A)$ defined by $i(a) = \{a\}$ for all $a \in A$. According to item 2, there does not exist a finitely supported injective mapping from $\wp_{fs}(A)$ onto one of its finitely supported proper subsets. Thus, there could not exist a bijection $f : \wp_{fs}(A) \to \wp_{fin}(A)$. Therefore, $|\wp_{fin}(A)| \neq |\wp_{fs}(A)| = 2|\wp_{fin}(A)|$.

24. This follows from the proof of Theorem 5.5(5).

25. This follows from the proof of Theorem 5.5(7).

26/27. This is actually Corollary 5.6.

28/29. As in the proof of Theorem 5.9, the fixed point of f is defined in each case as the least upper bound of the uniformly supported, stationary ascending sequence $\emptyset \subseteq f(\emptyset) \subseteq f^2(\emptyset) \subseteq \ldots \subseteq f^n(\emptyset) \subseteq \ldots$. Since f is not necessarily monotone, this fixed point of f is not necessarily a least one.

30/31. Let X be one of the sets A or $\wp_{fs}(A)$. Let $\mathscr{F}$ be finitely supported totally ordered subset of the invariant set $\wp_{fin}(X)$ equipped with the equivariant order relation $\subseteq$. Since $\mathscr{F}$ is totally ordered with respect to the inclusion relation on $\wp_{fin}(X)$, then there do not exist two different finite subsets of X of the same cardinality belonging to $\mathscr{F}$. Since $\mathscr{F}$ is finitely supported, then there exists a finite set $S \subseteq A$ such that $\pi \star Y \in \mathscr{F}$ for each $Y \in \mathscr{F}$ and each $\pi \in Fix(S)$. However, for each $Y \in \mathscr{F}$ and each $\pi \in Fix(S)$ we have that $|\pi \star Y| = |Y|$. Since there do not exist two distinct elements in $\mathscr{F}$ having the same cardinality, we conclude that $\pi \star Y = Y$ for all $Y \in \mathscr{F}$ and all $\pi \in Fix(S)$. Thus, $\mathscr{F}$ is uniformly supported by S. Since $\wp_{fin}(X)$ does not contain an infinite uniformly supported subset (see Lemma 9.1), $\mathscr{F}$ must be finite, and so there exists the (finite) union of the members of $\mathscr{F}$ which is an elements of $\wp_{fin}(X)$ and an upper bound for $\mathscr{F}$. Suppose by contradiction that there exists a maximal element X_0 of $\wp_{fin}(X)$. Then $X_0 = \{x_1, \dots x_n\}$, $x_1, \dots, x_n \in X$ for some $n \in \mathbb{N}$. Since X is infinite, there exists $y \in X \setminus X_0$. However, $\{x_1, \dots x_n\} \subsetneq \{x_1, \dots x_n, y\}$ with $\{x_1, \dots x_n, y\} \in \wp_{fin}(X)$ which contradicts the maximality of X_0.

32. This item follows from the proof of Lemma 5.3.

33. This item follows from the proof of Corollary 5.1. □

10.2 Translated Properties of Atoms

Theorem 10.2 *The following properties of atoms are naturally translated from the ZF framework.*

1. *There exists a finitely supported (more exactly, an equivariant) injective mapping $g : \wp_{fs}(\wp_{fs}(A)) \to \wp_{fs}(\wp_{fs}(A))$ that is not surjective.*
2. *There exists a finitely supported (more exactly, an equivariant) injective mapping $g : \wp_{fs}(\wp_{fin}(A)) \to \wp_{fs}(\wp_{fin}(A))$ that is not surjective (analyze this in comparison with Theorem 10.1(4)).*
3. *There exists a finitely supported injective mapping $g : T^{\delta}_{fin}(A) \to T^{\delta}_{fin}(A)$ that is not surjective.*
4. *There exists an equivariant bijective mapping from $\wp_{fs}(A)$ onto the family $\{0,1\}^A_{fs}$ of those finitely supported functions from A to the ordinary ZF set $\{0,1\}$, i.e. $|\wp_{fs}(A)| = |2^A_{fs}|$ in FSM.*
5. *Tarski fixed point theorem for $\wp_{fs}(A)$:*
   Any finitely supported monotone function $f : \wp_{fs}(A) \to \wp_{fs}(A)$ has least fixed point and a greatest fixed point which are both supported by $supp(f)$.

Proof **1/2**. These items follow directly from Corollary 9.3(1).

3. This item follows from Corollary 9.3(4).

4. This follows from Theorem 2.3.

5. Let us consider a finitely supported monotone function $f : \wp_{fs}(A) \to \wp_{fs}(A)$. From Theorem 6.2 we know that f has a greatest fixed point and a least fixed point, but we do not know that both of them are supported by the same set $supp(f)$. Thus, we provide an independent proof of the result.

We know that $(\wp_{fs}(A), \star)$ is an invariant lattice We prove that the set $S := \{X \mid X \in \wp_{fs}(A), X \subseteq f(X)\}$ is a non-empty, finitely supported subset of $(\wp_{fs}(A), \star)$. Obviously, $\emptyset \in S$. We claim that S is supported by $supp(f)$. Let $\pi \in Fix(supp(f))$.

Let $X \in S$. Then $X \subseteq f(X)$. From the definition of $\star$ we have $\pi \star X \subseteq \pi \star f(X)$. According to Proposition 2.8, because $supp(f)$ supports f, we have $\pi \star X \subseteq \pi \star f(X) = f(\pi \star X)$, and so $\pi \star X \in S$. It follows that S is finitely supported, and $supp(S) \subseteq supp(f)$. As in the proof of Theorem 7.1, $T := \underset{X \in S}{\cup} X$ is finitely supported by $supp(S) \subseteq supp(f)$. We prove that $f(T) = T$. Let $X \in S$ arbitrary. We have $X \subseteq f(X) \subseteq f(T)$. By taking the supremum (union) on S, this leads to $T \subseteq f(T)$. However, because $T \subseteq f(T)$ an f is monotone, we also have $f(T) \subseteq f(f(T))$. Furthermore, $f(T)$ is supported by $supp(f) \cup supp(T)$, and so $f(T) \in S$. According to the definition of T, we get $f(T) \subseteq T$. Furthermore, T is the greatest fixed point of f. Indeed, whenever T' is an element in $\wp_{fs}(A)$ such that $f(T') = T'$, it follows that $T' \in S$, and so $T' \subseteq T$. Similarly, we prove that the set $S' = \{X \in \wp_{fs}(A) \,|\, f(X) \subseteq X\}$ is finitely supported by $supp(f)$. Similarly as in the paragraphs above, there exists $\underset{X \in S'}{\cap} X' \in \wp_{fs}(A)$ (also supported by $supp(f)$) which is the least fixed point of f. □

10.3 Choice Properties of Atoms

A finitely supported prime ideal of a Boolean algebra $(L, \sqcup, \sqcap)$ is a finitely supported subset I of L with the properties that I is a lower set (for every $x \in I$, $y \leq x$ implies that $y \in I$), I is directed (for every $x, y \in I$ we have $x \sqcup y \in I$) and I is prime (for every $x, y \in L$ we have that $x \sqcap y \in I$ implies that $x \in I$ or $y \in I$).

Although most of the choice principles (including axiom of countable choice, prime ideal principle and ultrafilter principle) fail in FSM, we have:

Theorem 10.3 *For the set A of atoms, the following choice principles are valid:*

1. *$(\wp_{fs}(A), \subseteq, \star)$ is an invariant Boolean lattice and the subset $\wp_{fin}(A)$ is a finitely supported prime and maximal ideal of $\wp_{fs}(A)$ (prime ideal for atoms).*
2. *$(\wp_{fs}(A), \subseteq, \star)$ is an invariant Boolean lattice and $\wp_{cofin}(A) = \{X \subseteq A \,|\, A \setminus X \text{ is finite}\}$ is a finitely supported prime and maximal filter of $\wp_{fs}(A)$ (ultrafilter theorem for atoms).*
3. *The invariant partially ordered set $(\wp_{fs}(A), \subseteq)$ contains infinitely many equivariant maximal subsets of pairwise incomparable elements (Kurepa antichain theorem for atoms).*
4. *Ramsey theorem for atoms holds in FSM; moreover, it admits a constructive proof: Let $n \geq 1$. If there exists an FSM colouring on the elements of $\wp_n(A)$ with finitely many colours belonging to a certain invariant set $(Y, \diamond)$ (i.e. a finitely supported function from $\wp_n(A)$ to a finite subset of Y), then we can construct $M \subseteq A$ infinite and finitely supported such that all n-sized subsets of M are FSM coloured with the same colour.*

Proof 1. $(\wp_{fs}(A), \subseteq, \star)$ is an invariant Boolean lattice from Theorem 7.1. We have that $\wp_{fin}(A)$ is an equivariant subset of $\wp_{fs}(A)$. It is obviously a directed lower set. We prove now that $\wp_{fin}(A)$ is prime. Let X and Y be two finitely supported subsets of A such that $X \cap Y \in \wp_{fs}(A)$ is finite. Assume that neither X nor Y is finite. Then, because both X and Y are finitely supported subsets of A, and every finitely supported subset of A is either finite or cofinite, we have that both X and Y are cofinite, which means both $A \setminus X$ and $A \setminus Y$ are finite. By de Morgan laws we have $A \setminus (X \cap Y) = (A \setminus X) \cup (A \setminus Y)$, and so $A \setminus (X \cap Y)$ is finite (as a union of two finite subsets of A), which means $X \cap Y$ is infinite, a contradiction. Thus, at least one of the subsets X and Y should be finite. Now we prove that $\wp_{fin}(A)$ is a maximal ideal in $\wp_{fs}(A)$. Let I be an ideal of $\wp_{fs}(A)$ with $\wp_{fin}(A) \subsetneq I \subseteq \wp_{fs}(A)$. Then I contains a cofinite subset of A of form $A \setminus X$ with X finite. Since I is an ideal and $X \in I$, we have $A = X \cup (A \setminus X) \in I$. Since I is a lower set, we have that every $Y \in \wp_{fs}(A)$ (i.e. every $Y \subseteq_{fs} A$) is a member of I. Then $I = \wp_{fs}(A)$.

2. By duality with item 1.

3. For each $n \in \mathbb{N}$ we define $\wp_n(A)$. Since permutations of atoms are bijective functions, the effect of any permutation of atoms on an n-sized subset of A is another n-sized subset of A. Thus, each $\wp_n(A)$ is an equivariant subset of $\wp_{fs}(A)$. Obviously, for a fixed $n \in \mathbb{N}$, any two elements of $\wp_n(A)$ are incomparable via $\subseteq$ because they are different subsets of A having the same cardinality. Furthermore, if there was a finitely supported antichain U of $\wp_{fs}(A)$ with $\wp_n(A) \subsetneq U \subseteq \wp_{fs}(A)$, then U should contain at least a subset of A having the size different from n (since all the subsets of A of size n form $\wp_n(A)$). Such a subset of A is comparable with at least one n-sized subset of A, and so U would not be an antichain of $\wp_{fs}(A)$. Thus, $\wp_n(A)$ is a maximal antichain of $\wp_{fs}(A)$ for each $n \in \mathbb{N}$.

4. Firstly, we remark that $\wp_n(A)$ is an equivariant subset of $(\wp_{fs}(A), \star)$, and so it is itself an invariant set with the S_A-action $\star|_{\wp_n(A)}$ defined by $\pi \star \{x_1, \dots, x_n\} = \{\pi \cdot x_1, \dots, \pi \cdot x_n\} = \{\pi(x_1), \dots, \pi(x_n)\}$ for all $\pi \in S_A$ and all $\{x_1, \dots, x_n\} \in \wp_n(A)$. Suppose that the set of colours $\{c_1, \dots, c_m\}$ belong to the invariant set $(Y, \diamond)$. We also have that $\{c_1, \dots, c_m\}$ is subset of Y finitely supported by $S = supp(c_1) \cup \dots \cup supp(c_m)$. By hypothesis, there exists an FSM colouring on $\wp_n(A)$, which means there exists a finitely supported function $f : \wp_n(A) \to \{c_1, \dots, c_m\}$. From Proposition 2.8 we have that $f(\pi(X)) = f(\pi \star X) = \pi \diamond f(X)$ for all $X \in \wp_n(A)$ and $\pi \in Fix(supp(f))$ (1). Let us consider $M = A \setminus (supp(f) \cup supp(c_1) \cup \dots \cup supp(c_m))$. Since $supp(f) \cup supp(c_1) \cup \dots \cup supp(c_m)$ is finite, it follows that M is cofinite, and so it is finitely supported with $supp(M) = supp(f) \cup supp(c_1) \cup \dots \cup supp(c_m)$. Thus, M is an infinite, finitely supported subset of A. Furthermore, $\wp_n(M) = \{X \subseteq M \,|\, |X| = n\}$ is a subset of $\wp_n(A)$ supported by $supp(M)$; this follows because, whenever $\sigma \in Fix(supp(M))$ and $X \in \wp_n(M)$, we have $X \subseteq M$ and $|X| = n$, and so $\sigma \star X \subseteq \sigma \star M = M$ and $|\sigma \star X| = n$, from which $\sigma \star X \in \wp_n(M)$. Let us consider $X, Y \in \wp_n(M)$ with $X \neq Y$. Suppose that X and Y have $k < n$ identical elements. After o renumbering of the elements from X and Y (which is possible without requiring any form of choice because both X and Y are finite) we can suppose that the first k elements from X and Y coincide, while the other elements are all different. Thus, X is of from $\{x_1, \dots, x_k, x_{k+1}, \dots, x_n\}$ and Y is of form $\{x_1, \dots, x_k, y_{k+1}, \dots, y_n\}$ where $x_i \neq y_j$ for all $i \in \{1, \dots, n\}$ and $j \in \{k+1, \dots, n\}$. Define a function $\pi : A \to A$ by:

$$\pi(x) = \begin{cases} x, & \text{if}\, x \in \{x_1, \ldots, x_k\}\,; \\ y_l, & \text{if}\, x = x_l \in \{x_{k+1}, \ldots, x_n\}\,; \\ x_l, & \text{if}\, x = y_l \in \{y_{k+1}, \ldots, y_n\}\,; \\ x, & \text{if}\, x \in A \setminus \{x_1, \ldots, x_k, x_{k+1}, \ldots x_n, y_{k+1}, \ldots, y_n\}\,. \end{cases}$$

From the above definition we have that π is a finite permutation of A, and $\pi \star X = \{\pi(x_1), \ldots, \pi(x_k), \pi(x_{k+1}), \ldots, \pi(x_n)\} = \{x_1, \ldots, x_k, y_{k+1}, \ldots, y_n\} = Y$. Since π changes only elements belonging to M, it follows that π fixes $supp(f) \cup supp(c_1) \cup \ldots \cup supp(c_m)$ pointwise. Thus, since $\pi \in Fix(supp(f))$, by (1) we have $f(Y) = f(\pi \star X) = \pi \diamond f(X)$. However, since π fixes $supp(c_1) \cup \ldots \cup supp(c_m)$ pointwise and $f(X) \in \{c_1, \ldots, c_m\}$ we have $\pi \diamond f(X) = f(X)$, and so $f(Y) = f(X)$. Since X and Y were arbitrarily chosen from the family of n-sized subsets of M, it follows that $f(\wp_n(M))$ is a single-element set. Furthermore, $g = f|_{\wp_n(M)}$ is a finitely supported function (supported by $supp(f) \cup supp(M) = supp(f) \cup supp(c_1) \cup \ldots \cup supp(c_m)$) which is the required FSM colouring function on $\wp_n(M)$. □

Chapter 11
Freshness in Finitely Supported Mathematics

Abstract The goal of this chapter is to describe properties of elements placed outside the support of a given element. We present a specific quantifier (introduced initially by Gabbay and Pitts) that encodes 'for all but finitely many', and show that is placed between $\forall$ and $\exists$.

11.1 The NEW Quantifier

Definition 11.1 If $x \in A$ and $y \in Y$ where Y is an invariant set, we say that *x is fresh for y*, and denote this by $x\#y$ if $\{x\} \cap supp(y) = \emptyset$.

Since A is infinite, for any arbitrary element in an invariant set, we can always find a name outside its support, namely $\forall x.\exists a \in A.\ a\#x$.

We can generalize this definition to arbitrary invariant sets.

Definition 11.2 Let X and Y be invariant sets. If $x \in X$ and $y \in Y$, we say that *x is fresh for y* and denote this by $x\#y$ if $supp(x) \cap supp(y) = \emptyset$.

From Proposition 2.1 and from the bijectivity of each permutation of atoms, the next result follows immediately.

Proposition 11.1 *Let $(X, \cdot)$ and $(Y, \diamond)$ be invariant sets. If $x \in X$ and $y \in Y$ with $x\#y$, then $\pi \cdot x\#\pi \diamond y$ for all $\pi \in S_A$.*

Remark 11.1 The following properties are consequences of the definition of support:

1. If $a, b \in A$, then $(a \neq b \Longleftrightarrow a\#b)$.
2. If $a, b \in A$, $(X, \cdot)$ is an invariant set with $x \in X$, and $a, b\#x$, then $(a\,b) \cdot x = x$.

Definition 11.3 Let P be a predicate on A. We say that $\text{И}a.P(a)$ if $P(a)$ is true for all but finitely many elements of A. И is called the *freshness quantifier*.

A. Alexandru, G. Ciobanu, *Foundations of Finitely Supported Structures*,
https://doi.org/10.1007/978-3-030-52962-8_11

Definition 11.3 also makes sense if P is a predicate in the logic of ZFA. This quantifier was introduced in [29]; in that article, И is called 'the new quantifier'. It is denoted by a reflected sans-serif Roman letter by analogy with $\forall$, the reflected A of 'for all' and $\exists$, the reflected E of 'exists'.

Proposition 11.2 *[44]*

Let $(X,\cdot)$ be an invariant set and $R \subseteq A \times X$ a finitely supported subset. For each $x \in X$ the following are equivalent:

1. $\forall a \in A.(a\#x, R \Rightarrow (a,x) \in R)$;
2. $\text{И} a.(a,x) \in R$;
3. $\exists a \in A.(a\#x, R \wedge (a,x) \in R)$.

Proof We prove that $[\forall a.(a\#x, R \Rightarrow (a,x) \in R)] \Rightarrow [\text{И} a.(a,x) \in R] \Rightarrow [\exists a.(a\#x, R \wedge (a,x) \in R)] \Rightarrow [\forall a.(a, R\#x \Rightarrow (a,x) \in R)]$.

We know that the set $\{a \in A \,|\, a\#x, R\}$ is cofinite. If we suppose that $\forall a.(a\#x, R \Rightarrow (a,x) \in R)$, we have $\{a \in A \,|\, a\#x, R\} \subseteq \{a \in A \,|\, (a,x) \in R\}$, and so the set $\{a \in A \,|\, (a,x) \in R\}$ is cofinite, which means $\text{И} a.(a,x) \in R$.

Now, if we suppose $\text{И} a.(a,x) \in R$, then $\{a \in A \,|\, (a,x) \in R\}$ is cofinite, and so $\{a \in A \,|\, a\#x, R\} \cap \{a \in A \,|\, (a,x) \in R\}$ is the intersection of two cofinite subsets of the infinite set A and so cannot be empty. Indeed, if we assume that the intersection $B \cap C$ of two cofinite subsets of A is empty, we have $C_{B \cap C} = A$, and so $C_B \cup C_C = A$; since C_B and C_C are finite, we obtain A is finite which contradicts the infiniteness of A. Therefore, $\exists a.(a\#x, R \wedge (a,x) \in R)$.

Let us suppose now that $\exists a.(a\#x, R \wedge (a,x) \in R)$ and let b be an arbitrary atom such that $b\#x, R$. We have $(b\,a) \star R = R$ because $a, b\#R$ and so $(b\,a)$ fixes $supp(R)$ pointwise. Since $(a,x) \in R$, we have $(b\,a) \otimes (a,x) \in R$. However, $(b\,a) \otimes (a,x) = ((b\,a)(a), (b\,a) \cdot x) = (b, (b\,a) \cdot x)$. Since $a, b\#x$, we also have $(b\,a) \cdot x = x$, and hence $(b,x) = (b\,a) \otimes (a,x) \in R$. □

Corollary 11.1 *Let $(X,\cdot)$ be an invariant set and $R \subseteq A \times X$ an equivariant subset. For each $x \in X$ the following are equivalent:*

1. $\forall a \in A.(a\#x \Rightarrow (a,x) \in R)$;
2. $\text{И} a.(a,x) \in R$;
3. $\exists a \in A.(a\#x \wedge (a,x) \in R)$.

Proposition 11.3 *Let p and p' be properties of atoms for which $\{a \in A \,|\, p\}$ and $\{a \in A \,|\, p'\}$ are finitely supported subsets of A. The following properties hold:*

1. $\neg(\text{И} a.p)$ *if and only if* $\text{И} a.(\neg p)$;
2. $\text{И} a.(p \wedge p')$ *if and only if* $(\text{И} a.p) \wedge (\text{И} a.p')$.

Proof Let $S = \{a \in A \,|\, p(a)\}$ and $S' = \{a \in A \,|\, p'(a)\}$. Since $S, S' \in \wp_{fs}(A)$, it follows that S, S' have to be either finite or cofinite. Thus, S is not cofinite if and only if S is finite. However, S is finite if and only if $C_S = A \setminus S$ is cofinite. The first item of the proposition holds. Since the union of two sets is finite if and only if they are both finite, we have that $S \cap S'$ is cofinite if and only if both S and S' are cofinite. The second item of the proposition follows. □

The next corollary follows immediately by using the propositional logic.

Corollary 11.2 *Let p and p' be properties of atoms for which $\{a \in A \,|\, p\}$ and $\{a \in A \,|\, p'\}$ are finitely supported subsets of A. The following properties hold:*

1. *$\text{И}a.(p \vee p')$ if and only if $(\text{И}a.p) \vee (\text{И}a.p')$;*
2. *$(\text{И}a.p) \Rightarrow (\text{И}a.p')$ if and only if $\text{И}a.(p \Rightarrow p')$.*

11.2 The NEW Quantifier in the Logic of ZFA and FM

Similar freshness properties hold in the frameworks of ZFA and FM set theories.

Proposition 11.4 *Suppose that $p((x_i)_i)$ is a formula in the logic of ZFA or FM, where the free variables of p are listed in the set $(x_i)_i$, and $(x_i)_i$ are arbitrary distinct variables. Then $\forall a, b \in A.(p((x_i)_i) \Leftrightarrow p((a\,b) \cdot (x_i)_i))$, where $p((a\,b) \cdot (x_i)_i)$ denotes the result of substituting $(a\,b) \cdot x_j$ for all free occurrences x_j from the set $(x_i)_i$ in p.*

Proposition 11.4 is an equivariance property. A complete proof of this property can be found in [29]. The idea of proving this is to proceed by induction on the structure of the formula p, using the properties: $x = y \Rightarrow (a\,b) \cdot x = (a\,b) \cdot y$, $x \in y \Rightarrow (a\,b) \cdot x \in (a\,b) \cdot y$ and $(a\,b) \cdot A = A$, where the last two properties follow from the definition of the permutation action hierarchically constructed on $\nu(A)$.

Proposition 11.5 *Let $(x_i)_i$ be a set of distinct variables, and p a formula in the logic of FM. We have the following implications:*

$$[\forall a.(a\#(x_i)_i \Rightarrow p)] \Rightarrow [\text{И}a.p] \Rightarrow [\exists a \in A(a\#(x_i)_i \wedge p)].$$

Proof We know the set $\{a \in A \,|\, a\#(x_i)_i\}$ is cofinite. If we assume $\forall a.(a\#(x_i)_i \Rightarrow p)$, we have $\{a \in A \,|\, a\#(x_i)_i\} \subset \{a \in A \,|\, p\}$, and so $\text{И}a.p$. Now, if we assume $\text{И}a.p$, then $\{a \in A \,|\, p\}$ is cofinite, and so $\{a \in A \,|\, a\#(x_i)_i\} \cap \{a \in A \,|\, p\}$ is the intersection of two cofinite subsets of the infinite set A, which cannot be empty. □

Proposition 11.6 *Let $(x_i)_i$ be a set of distinct variables, and p a formula in the logic of FM. If the free variables of the formula p are contained in the set of distinct variables $\{a, (x_i)_i\}$, we have $[\exists a \in A(a\#(x_i)_i \wedge p)] \Rightarrow [\forall a.(a\#(x_i)_i \Rightarrow p)]$.*

Proof Let us suppose that for some $a \in A$ we have $a\#(x_i)_i \wedge p$. Then, by Proposition 11.4, we obtain that for every atom b we have $p(b, (a\,b) \cdot (x_i)_i)$, where $p(b, (a\,b) \cdot (x_i)_i)$ denotes the formula obtained when we substitute b for all free occurrences of a in p, and $(a\,b) \cdot x_j$ for all free occurrences of x_j in p. Now if we suppose $b\#(x_i)_i$, we get $(a\,b) \cdot (x_i)_i = (x_i)_i$ (because we also have $a\#(x_i)_i$), and so we obtain $p(b, (x_i)_i)$. □

Propositions 11.5 and 11.6 actually represent the correspondent of Corollary 11.1 in the FM framework.

Chapter 12
Abstraction in Finitely Supported Mathematics

Abstract The notion of abstraction appearing in the theory of nominal sets is used in order to model basic concepts in computer science such as renaming, binding and fresh name. We provide a uniform presentation of the existing results involving abstraction, and provide connections with the theory of finitely supported partially ordered sets.

12.1 Generalizing α-abstraction from λ-calculus

Abstraction in FSM is defined by generalizing the notion of abstraction in the λ-calculus. Let us consider the terms t of the untyped λ-calculus [18]:

$$t ::= a \,|\, tt \,|\, \lambda a.t \,,$$

where a ranges over an infinite set U of names of variables.
For $a, b \in U$ we consider the following three versions of the notion of variable renaming for λ-terms t:

- $[b|a]t$, the textual substitution of b for all free occurrences of a in t;
- $\{b|a\}t$, the capture-avoiding substitution of b for all free occurrences of a in t;
- $(b\,a) \cdot t$, the transposition of all occurrences (be they free, bound or binding) of a and b in t.

The α-equivalence in the λ-calculus (denoted by $=_\alpha$) is defined as the least congruence on the set of λ-terms that identifies $\lambda a.t$ with $\lambda b.\{b|a\}t$. The notion of renaming defined by using transpositions can be used to characterize the α-equivalence, as Theorem 12.1 shows. Moreover, when we work with this notion of renaming, it is not necessary to know whether any of the operations in the underlying signature for λ-terms are supposed to be variable-binders in order to define it.

Theorem 12.1 *The relation $=_\alpha$ of α-equivalence between λ-terms coincides with the binary relation $\sim$ inductively generated by the following axioms and rules:*

A. Alexandru, G. Ciobanu, *Foundations of Finitely Supported Structures*,
https://doi.org/10.1007/978-3-030-52962-8_12

$$\frac{a \in U}{a \sim a}, \quad \frac{t_1 \sim t_1', t_2 \sim t_2'}{t_1 t_2 \sim t_1' t_2'},$$
$$\frac{(ba) \cdot t \sim (ba') \cdot t', b \neq a, a' \text{ and } b \text{ do not occur in } t, t'}{\lambda a.t \sim \lambda a' t'}.$$

Proof It is easy to check that $(ba) \cdot (-)$ preserves $=_\alpha$, and hence $=_\alpha$ is closed under the axioms and rules defining $\sim$. So, $\sim$ is contained in $=_\alpha$. The converse inclusion follows by proving that $\sim$ is a congruence relating $\lambda a.t$ and $\lambda b.\{b|a\}t$. This follows because $(ba) \cdot (-)$ preserves $\sim$, and if b does not occur in t, then $(ba) \cdot t \sim \{b|a\}t$. The last result can easily be proved by induction on term size. □

This theorem shows that some aspects related to variable binding can be phrased in terms of the elementary operation of variable-transposition $(ba) \cdot (-)$ rather than the more complicated operation of variable-substitution. Generally, we can define $\pi \cdot t$ to be the result of permuting the atoms in t under the effect of a one-to-one transformation $\pi : U \to U$ of U onto itself. As it is presented in [29], this 'permutation action' permits one to formalize an important abstractness property of meta-theoretic assertions involving the notion of 'variable', namely that the validity of assertions about syntactical objects should be sensitive only to distinctions between variable names, rather than to the particular names themselves. Formally, this represents the equivariance property of an assertion $p(t)$ about terms t claiming that for all π and t we have $p(t) \Leftrightarrow p(\pi \cdot t)$. The validity of the previous statement depends on the nature of the assertion $p(t)$.

Consider that the set U of names in λ-calculus corresponds to the set A in FSM. By employing the freshness quantifier, the rule for α-equivalence of λ-terms has the following form.

Proposition 12.1 *The rule for α-equivalence of λ-abstractions in Theorem 12.1*

$$\frac{(ba) \cdot t \sim (ba') \cdot t', b \neq a, a' \text{ and } b \text{ do not occur in } t, t'}{\lambda a.t \sim \lambda a' t'}$$

is the same as the rule

$$\frac{\text{И} b.((ba) \cdot t \sim (ba') \cdot t')}{\lambda a.t \sim \lambda a'.t'}.$$

Proof Let Λ be the set of λ-terms t presented in Example 2.1(1). Let us define $R = \{(b,(a,t,a',t')) \mid (ba) \cdot t \sim (ba') \cdot t'\}$. We know that both A and Λ are invariant set under the canonical with the S_A-actions defined in Proposition 2.2(1) and in Example 2.1(1), respectively. Therefore, $(A \times \Lambda \times A \times \Lambda)$ and $A \times (A \times \Lambda \times A \times \Lambda)$ are invariant sets. For simplicity, we denote all the S_A-actions by $\cdot$. It is easy to prove that R is an equivariant subset of $A \times (A \times \Lambda \times A \times \Lambda)$. Indeed, let $(b,(a,t,a',t'))$ be an arbitrary element of R, and π an arbitrary element of S_A. We have $\pi \cdot (b,(a,t,a',t')) = (\pi(b),(\pi(a),\pi \cdot t,\pi(a'),\pi \cdot t'))$. Moreover, because $\sim$ coincides with α-equivalence (Theorem 12.1) and the α-equivalence is an equivariant relation (see Example 2.1(2)), we have $\pi \cdot (ba) \cdot t \sim \pi \cdot (ba') \cdot t'$, which means

$(\pi\circ(b\,a))\cdot t\sim(\pi\circ(b\,a'))\cdot t'$ from which $((\pi(b)\,\pi(a))\circ\pi)\cdot t\sim((\pi(b)\,\pi(a'))\circ\pi)\cdot t'$, and finally $(\pi(b)\,\pi(a))\cdot\pi\cdot t\sim(\pi(b)\,\pi(a'))\cdot\pi\cdot t'$. This means $\pi\cdot(b,(a,t,a',t'))\in R$, and so R is equivariant. The assertion "$b\neq a$, a' and b does not occur in t,t'" is the same as "$b\#a,a',t,t'$". The desired result follows from Corollary 11.1. □

Proposition 12.1 suggests how to generalize the notion of α-equivalence from the syntax trees of λ-terms to arbitrary objects in FSM. Let X be an invariant set. Let $\sim_A$ be the binary relation on $A\times X$ defined by:

$$(a,x)\sim_A(b,y) \text{ if and only if } \text{И}c.((c\,a)\cdot x=(c\,b)\cdot y).$$

It is easy to prove that $\sim_A$ is an equivalence relation on $A\times X$. The reflexivity and symmetry result directly from the definition of $\sim_A$. The transitivity of $\sim_A$ is obtained from Proposition 11.3(2).

Definition 12.1 For an invariant set X and arbitrary elements $x\in X$ and $a\in A$, the *abstraction of* a *in* x is the element $[a]x$ defined as the equivalence class of (a,x) modulo $\sim_A$.

Proposition 12.2 *Let X be an invariant set, $x,y\in X$ and $a,b\in A$. Then $(a,x)\sim_A(b,y)$ if and only if either $a=b$ and $x=y$, or $b\neq a$, $b\#x$ and $y=(b\,a)\cdot x$.*

Proof Let $x,y\in X$ and $a,b\in A$ with $(a,x)\sim_A(b,y)$. If $b=a$, then $x=y$. Let us suppose $b\neq a$. If $(a,x)\sim_A(b,y)$, we have that $(d\,a)\cdot x=(d\,b)\cdot y$ for all but finitely many atoms d. Let U be a finite set of atoms such that $(d\,a)\cdot x=(d\,b)\cdot y$ for all $d\in A\setminus U$. Since $S=supp(a)\cup supp(x)\cup supp(b)\cup supp(y)$ is finite, we can choose an element $c\in A\setminus(U\cup S)$. Thus, there is $c\#a,x,b,y$ and $(c\,a)\cdot x=(c\,b)\cdot y$ (1).

Since $c\#y$ and $b=(c\,b)\cdot c$, by Proposition 11.1, we also have $b\#(c\,b)\cdot y$ and hence $b\#(c\,a)\cdot x$. From Proposition 11.1, we have $(c\,a)\cdot b\#(c\,a)\cdot((c\,a)\cdot x)$, and so $(c\,a)\cdot b\#x$. However, $b\#a,c$, and we obtain $(c\,a)\cdot b=b$ and $b\#x$.

We must prove that $y=(b\,a)\cdot x$. Firstly, we remark that $(c\,b)\circ(c\,a)=(b\,a)\circ(c\,b)$ (2) which follows by easy calculation.

We have $y=(c\,b)\cdot((c\,b)\cdot y)\overset{(1)}{=}(c\,b)\cdot((c\,a)\cdot x)\overset{(2)}{=}(b\,a)\cdot((c\,b)\cdot x)\overset{b,c\#x}{=}(b\,a)\cdot x$.

Conversely, let $y=(b\,a)\cdot x$ (3) and $b\#x$.

Since A is not finite, the set of atoms c for which $c\#a,x,b,y$ is cofinite. Firstly, we remark that $(c\,b)\circ(b\,a)=(c\,a)\circ(c\,b)$ (4) which follows by easy calculation. Moreover, when $c\#a,x,b,y$ we also have $(c\,b)\cdot y\overset{(3)}{=}(c\,b)\cdot((b\,a)\cdot x)\overset{(4)}{=}(c\,a)\cdot((c\,b)\cdot x)\overset{c,b\#x}{=}(c\,a)\cdot x$. Therefore, $\text{И}c.((c\,a)\cdot x=(c\,b)\cdot y)$, which means $(a,x)\sim_A(b,y)$. □

From Proposition 12.2, we have $(a,x)/\sim_A=\{(b,(b\,a)\cdot x)\,|\,b\in A\wedge(b=a\vee b\#x)\}$.

Example 12.1 $[a](A\setminus\{a\})=\{(a,A\setminus\{a\}),(b,A\setminus\{b\}),\dots\}$
$[a]\{a,b\}=\{(a,\{a,b\}),(c,\{c,b\}),(d,\{d,b\}),(e,\{e,b\}),\dots\}$.

From Proposition 12.2, we obtain the following result.

Proposition 12.3 *Let X be an invariant set. Let $a,b \in A$ and $x,y \in X$. We have $[a]x = [b]y$ if and only if one of the following statements is true:*

- *$a = b$ and $x = y$;*
- *$a \neq b$, $b\#x$ and $y = (b\,a) \cdot x$.*

Corollary 12.1 *Let X be an invariant set. Let $a,b \in A$ and $x,y \in X$. If we have $c\#a,b,x,y$ and $[a]x = [b]y$, then $(a\,c) \cdot x = (b\,c) \cdot y$.*

12.2 An Invariant Partially Ordered Set of Abstractions

The results in this section slightly extend those from [48].

Proposition 12.4 *Let X be an invariant set. Then the set $[A]X = \{[a]x \,|\, a \in A, x \in X\}$ is also an invariant set with the S_A-action $\cdot_{[A]X} : S_A \times A[X] \to A[X]$ defined by $\pi \cdot_{A[X]} a[x] = [\pi(a)]\pi \cdot x$ for all $\pi \in S_A$ and all $[a]x \in A[X]$. Moreover, we have $supp([a]x) = supp(x) \setminus \{a\}$ for all $a \in A$ and $x \in X$.*

Proof We firstly prove that the function $\cdot_{[A]X} : S_A \times A[X] \to A[X]$ defined by $\pi \cdot_{A[X]} a[x] = [\pi(a)]\pi \cdot x$ for all $\pi \in S_A$ and all $[a]x \in A[X]$ is well-defined, i.e. it does not depend on the chosen representatives for the equivalence classes modulo $\sim_A$. Let (a,x) and (b,y) be two elements in the same equivalence class modulo $\sim_A$, i.e. $[a]x = [b]y$. Then either $a = b$ and $x = y$, or $b \neq a$, $b\#x$ and $y = (b\,a) \cdot x$. If $a = b$ and $x = y$, then clearly $\pi(a) = \pi(b)$ and $\pi \cdot x = \pi \cdot y$. So suppose $b \neq a$, $b\#x$ and $y = (b\,a) \cdot x$. Since π is one-to-one, it is clear that $\pi(b) \neq \pi(a)$. According to Proposition 11.1, we also have $\pi(b)\#\pi \cdot x$. Now $\pi \cdot y = \pi \cdot ((b\,a) \cdot x) = (\pi \circ (b\,a)) \cdot x = ((\pi(b)\,\pi(a)) \circ \pi) \cdot x = (\pi(b)\,\pi(a)) \cdot (\pi \cdot x)$. From Proposition 12.2, we obtain $[\pi(a)]\pi \cdot x = [\pi(b)]\pi \cdot y$. Since the function $\cdot$ is an S_A-action on X, it is clear that $\cdot_{[A]X}$ is also an S_A-action on $[A]X$.

Let $V \subset A$ be a finite set which supports the element (a,x) from $A \times X$. Let $\pi \in Fix(V)$. Therefore, $\pi \cdot (a,x) = (a,x)$, which means $\pi(a) = a$ and $\pi \cdot x = x$. We have $\pi \cdot a[x] = [\pi(a)]\pi \cdot x = [a]x$ which means V supports $[a]x$ in $[A]X$. However, $supp([a]x)$ is the least finite set supporting $[a]x$. Whenever V supports (a,x) we have that V supports $[a]x$, and so $supp([a]x) \subseteq V$. Since $supp((a,x)) = \cap\{V \,|\, V\, supports\, (a,x)\}$, we have $supp([a]x) \subseteq supp((a,x)) = supp(x) \cup \{a\}$ (1).

Let $S \subset A$ be a finite set which supports $[a]x$ in $[A]X$. Let $\pi \in Fix(S \cup \{a\})$. Since S supports $[a]x$, we have $\pi \cdot [a]x = [a]x$. We also have $\pi(a) = a$. Furthermore, we obtain $[a]x = \pi \cdot [a]x = [\pi(a)]\pi \cdot x = [a]\pi \cdot x$. According to Proposition 12.2, we have $\pi \cdot x = x$. Therefore, $S \cup \{a\}$ supports x in X. However, $supp(x)$ is the least finite set supporting x. Whenever S supports $[a]x$, we have that $S \cup \{a\}$ supports x, and so $supp(x) \subseteq \{a\} \cup supp([a]x)$. Since we have already proved that $supp([a]x) \subseteq supp(x) \cup \{a\}$, it remains to show that $a \notin supp([a]x)$, that is $a\#[a]x$. Since A is not finite, we can choose an atom b such that $b\#a,x$. According to Proposition 12.2, we have $(b,(b\,a) \cdot x) \sim_A (a,x)$, and so $(b\,a) \cdot a[x] = [(b\,a)(a)]((b\,a) \cdot x) = [b]((b\,a) \cdot x) = [a]x$. However, $b\#a,x$, and hence, by (1), we have $b\#[a]x$. According to Proposition 11.1, we have $(b\,a) \cdot b\#(b\,a) \cdot [a]x$, and so $a\#a[x]$. □

Definition 12.2 For each invariant poset $(X,\sqsubseteq,\cdot)$ we define the binary relation $\preceq_A$ on $A \times X$ by: $(a,x) \preceq_A (b,y)$ if and only if there is $c \in A \setminus supp(a,b,x,y)$ such that $(a\,c)\cdot x \sqsubseteq (b\,c)\cdot y$.

Proposition 12.5 *For each invariant poset $(X,\sqsubseteq,\cdot)$, we have that $(a,x) \preceq_A (b,y)$ if and only if for all $c \in A \setminus supp(a,b,x,y)$ we have $(a\,c)\cdot x \sqsubseteq (b\,c)\cdot y$.*

Proof The reverse direction is trivial because $A \setminus supp(a,b,x,y)$ is non-empty. Let us suppose now that $(a,x) \preceq_A (b,y)$. Therefore, $\exists c \in A \setminus supp(a,b,x,y)$ such that $(a\,c)\cdot x \sqsubseteq (b\,c)\cdot y$. Let $d \in A \setminus supp(a,b,x,y)$. Since $\sqsubseteq$ is equivariant, we have $(c\,d)\cdot(a\,c)\cdot x \sqsubseteq (c\,d)\cdot(b\,c)\cdot y$. It follows by easy calculation that $(a\,d)\cdot(c\,d)\cdot x \sqsubseteq (b\,d)\cdot(c\,d)\cdot y$. Since $c,d\#x,y$, we get $(a\,d)\cdot x \sqsubseteq (b\,d)\cdot y$. □

Following Corollary 11.1, we obtain the next result.

Corollary 12.2 *For each invariant poset $(X,\sqsubseteq,\cdot)$, we have that $(a,x) \preceq_A (b,y)$ if and only if $\mathsf{N}c.((a\,c)\cdot x \sqsubseteq (b\,c)\cdot y)$.*

Proposition 12.6 *For each invariant poset $(X,\sqsubseteq,\cdot)$, the binary relation $\preceq_A$ is a preorder on $A \times X$.*

Proof Reflexivity is trivial. Let us suppose $(a,x) \preceq_A (b,y)$ and $(b,y) \preceq_A (c,z)$. Then there exists $d \in A \setminus supp(a,b,x,y)$ such that $(a\,d)\cdot x \sqsubseteq (b\,d)\cdot y$ and $e \in A \setminus supp(b,c,y,z)$ such that $(b\,e)\cdot y \sqsubseteq (c\,e)\cdot z$. According to Proposition 12.5, for $f \in A \setminus supp(a,b,c,x,y,z)$ we have that $(a\,f)\cdot x \sqsubseteq (b\,f)\cdot y \sqsubseteq (c\,f)\cdot z$. By transitivity of $\sqsubseteq$, we obtain $(a,x) \preceq_A (c,z)$. □

Theorem 12.2 *For each invariant poset $(X,\sqsubseteq,\cdot)$, $[A]X$ is an invariant poset with the partial order defined by $[a]x \sqsubseteq_{[A]X} [b]y$ if and only if $(a,x) \preceq_A (b,y)$.*

Proof It is trivial to prove that $\sqsubseteq_{[A]X}$ is a well-defined partial order on $[A]X$. It remains to prove that $\sqsubseteq_{[A]X}$ is equivariant. Therefore, we have to prove that $(a,x) \preceq_A (b,y)$ implies $(\pi(a),\pi\cdot x) \preceq_A (\pi(b),\pi\cdot y)$ for each $\pi \in S_A$. Let $(a,x) \preceq_A (b,y)$. Then there exists $c \in A \setminus supp(a,b,x,y)$ such that $(a\,c)\cdot x \sqsubseteq (b\,c)\cdot y$. Let $\pi \in S_A$. Since $\sqsubseteq$ is equivariant, we obtain $(\pi(a)\,\pi(c))\cdot(\pi\cdot x) \sqsubseteq (\pi(b)\,\pi(c))\cdot(\pi\cdot y)$. Since $c \in A \setminus supp(a,b,x,y)$, we obtain $\pi(c) \in A \setminus supp(\pi(a),\pi(b),\pi\cdot x,\pi\cdot y)$ (see Proposition 11.1). Therefore, $(\pi(a),\pi\cdot x) \preceq_A (\pi(b),\pi\cdot y)$. □

From the equivariance of $\sqsubseteq$ and $\sqsubseteq_{[A]X}$, we obtain the following result.

Proposition 12.7 *Let $(X,\sqsubseteq,\cdot)$ be an invariant poset and $x,y \in X$.*
Then $x \sqsubseteq y$ if and only if $[a]x \sqsubseteq_{[A]X} [a]y$ for each $a \in A$.

Proposition 12.8 *Let X be an invariant set. For each $[a]x$ in $[A]X$ and $b \in A \setminus supp([a]x)$, there exists a unique $y = (a\,b)\cdot x$ such that $[a]x = [b]y$. We call this y the* concretion of $[a]x$ and b, *and we denote it by $([a]x)@b$.*

Proof Let $b \in A \setminus supp([a]x)$. We have either $b = a$, or $b \neq a$ and $b \notin supp(x)$. The required result follows easily from Proposition 12.2 □

By easy calculation, we obtain the following result.

Proposition 12.9 *Let X be an invariant set. For each* $[a]x \in [A]X$ *and* $b, c \in A$, $b \neq c$ *with* $b, c \notin supp([a]x)$ *we have* $(bc) \cdot (([a]x)@b) = ([a]x)@c$.

Proposition 12.10 *Let X be an invariant set. For each* $b \in A \setminus supp([a]x)$ *we have* $supp(([a]x)@b) \subseteq supp([a]x) \cup \{b\}$.

Proof Let $b \in A \setminus supp([a]x)$. We have either $b = a$, or $b \neq a$ and $b \notin supp(x)$. If $b = a$ we have $supp(([a]x)@b) = supp((ab) \cdot x) = supp(x) = supp([a]x) \cup \{a\}$. If $b \neq a$ and $b \notin supp(x)$, by Proposition 2.1, we obtain $supp((ab) \cdot x) = (ab) \cdot supp(x) \subseteq supp(x) \setminus \{a\} \cup \{b\} = supp([a]x) \cup \{b\}$. □

Proposition 12.11 *Let* $(X, \sqsubseteq, \cdot)$ *be an invariant poset and* $x, y \in X$. *If* $[a]x \sqsubseteq_{[A]X} [b]y$ *for some* $a, b \in A$, *then* $([a]x)@c \sqsubseteq ([b]y)@c$ *for any* $c \in A \setminus supp([a]x, [b]y)$.

Proof From the definition of concretion, we obtain that $[a]x = [c](([a]x@c) = [c](ac) \cdot x$ and $[b]y = [c](([b]y@c) = [c](bc) \cdot y$. Therefore, $[c](ac) \cdot x \sqsubseteq_{[A]X} [c](bc) \cdot y$. According to Proposition 12.7, we obtain $(ac) \cdot x \sqsubseteq (bc) \cdot y$ as required. □

Theorem 12.3 *Let* $(X, \sqsubseteq, \cdot)$ *be an invariant cpo. Then* $[A]X$ *is an invariant cpo.*

Proof We must prove that any finitely supported countable totally ordered family (chain) $([a_n]x_n)_{n \in \mathbb{N}} \in [A]X$ has a least upper bound. Let $([a_n]x_n)_{n \in \mathbb{N}} \in [A]X$ be a finitely supported chain which means that all the elements $[a_n]x_n$ of the chain are supported by the same set. Let us consider $a \notin supp([a_n]x_n)$ for each $n \in \mathbb{N}$. According to Proposition 12.11, we obtain another countable chain $(([a_n]x_n)@a)_{n \in \mathbb{N}} \in X$. Since by Proposition 12.10 it holds $supp(([a_n]x_n)@a) \subseteq supp([a_n]x_n) \cup \{a\}$ for all $n \in \mathbb{N}$, we have that $supp([a_n]x_n)_{n \in \mathbb{N}} \cup \{a\}$ supports the chain $(([a_n]x_n)@a)_{n \in \mathbb{N}}$. Since X is an invariant cpo, we obtain that $(([a_n]x_n)@a)_{n \in \mathbb{N}}$ has a least upper bound denoted by $\bigsqcup_{n \in \mathbb{N}}(([a_n]x_n)@a)$. Hence $([a_n]x_n)@a \sqsubseteq \bigsqcup_{n \in \mathbb{N}}(([a_n]x_n)@a)$ for each $n \in \mathbb{N}$. Since $\sqsubseteq$ is equivariant, $[a](([a_n]x_n)@a) \sqsubseteq_{[A]X} [a]\bigsqcup_{n}(([a_n]x_n)@a)$ for each $n \in \mathbb{N}$. Thus, $[a]\bigsqcup_{n \in \mathbb{N}}(([a_n]x_n)@a)$ is an upper bound for the chain $([a_n]x_n)_{n \in \mathbb{N}} \in [A]X$. It remains to prove that $[a]\bigsqcup_{n \in \mathbb{N}}(([a_n]x_n)@a)$ is the least such upper bound.

Suppose that there is another upper bound $[b]y$ such that $[a_n]x_n \sqsubseteq_{A[X]} [b]y$ for all $n \in \mathbb{N}$. We have to prove that $[a]\bigsqcup_{n \in \mathbb{N}}(([a_n]x_n)@a) \sqsubseteq_{[A]X} [b]y$. Let $c \in A \setminus supp(a, b, y, ([a_n]x_n)_{n \in \mathbb{N}})$. We claim that $(ac) \cdot \bigsqcup_{n \in \mathbb{N}}(([a_n]x_n)@a) \sqsubseteq (bc) \cdot y$. Indeed, from the equivariance of $\sqcup$ and Proposition 12.9, we have $(ac) \cdot \bigsqcup_{n}(([a_n]x_n)@a) = \bigsqcup_{n}((ac) \cdot ([a_n]x_n)@a)) = \bigsqcup_{n}(([a_n]x_n)@c)$. Since $[b]y$ is an upper bound and because of Proposition 12.11, we obtain $([a_n]x_n)@c \sqsubseteq ([b]y)@c$. Therefore, $\bigsqcup_{n \in \mathbb{N}}(([a_n]x_n)@c) \sqsubseteq ([b]y)@c = (bc) \cdot y$. Therefore, $[a]\bigsqcup_{n \in \mathbb{N}}(([a_n]x_n)@a) \sqsubseteq_{[A]X} [b]y$. □

Corollary 12.3 *Let* $(X, \sqsubseteq, \cdot, \bot)$ *be an invariant cppo. Then* $[A]X$ *is an invariant cppo.*

Proof According to Theorem 12.3, we have that $[A]X$ is an invariant cpo. Since $\bot$ is the least element in X, we obtain that for each $a \in A$ we have $[a] \bot$ is less than any element in $[A]X$ (see Proposition 12.7). Also, by Proposition 12.2, we obtain that $[a] \bot = [b] \bot$ for any $b \in A$. Thus, $[a] \bot$ is the least element in $[A]X$ for each $a \in A$.□

Chapter 13
Relaxing the Finite Support Requirement

Abstract P_A-sets are defined as classical sets equipped with actions of the group of all bijections of an amorphous set A. They are constructed in the same way as S_A-sets, except that for defining P_A-sets we consider all the bijections of A, not only the finitary ones. Furthermore, in contrast to the invariant sets, P_A-sets do not necessarily satisfy the finite support requirement. The Relaxed Fraenkel-Mostowski axiomatic set theory represents a refinement of Fraenkel-Mostowski set theory obtained by replacing the finite support axiom with a consequence of it which states that any subset of the set A of atoms is either finite or cofinite. More exactly, the aim of the Relaxed Fraenkel-Mostowski set theory is to replace the requirement 'finite support for all sets (built on a cumulative hierarchy from a family of basic elements)' with 'finite support only for set of basic elements' in order to obtain similar results as in the Fraenkel-Mostowski set theory. In this sense, although we do not require the existence of a finite support for any hierarchically defined structure, we prove that several properties of the set of atoms and of the group of all bijections of basic elements (particularly, local finiteness) are preserved. Similarly, we prove that P_A-sets have some properties that are similar to those of invariant sets. We also introduce a mathematics where each set is either finite or cofinite, and we relate it to the Relaxed Fraenkel-Mostowski and to the Fraenkel-Mostowski set theories. This chapter is based on the article [16].

13.1 P_A-sets

Adjoin to ZF an infinite set A (formed by elements whose internal structure is irrelevant, the only relevant attribute being their identity) which is called 'the set of atoms (basic elements)' and is emphasized distinctly from the ordinary ZF sets. We refer to A as a fixed ZF set, while noting that the internal structure of the elements of A is ignored. Additionally, we assume that A is amorphous, meaning that any subset of A is either finite or cofinite.

A. Alexandru, G. Ciobanu, *Foundations of Finitely Supported Structures*,
https://doi.org/10.1007/978-3-030-52962-8_13

Similarly to the case of nominal sets, P_A-sets are defined as Zermelo-Fraenkel sets equipped with actions of the group of all bijections of the fixed Zermelo-Fraenkel amorphous set A; however, despite nominal sets, we do not impose a finite support requirement. Let P_A be the group of all bijections of A. Despite S_A, the bijections belonging to P_A do not necessary leave unchanged all but finitely many atoms because they are not necessarily finitely supported.

Definition 13.1 • Assume A is a fixed infinite amorphous set. Let X be a ZF set. A P_A-*action* on X is a function $\cdot : P_A \times X \to X$ having the properties that $Id \cdot x = x$ and $\pi \cdot (\pi' \cdot x) = (\pi \circ \pi') \cdot x$ for all $\pi, \pi' \in P_A$ and $x \in X$.

- A P_A-*set* is a pair $(X, \cdot)$ where X is a ZF set, and $\cdot : P_A \times X \to X$ is a P_A-action on X.

Example 13.1 1. The set A of atoms is a P_A-set with the P_A-action $\cdot : P_A \times A \to A$ defined by $\pi \cdot a := \pi(a)$ for all $\pi \in P_A$ and $a \in A$.
2. The set P_A is a P_A-set with the P_A-action $\cdot : P_A \times P_A \to P_A$ defined by $\pi \cdot \sigma := \pi \circ \sigma \circ \pi^{-1}$ for all $\pi, \sigma \in P_A$.
3. Any ordinary ZF set X is a P_A-set with the (only possible) P_A-action $\cdot : P_A \times X \to X$ defined by $\pi \cdot x := x$ for all $\pi \in P_A$ and $x \in X$.
4. The set $U = \wp_{fin}(A) \cup \wp_{cofin}(A)$ is a P_A-set with the P_A-action $\star : P_A \times U \to U$ defined by $\pi \star Z := \{\pi \cdot z \,|\, z \in Z\}$ for all $\pi \in P_A$, and all $Z \in U$.

So, in Chapter 2 we considered S_A to be the set of all finitary bijections (i.e. bijections that can be expressed as a finite composition of transpositions) of a fixed ZF set A, while in this chapter we consider P_A to be the set of *all* bijections of a fixed *amorphous* set A. In FSM, A is amorphous (a consequence of the finite support requirement), and furthermore, *any higher-order construction is finitely supported*, while in the world of P_A-sets we require only the amorphous structure of A, without imposing a finite support requirement for higher-order constructions. Our goal is to prove that particular P_A-sets have some similar properties as finitely supported sets.

13.2 Properties of P_A-sets

Assume that the fixed infinite ZF set A is amorphous, i.e. any subset of A is either finite or cofinite. The properties presented below are also properties of atomic sets in FSM, but they can be proved without imposing a finite support requirement.

Proposition 13.1 *Every one-to-one mapping $f : A \to A$ is also onto.*

Proof Suppose by contradiction that $f : A \to A$ is an injection, with the property that $Im(f) \subsetneq A$. This means that there exists $x_0 \in A$ such that $x_0 \notin Im(f)$. We define an injective sequence $\mathscr{F}$ of elements from A which has the first term x_0 and the general term $x_{n+1} = f(x_n)$ for all $n \in \mathbb{N}$. The function $g : \mathbb{N} \to A$, defined by $g(n) = x_n$, is injective. Thus, $g(2\mathbb{N})$ and $g(2\mathbb{N}+1)$ are disjoint, infinite subsets of A. Therefore, $g(2\mathbb{N})$ is infinite and coinfinite, and this contradicts the structure of A. □

The finite support requirement in FSM is not mandatory in order to prove the inconsistency of Tarski theorem about choice, i.e, we are able to prove in the framework of P_A-sets that there is no injection from $A \times A$ to A.

Proposition 13.2 *There does not exist a one-to-one mapping from $A \times A$ to A.*

Proof Suppose by contradiction that there is an injective mapping $i : A \times A \to A$. Let us fix two atoms x and y with $x \neq y$. The sets $\{i(a,x) \,|\, a \in A\}$ and $\{i(a,y) \,|\, a \in A\}$ are disjoint and infinite. Thus, $\{i(a,x) \,|\, a \in A\}$ is an infinite and coinfinite subset of A, which contradicts the structure of A. □

The property of 'having a finite image' for functions from A to $\mathbb{N}$ (where the $\mathbb{N}$ can be replaced by any ZF set having the property that all its infinite subsets are non-amorphous) can be proved without involving the finite support axiom.

Proposition 13.3 *Any function $f : A \to \mathbb{N}$ has the property that $Im(f)$ is finite.*

Proof Assume by contradiction that $Im(f)$ is infinite. Since $Im(f) \subseteq \mathbb{N}$ we can define a partition of $Im(f)$ into two infinite disjoint subsets X and Y. Therefore $f^{-1}(X)$ and $f^{-1}(Y)$ are infinite and disjoint, and so they are cofinite subsets of A. However, there do not exist two disjoint cofinite subsets of A. □

Proposition 13.4 *The set A cannot be totally ordered.*

Proof The proof follows the same steps as the proof of Proposition 9.9.

Assume by contradiction there is a total ordered $\leq$ on A. For $z \in A$ we define the subsets $\downarrow z = \{y \in A \,|\, y \leq z\}$ and $\uparrow z = \{y \in A \,|\, z \leq y\}$ for all $z \in A$. We define the mapping $z \mapsto \downarrow z$ from A to $T = \{\downarrow z \,|\, z \in A\}$. The mapping is bijective, and so T is amorphous. Thus, any subset Z of T is either finite or cofinite. Similarly, the mapping $z \mapsto \uparrow z$ from A to $T' = \{\uparrow z \,|\, z \in A\}$ is bijective, which means that any subset of T' is either finite or cofinite.

We distinguish the following two cases:

1. There are only finitely many elements $x_1, \ldots, x_n \in A$ such that $\downarrow x_1, \ldots, \downarrow x_n$ are finite. Thus, for $y \in U = A \setminus \{x_1, \ldots, x_n\}$ we have $\downarrow y$ infinite. Since $\downarrow y$ is a subset of A, it should be cofinite, and so $\uparrow y$ is finite (because $\leq$ is a total order relation). Let $M = \{\uparrow y \,|\, y \in U\}$. We have that M is totally ordered with respect to sets inclusion. Furthermore, for an arbitrary $y \in U$ we cannot have $y \leq x_k$ for some $k \in \{1, \ldots, n\}$ because $\downarrow y$ is infinite, while $\downarrow x_k$ is finite, and so $\uparrow y$ is a subset of U. Thus, M is an infinite, totally ordered family formed by finite subsets of U. Since any element of U has only a finite number of successors (leading to the conclusion that $\geq$ is an well-ordering on U), we can define an order monomorphism between $\mathbb{N}$ and U. For example, choose $u_0 \neq u_1 \in U$, then let u_2 be *the greatest element* (with respect to $\leq$) in $U \setminus \{u_0, u_1\}$, u_3 be *the greatest element* in $U \setminus \{u_0, u_1, u_2\}$ (no choice principle is used since $\geq$ is an well-ordering, and so such a *greatest* element is precisely defined), and so on, and find an infinite, countable sequence $u_0, u_1, u_2, \ldots$. Since $\mathbb{N}$ is non-amorphous (being expressed as the union between the even elements and the odd elements), we conclude that U is non-amorphous.

2. We have cofinitely many elements z such that $\downarrow z$ is finite. Thus, there are only finitely many elements $y_1, \dots, y_m \in A$ such that $\downarrow y_1, \dots, \downarrow y_m$ are infinite. Since every infinite subset of A is cofinite, only $\uparrow y_1, \dots, \uparrow y_m$ are finite. Let $z \in A \setminus \{y_1, \dots, y_m\}$ which means $\uparrow z$ infinite. Since $\uparrow z$ is a subset of A it should be cofinite, and so $\downarrow z$ is finite. As in the above item, the set $M' = \{\downarrow z \,|\, z \in A \setminus \{y_1, \dots, y_m\}\}$ is an infinite, totally ordered (by inclusion) family of finite sets, and so $\leq$ is an well-ordering on $A \setminus \{y_1, \dots, y_m\}$. Therefore, $A \setminus \{y_1, \dots, y_m\}$ has an infinite, countable subset, which is a contradiction. □

In the world of P_A-sets, a bijection of A onto A is not necessarily a permutation of A in the sense of Definition 2.1. This is because, despite the FSM approach, in the framework of P_A-sets we do not require any bijection of A onto A to be a finitely supported function. This means that a bijection of A could be non-finitary, i.e. it could permute infinitely many atoms. However, we are able to prove that the set P_A of all bijections of A onto A satisfies some similar properties as the set S_A of all (finite) permutations of A in FSM.

Theorem 13.1 *The group P_A of all one-to-one transformations of A onto itself is a torsion group.*

Proof We prove that P_A is a torsion group, i.e. every element of P_A has finite order. Firstly, we prove that the cycles of an arbitrary $\sigma \in P_A$ are finite. Moreover, there is $m \in \mathbb{N}$ such that all but finitely many cycles of σ have length m. Let us suppose that σ has an infinite cycle. If we assume that σ has at least two infinite cycles, then the set of points of one of these cycles is infinite and coinfinite. However, every subset of A is finite or cofinite, and this means that every cycle of σ is finite. If we suppose now that σ has only one infinite cycle, we obtain that $\sigma \circ \sigma$ is a bijection of A with at least two infinite cycles, and so we get a contradiction. Now, for every $n \in \mathbb{N}$, the set of points in the cycles of σ which have length n is also finite or cofinite. If there is n such that this set is cofinite, then the proof is finished. If not, then there are infinitely many different cycle lengths. We can define a partition of the set of cycle lengths into two infinite sets X and Y. Then the set of points from cycles with length in X is infinite and coinfinite, and so we get a subset of A (the set of points from cycles with length in X) which is neither finite nor cofinite. Again we contradict the relation $\wp(A) = \wp_{fin}(A) \cup \wp_{cofin}(A)$. We obtained that all the cycles of σ are finite, and we have a finite number of cycle lengths. According to Theorem 5.1.2 in [33], the order of σ is the least common multiple of these cycle lengths. Thus, P_A is a torsion group. □

Remark 13.1 Since the set of cycle lengths is a subset of $\mathbb{N}$, we could define a partition of the set of cycle lengths into two infinite sets X and Y. The construction of $\mathbb{N}$ makes sense within P_A-sets framework. However, the set A of atoms cannot be partitioned into two infinite subsets A_1 and A_2.

Classes of amorphous sets in ZF were presented in [50]. For the so called 'bounded' amorphous sets, it is proved that the group of those bijections of a bounded amorphous set is locally finite (Corollary 2.2 from [50]); a locally finite group is a group for which every finitely generated subgroup is finite. We prove below that we do not require such a boundedness restriction for the set of atoms in the world of P_A-sets, that is the group of all bijections of the set of atoms is locally finite in the world of P_A-sets; the only requirement is the amorphous structure of the set of atoms.

Theorem 13.2 *P_A is a locally finite group, namely every finitely generated subgroup of P_A is finite.*

Proof We prove by contradiction that, if $G \leq P_A$ is a finitely generated group, then G is finite. Let $\sigma_1, \sigma_2, \dots, \sigma_m \in P_A$ and $G = < \sigma_1, \dots, \sigma_m >$. A G-renaming is the orbit of an element $\alpha \in A$ under the canonical action of G on A defined by $(\pi, a) \mapsto \pi(a)$. Let us suppose that G is infinite. We prove that there exists r (depending on m) such that all but finitely many G-renamings have size r. Let us assume by contradiction that there is an infinite G-renaming under the canonical action of G on A. We define *a word with k letters in $\sigma_1, \sigma_2, \dots, \sigma_m$* to be a finite composition of k bijections and inverses from the set $\{\sigma_1, \sigma_2, \dots, \sigma_m\}$. This terminology comes from the theory of free groups. Of course, the set of words with k letters is finite for each k. We consider an infinite family of words in $\sigma_1, \sigma_2, \dots, \sigma_m$. If there is $a \in A$ with an infinite orbit, we can define the image $\{a_1, a_2, \dots, a_m\}$ of a under the words of the family by $a_1 = \sigma_1(a), \dots, a_m = \sigma_m(a)$. Let $a_{m+1} = \sigma_{m+1}(a)$, where σ_{m+1} is the first word in the family such that $\sigma_{m+1}(a) \notin \{a_1, a_2, \dots, a_m\}$. Such a σ_{m+1} exists because we suppose that the orbit of a is infinite. Indeed, we can define a method of covering the family of words in $\sigma_1, \sigma_2, \dots, \sigma_m$ in the following way: firstly, we cover the words with two letters (in alphabetical order, in the same way as we 'read a dictionary'); because the set of letters is finite, and so well-ordered. Notice that each finite set can be well-ordered, and to prove this we do not need the axiom of choice. Then we cover the words with three letters, and so on. We pick the first word we find with the required property that the image of a under it is not a member of $\{a_1, a_2, \dots, a_m\}$. We present a constructive method to choose the first element in the family of words in $\sigma_1, \sigma_2, \dots, \sigma_m$ with the required property. The method of covering the related family presented before induce a well-order relation on that family if we consider only the distinct words in the family (note that if we form all the possible words with the letters $\sigma_1, \sigma_2, \dots, \sigma_m$ we could have words in $\sigma_1, \sigma_2, \dots, \sigma_m$ with different numbers of letters, but which are equal); this order is called 'lexicographic order'. With a_{m+1} already found, we repeat the procedure described before to find a_{m+2}, and so on. More precisely, since we can define a well-ordering on the family of words in $\sigma_1, \sigma_2, \dots, \sigma_m$ (by ordering lexicographically the words with two letters, then with three letters, etc, deleting then those terms that are equal with a previously presented term), we can form the infinite sequence $a_{m+1}, a_{m+2}, \dots$ without requiring any form of choice (i.e. we formed this sequence in a constructive manner). Thus, we obtain an infinite countable subset of A denoted by B. This contradicts the fact that all subsets of A are finite or cofinite because B has both infinite and coinfinite subsets, and so there is no infinite G-renaming. Let us suppose by contradiction that we have infinitely many G-renamings with size k and infinitely many G-renamings with size l. Since the G-renamings which are different are also disjoint, we conclude that elements in the G-renamings with size k form a set which is both infinite and coinfinite, which represents a contradiction. If the G-renamings are arbitrarily large, we again obtain a contradiction. The proof of this fact is similar to that of Theorem 13.1: we use a partition of the G-renaming sizes into two infinite sets X and Y, and see that the set of elements with G-renaming sizes in X is infinite and coinfinite. We conclude that all but finitely many G-renamings have size r. Thus, there are infinitely many G-renamings, and all but finitely many of these G-renamings have size r (otherwise, if we assume that there are only a finite number of G-renamings, because each G-renaming is finite, it follows that A is finite, which is a contradiction).

We define an equivalence relation on the set of G-renamings saying that two G-renamings are equivalent if and only if the actions of G on them are isomorphic. Let U and V be two equivalent G-renamings of size r. Let $\phi : G \to S(U)$ be the associated representation of the action of G on U (by $S(U)$ we denoted the one-to-one mappings of U onto itself), and $\phi' : G \to S(V)$ be the associated representation of the action of G on V. Thus, there must exist a group isomorphism $\nu : S(U) \to S(V)$ such that $\phi' = \nu \circ \phi$. Since $|U| = r$ there is an isomorphism $i : S(U) \to S_r$ (where S_r is the group of one-to-one mappings of the set $\{1,\dots,r\}$ onto itself). If we define $i' : S(V) \to S_r$ by $i' = i \circ \nu^{-1}$, we have $i \circ \phi = i' \circ \phi'$, and so the actions of G on U and V have associated the same homomorphism from G to S_r.

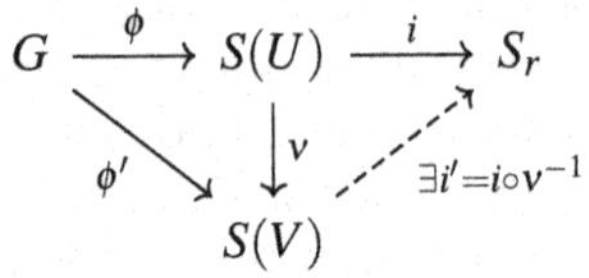

Conversely, let U and V be two G-renamings of size r that have associated the same homomorphism from G to S_r. Let $\phi : G \to S(U)$ be the associated representation of the action of G on U, and $\phi' : G \to S(V)$ be the associated representation of the action of G on V. Let $i : S(U) \to S_r$ and $i' : S(V) \to S_r$ be isomorphisms such that $i \circ \phi = i' \circ \phi'$. Then there exists an isomorphism $\nu : S(U) \to S(V)$ defined as $\nu = i'^{-1} \circ i$ such that $\phi' = \nu \circ \phi$, that is the actions of G on U and V are isomorphic.

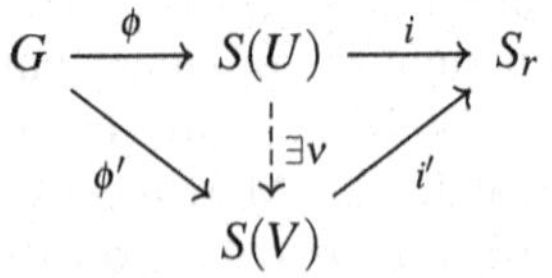

Thus, to an equivalence class of G-renamings we can associate (it corresponds bijectively) one homomorphism from G to S_r. Since G is finitely generated, it follows that there is only a finite number of homomorphisms from G to S_r, and so there is only a finite number of equivalence classes of G-renamings under the previously defined equivalence relation on the family of G-all renamings. It follows that one equivalence class (denoted by $\mathscr{O}$) is infinite; otherwise, if all the equivalence classes are finite, because there are only finitely many equivalence classes, then there is only a finite number of G-renamings, which is a contradiction.

Let X_0 be a G-renaming that is a representative of $\mathscr{O}$. The action of G on X_0 can be seen as a homomorphism f from G to S_r. In fact this action can be viewed as the associated representation by permutations $\psi : G \to S(X_0)$ defined by $\psi(\sigma)(x) = \sigma(x)$. Since $|X_0| = r$, there is an isomorphism $\varphi : S(X_0) \to S_r$, and $f = \varphi \circ \psi$. If $\sigma \in Ker f$, then $\varphi(\psi(\sigma)) = 1_r$, and so $\psi(\sigma) = \varphi^{-1}(1_r) = 1_{X_0}$. Thus, $\sigma(x) = \psi(\sigma)(x) = 1_{X_0}(x) = x$ for all $x \in X_0$, and so σ fixes all the elements in X_0, which means σ fixes all the elements whose G-renamings are in $\mathscr{O}$ (Definition 4.19 of [45] is useful to show how isomorphic actions look like). Indeed, let Y_0 be another G-renaming from $\mathscr{O}$, and $\eta : G \to S(Y_0)$ the representation by permutations of the action of G on Y_0. Since the actions of G on X_0 and Y_0 are isomorphic, there must exist a group isomorphism $\mu : S(X_0) \to S(Y_0)$ such that $\eta = \mu \circ \psi$. Thus, $\eta(\sigma) = \mu(\psi(\sigma)) = \mu(1_{X_0}) = 1_{Y_0}$, and so $\sigma(y) = \eta(\sigma)(y) = 1_{Y_0}(y) = y$ for all $y \in Y_0$. The number of elements of A fixed by σ is infinite (because $\mathscr{O}$ is infinite), and because $\wp(A) = \wp_{fin}(A) \cup \wp_{cofin}(A)$, the number of elements of A not fixed by σ is finite. Therefore, $Ker f$ is formed by bijections of A which keep fixed all but finitely many atoms.

We also have that S_A is locally finite. Indeed, let $\pi_1, \ldots, \pi_k \in S_A$ such that π_1 permutes the atoms from a finite subset of A named $U_1, \ldots$, π_k permutes the atoms from a finite subset of A named U_k. Let $U = U_1 \cup \ldots \cup U_k$. Then each of $\pi_1, \ldots, \pi_k$ is a permutation of U. If u is the finite cardinality of U, then we obtain that $[\{\pi_1, \ldots, \pi_k\}] \leq S(U) \cong S_u$, and of course, $[\{\pi_1, \ldots, \pi_k\}]$ is finite.

Now, since $Ker f \leq S_A$ and S_A is locally finite, we have that $Ker f$ is locally finite (see [45]). There is an isomorphism from $G/Ker f$ to $Im(f) \leq S_r$; the proof is similar to the proof of the fundamental isomorphism theorem for groups. We know that for a finitely generated group G, every subgroup of finite index in G is finitely generated (see [45]). Since $Ker f$ is a subgroup of finite index in G, we have that $Ker f$ is finitely generated. Since $Ker f$ is also a locally finite group, then $Ker f$ is finite. Since $Ker f$ and $G/Ker f$ are both finite, it follows that G is finite; the proof is similar to the proof of Lagrange Theorem (we can choose a system of representatives for the set of left cosets modulo $Ker f$ since $G/Ker f$ is finite). □

13.3 Axiomatic Construction of RFM

In Section 13.2, to define P_A-sets, we assumed that A is a fixed ZF set formed by elements whose internal structure is irrelevant (it is not taken into consideration). Actually, such elements have the same property as atoms in ZFA (having no internal structure). In the following sections we will consider A to be the set of atoms in ZFA. This does not change at all the proofs of the results in Section 13.2, and so these results are preserved in the framework described below over ZFA.

According to the finite support axiom in FM set theory, each subset of A has to be finitely supported, and so only finite or cofinite sets of atoms are allowed in the FM universe. The finite support axiom of FM set theory is very strong, meaning it requires the existence of finite supports for any FM construction. One can think to study the consequences of replacing this strong axiom with a more relaxed one stating that each subset of A is either finite or cofinite. Obviously, 'finite-cofinite' is easier to manipulate than 'finitely supported'. The aim of this approach is to replace the requirement 'finite support for all sets (built on a cumulative hierarchy)' with 'finite support only for subsets of atoms' in order to obtain some similar results as in the FM case. In this sense, although the finite support axiom from FM set theory is relaxed in this new theory (i.e. we no longer require that each bijection of A is finitely supported), several properties of the group of all bijections of A, such as torsioness or local finiteness are preserved. In the same way as FM is the correspondent of FSM over ZFA, RFM is the correspondent of the universe of P_A-sets when A is assumed to be the set of atoms in ZFA, and additionally, A is assumed to be amorphous.

Definition 13.2 The following axioms give a complete characterization of Relaxed Fraenkel-Mostowski set theory:

1. $\forall x.(\exists y.y \in x) \Rightarrow x \notin A.$ (only non-atoms can have elements)
2. $\forall x,y.(x \notin A\, and\; y \notin A\; and\; \forall z.(z \in x \Leftrightarrow z \in y)) \Rightarrow x = y.$ (axiom of extensionality)
3. $\forall x,y.\exists z.z = \{x,y\}.$ (axiom of pairing)
4. $\forall x.\exists y.y = \{z \,|\, z \subseteq x\}.$ (axiom of powerset)
5. $\forall x.\exists y.y \notin A\, and\, y = \{z \,|\, \exists w.(z \in w\, and\, w \in x)\}.$ (axiom of union)
6. $\forall x.\exists y.(y \notin A\, and\, y = \{f(z) \,|\, z \in x\}),$
   for each functional formula $f(z)$ (axiom of replacement)
7. $\forall x.\exists y.(y \notin A\, and\, y = \{z \,|\, z \in x\, and\, p(z)\}),$
   for each formula $p(z)$ (axiom of separation)
8. $(\forall x.(\forall y \in x.p(y)) \Rightarrow p(x)) \Rightarrow \forall x.p(x).$ (induction principle)
9. $\exists x.(\emptyset \in x\, and\, (\forall y.y \in x \Rightarrow y \cup \{y\} \in x)).$ (axiom of infinite)
10. *A is not finite.*

11' *Each subset of A is either finite or cofinite.* (axiom of the structure of A)

The RFM axiomatic description means that RFM set theoretical constructions (obtained by applying axioms 1-9) are added only respecting the general requirement that *each construction is only added if this construction coupled with everything that has already been added does not contradict axiom 11'.*

Actually, the reader can remark that axiom 11' presented above replaces axiom 11 in the description of the FM set theory, the rest of the axioms being unchanged (they are also axioms of ZFA). More precisely, by introducing axiom 11', we do not require in RFM the existence of a finite support for all higher-order constructions as in the FM case, but only for the elements of the powerset of A. Thus, RFM axioms are ZFA axioms together with axiom 11' (which is a 'finite support requirement for $\wp(A)$', and is weaker that axiom 11 in the FM set theory), while FM axioms are ZFA axioms together with axiom 11 (which is 'finite support requirement for any set theoretical construction'). *We conclude that RFM set theory is actually ZFA set theory where the set of atoms is assumed to be amorphous.* We also note that in RFM we do not require any set theoretical construction to be either finite or cofinite, but only the subsets of atoms (i.e. only the elements in $\wp(A)$) should satisfy this requirement; higher-order constructions in RFM could be simultaneously infinite and coinfinite, as long as their construction does not contradict axiom 11'. Thus, the universe of RFM is obtained by excluding from the ZFA universe those subsets of A which are simultaneously infinite and coinfinite (such subsets of A cannot be involved in a higher-order construction in RFM). Moreover, the RFM sets are not necessarily finitely supported; only the powerset of A is finitely supported.

13.4 The World of Either Finite or Cofinite Atomic Sets

It is important to remark that in RFM set theory the requirement 'to be either finite or cofinite' is applied only for the elements of $\wp(A)$ (i.e. only for the subsets of A) and not for all sets with atoms. If we involved such a requirement for all atomic sets, we would obtain the Finite-Cofinite Mathematics over atoms (FCM), namely a refinement of ZFA formed only by either finite or cofinite atomic sets and whose model $FCM(A)$ is defined as below.

- $FCM_0(A) = \emptyset$;
- $FCM_{\alpha+1}(A) = A + \wp_{fc}(FCM_\alpha(A))$ for every non-limit ordinal α;
- $FCM_\lambda(A) = \bigcup_{\alpha<\lambda} FCM_\alpha(A)$ (λ a limit ordinal);
- $FCM(A) = \bigcup_{\alpha} FCM_\alpha(A) \cup \mathcal{V}$.

where $\mathcal{V}$ is the model of ZF set theory described in Section 2.5 and $\wp_{fc}(X)$ represents the family of those subsets of X that are either finite or cofinite (i.e. $\wp_{fc}(X) = \wp_{fin}(X) \cup \wp_{cofin}(X)$). The requirement 'to be either finite or cofinite' is applied only to atomic sets (i.e. only to those sets that contain atoms in their construction), and so the non-atomic (trivial invariant) ZF sets do not necessarily be either finite or cofinite since the entire universe $\mathcal{V}$ of all ZF sets is contained in $FCM(A)$. Thus, all ZF sets belong to $FCM(A)$. An element of $FCM(A)$ is either a classical ZF set, an atom or an hereditary finite or cofinite atomic set. We know that $\wp_{fc}(A) = \wp_{fs}(A)$. Clearly, $FCM(A) \subset \mathcal{V}(A)$. According to Proposition 2.5 and Corollary 2.3, whenever the elements $x_1,\dots,x_{n-1},x_n$ from a finitely supported set $X \in \mathcal{V}(A)$ are finitely supported by $S_1,\dots,S_{n-1}$ and S_n, respectively, under the action $\cdot$ on $\mathcal{V}(A)$, then $\{x_1,\dots,x_n\}$ is supported by the finite set of atoms $S_1 \cup \dots \cup S_n$, and the complement of $\{x_1,\dots,x_n\}$ is supported by $supp(X) \cup S_1 \cup \dots \cup S_n$. Since any finite or cofinite family of atoms is finitely supported, meaning that $\wp_{fc}(A)$ is an invariant set (because permutations of atoms are injective functions, and so for all permutations of atoms π we have that $\pi \cdot Y$ is finite whenever Y is a finite set of atoms and $\pi \cdot Z$ is cofinite whenever Z is a cofinite set of atoms), by structural induction it follows that any element in $FCM(A)$ is finitely supported. Furthermore, for any ordinal α, $FCM_\alpha(A)$ is an invariant set. Thus, FCM is contained in FM, meaning that any FCM set is hereditary finitely supported.

13.5 Properties of Atomic Sets in RFM

The ZF choice principles described in the first part of Chapter 3 can be directly reformulated in RFM by replacing 'ZF' with 'ZFA'.

Theorem 13.3 *The choice principles **AC, ZL, HP, DC, CC, PCC, Fin, PIT, UFT, OP, KW, RKW, OEP, SIP, FPE, AC(fin)** and **GCH** are inconsistent with the axioms of RFM set theory, meaning that the negation of each of the above principle is a logical consequence of RFM axioms.*

Proof (Sketch) RFM set theory is the ZFA set theory where amorphous sets are allowed. According to axiom 11' in RFM set theory, for each subset X of the infinite set A we have that either X is finite or $A \setminus X$ is finite. Therefore, the statement "Every infinite set X has an infinite subset Y such that $X \setminus Y$ is also infinite" is false in the RFM framework. The existence of ZF amorphous sets is in contradiction with **AC(fin)** [25]. This result is preserved in ZFA (proof omitted). Also, amorphous sets cannot be totally ordered. Let us assume now that **Fin** is a valid statement in the RFM framework. Then we can find an injection $f : \mathbb{N} \to A$. Obviously, $f(2\mathbb{N})$ and $f(2\mathbb{N}+1)$ are disjoint, infinite subsets of A. Therefore, $f(2\mathbb{N})$ is infinite and coinfinite, and this contradicts the structure of A. Thus, **Fin** fails in the RFM framework. The rest of the theorem remains as an exercise. □

Remark 13.2 Note that the proof of Theorem 13.3 is consistent with the RFM axioms. However, such a proof cannot be made in FSM because in FSM only finitely supported objects are allowed. Although axiom 11' is a direct consequence of axiom 11 of FM set theory, a result obtained in the RFM framework remains valid in the FM framework only if all the objects appearing in its proof are finitely supported according to canonical hierarchically defined S_A-actions. More exactly, the required implications in the proof of Theorem 13.3 are not necessarily valid in FSM, unless we can reformulate them in terms of finitely supported objects. A proof of an FSM result (and of an FM result) should involve only finitely supported constructions, i.e. it should be internally consistent in the world of finitely supported structures and not retrieved from ZF or ZFA. This is the reason why we do not present Theorem 3.1 as a consequence of Theorem 13.3. More details are in Remark 3.1. Actually, we want to say that although A is amorphous in ZF (or in ZFA if A is associated with the set of atoms from ZFA), regarding the inconsistency of **AC(fin)**, we would obtain that there exists an atomic ZF set X having an infinite family $\mathscr{F}$ of finite subsets with no choice function, but we would not be able to establish directly whether X and $\mathscr{F}$ are finitely supported under the canonical permutation action in order to obtain the inconsistency of **AC(fin)** with FSM. However, this reasoning works for establishing the inconsistency of **AC(fin)** with RFM, where the finite support principle is no longer required.

Remark 13.3 Although the finite support axiom is relaxed in RFM, we are able to prove that some specific FM properties of (structures involving) atoms are preserved by replacing 'FM' with 'RFM'. In this sense, the following FM properties of atoms remain valid in RFM although the finite support axiom of FM set theory is replaced by axiom 11' of RFM set theory.

Let us consider the following FM properties:

1. Every FM one-to-one mapping $f : A \to A$ is also onto.
2. There does not exist an FM one-to-one mapping from $A \times A$ to A.
3. Any FM function $f : A \to \mathbb{N}$ has the property that $Im(f)$ is finite.
4. There does not exist an FM total order relation on A.
5. Each subgroup of P_A which is finitely generated is also finite, where P_A is the set of all FM bijections of A onto A.
6. Choice principles from Chapter 3 are inconsistent with the FM axioms.

The above properties have an RFM correspondent. This assertion can be proved by following the proofs of the properties of P_A-sets in Section 13.2. Adapting the related proofs to RFM does not imply any technical difficulty.

1. In RFM, every one-to-one mapping $f : A \to A$ is also onto.
2. In RFM, there does not exist a one-to-one mapping from $A \times A$ to A.
3. In RFM, any function $f : A \to \mathbb{N}$ has the property that $Im(f)$ is finite.
4. There does not exist a total order relation on A in RFM.
5. In RFM, each subgroup of P_A which is finitely generated is also finite, where P_A is the set of all bijections of A onto A.
6. The choice principles proved to be inconsistent in FM are inconsistent in RFM.

However, there exist some properties of atoms that are valid in FM, but they do not necessarily remain valid in RFM. In this sense, the following results are valid in FM according to the finite support requirement. If the finite support requirement is no longer involved, their validity is no longer preserved.

1. An FM surjective function $f : A \to A$ is also injective.
2. Every FM injective mapping $f : A \to A$ is a finite permutation of A.
3. Any FM injection $f : \wp_{fs}(A) \to \wp_{fs}(A)$ is also surjective.
4. There is no FM injection $f : \wp_{fs}(A) \times \wp_{fs}(A) \to \wp_{fs}(A)$.
5. Every FM injection $f : P_A \to P_A$ is also surjective.
6. There does not exist an FM injection $f : A \times A \to \wp_{fs}(A)$.
7. There does not exist an FM total order relation on $\wp_{fs}(A)$.
8. A function $f : A \to X$ between A and an arbitrary infinite ZF set X constructed without involving atoms is well-defined in FM if and only if there exists $x \in X$ such that $\{y \in A \mid f(y) \neq x\}$ is finite.
9. $\wp_{fin}(A)$ is infinite, but $|\wp_{fin}(A)| \neq 2|\wp_{fin}(A)|$, where $|Y|$ generally denotes the cardinality of Y.
10. Let $f : \wp_{fin}(A) \to \wp_{fin}(A)$ be an FM monotone function. Then there exists a least $X \in \wp_{fin}(A)$ such that $f(X) = X$.

13.6 Pictorial Summary

With the notations used in this book (where by ZF we actually understand the non-atomic ZF framework, i.e. all those hierarchical constructions over empty set that do not involve atoms), we have the following diagram that presents the relationships between the structures constructed in various set theories presented here. In the diagram below, A is considered to be the set of atoms in ZFA obtained by weakening the ZF axiom of extensionality. A ZFA set containing no atoms is called 'non-atomic' (such a set is a ZF set, i.e. a set hierarchically defined over $\emptyset$), while a set containing atoms is called 'atomic' (such a set is hierarchically defined over $\emptyset$ and over the elements of A, and contains at least one atom somewhere in its structure).

In the concentric circles presented below we have:

- **ZF** is represented by the empty set and all non-atomic sets (constructed over the empty set).
- **FCM** is formed by all ZF sets, all atoms and all hereditary finite or cofinite atomic sets.
- **FM** is formed by all ZF sets, all atoms and all hereditary finitely supported atomic sets.
- **RFM** is formed by all ZF sets, all atoms and all atomic sets constructed by respecting the requirement that any subset of A is either finite or cofinite.
- **ZFA** is formed by all ZF sets, all atoms and all atomic sets, i.e. ZFA is formed by all atoms and all sets hierarchically constructed over empty set and over atoms.

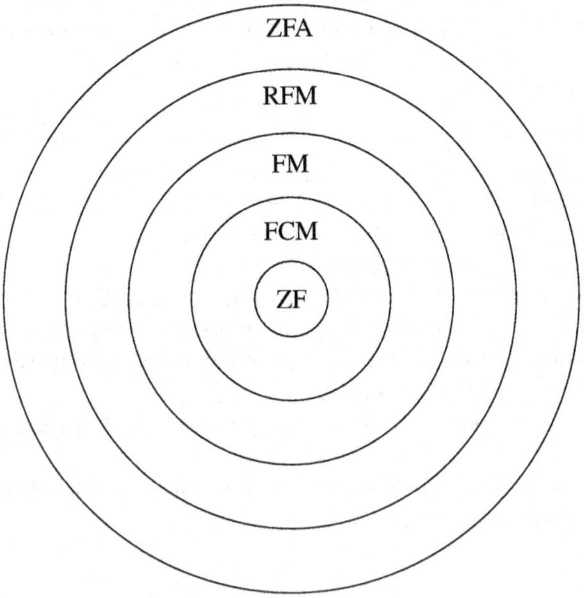

Looking to the above diagram:

- ZF is formed by those trivial invariant sets (i.e. non-atomic sets) that are equipped with the permutation action $(\pi,x) \mapsto x$.
- In FCM\ZF we find the atoms and the atomic constructions which are either finite or cofinite subsets of the atomic constructions in the upper theories of this diagram.
- In FM\FCM we find the atomic structures with (hereditary) finite supports which are neither finite nor cofinite subsets of the upper-order structures in the FCM hierarchical construction. For example, if we consider $\wp_{fs}(\wp_{fs}(A)) = \wp_{fs}(\wp_{fc}(A))$ and fix an element $x_0 \in A$, the family $\{\{x_0,a\} \,|\, a \in A, a \neq x_0\}$ is finitely supported (it is an element of $\wp_{fs}(\wp_{fc}(A))$ supported by $\{x_0\}$), but it does not belong to $\wp_{fc}(\wp_{fc}(A))$ since it is a subset of $\wp_{fc}(A)$ which is simultaneously infinite and coinfinite.
- In RFM\FM we can find higher-order non-finitely supported atomic structures. More precisely, in RFM only the elements of $\wp(A)$ should necessarily be finitely supported (and so only finite and cofinite subsets of A are allowed in RFM), while higher-order constructions could be non-finitely supported (as long as axiom 11' is not contradicted) meaning that there may exist RFM sets that are not FM sets.
- In ZFA\RFM we have those subsets of A which are simultaneously infinite and coinfinite, as well as the higher-order constructions involving them.

According to the approach presented in this book, FSM is assimilated to FM with the mention that we actually do not need an alternative set theory in order to describe FSM, because we can choose A as fixed ZF set formed by elements whose internal structure is not taken into account (similarly as the set of atoms in ZFA obtained by weakening the ZF Axiom of Extensionality). In the diagram above FSM and FM should describe the same circle when the set of atoms in ZFA is replaced by the fixed set A (adjoined to ZF) used in the description of FSM (in this case FCM, RFM and ZFA being adapted over such A). Similarly as FM, FSM involves only constructions that are internally consistent with respect to the finite support requirement (meaning that every sub-structure of an FSM structure should be itself finitely supported). Non-finitely supported structures are not allowed in FSM and cannot be involved nor even in intermediate steps of a proof. The set A used in order to define FSM is actually a separately emphasized ZF set. More exactly, the internal structure of the elements of A is irrelevant (it is not taken into consideration), the only relevant attribute being their identity, and this fact allows us to emphasize the elements of A (called atoms) distinctly from the ZF constructions (that are non-atomic). Actually, any ordinary ZF set (except $\emptyset$) has an internal structure, while the internal structure of the elements of the fixed set A is no longer taken into consideration, meaning that the elements of A become basic elements (similarly as $\emptyset$) in a hierarchical construction of sets with atoms. In the scheme above, in the first internal circle, ZF represents the family of all non-atomic constructions, while the hierarchical atomic constructions are emphasized in the larger circles. As claimed in this book, FSM is divided into two parts: it contains both the family of non-atomic structures (identified with classical ZF structures), and the family of atomic structures with (hereditary) finite support (containing sub-structures hierarchically constructed from A under the rules in Proposition 2.2).

Finally, it is worth noting that it is not necessary to consider a different notion of set theory in order to study finitely supported subsets of the invariant sets. Thus, when dealing with FSM, actually we formally work on standard ZF set theory (by considering a fixed infinite set A formed by elements whose internal structure is irrelevant) rather than on non-standard FM set theory. In this sense FSM sets are subsets of classical ZF sets equipped with canonical actions of the group of all permutations of A satisfying a finiteness requirement. However, all the results in FSM can be immediately transferred in FM, just by considering A as the set of atoms in ZFA, because, exactly as in the FM framework, the constructions in FSM should be made according to the finite support requirement. To summarize, regarding FSM we followed Pitts' approach on nominal sets. In fact, invariant sets are nominal sets (when the set A is assumed to be countable), the term 'invariant' being used instead of 'nominal' as motivated by Tarski's approach on logicality. The original part of the book comes from the foundations of finitely supported algebraic constructions over invariant sets. Furthermore, we intended to provide a set theoretical approach accessible to a broad audience rather than a categorical approach, and we focused on the mathematical foundations of FSM by studying if results in standard set theory (regarding cardinality, infinity, choice, fixed points, etc) remain valid for atomic sets with finite support. We also emphasized specific properties of atomic sets induced by the finite support requirement (such as the existence of fixed points for monotone self-mappings over $\wp_{fin}(A)$, the existence of an iterative calculation of the least and of the greatest fixed points of a monotone self-mapping over $\wp_{fs}(A)$ by using the values of the mapping on $\emptyset$ and A, etc).

Chapter 14
Conclusion

Abstract It is presented a summary of the main ideas and results.

FSM is an abbreviation for the theory of finitely supported algebraic structures, related to the theory of nominal sets and to the theory of FM sets. FSM represents a recently developed research line that introduces an alternative definition of sets (more exactly, sets are equipped with permutation actions), and provides a first step in dealing with infinite atomic algebraic structures (that are finitely supported modulo certain canonical permutation actions). More exactly, the elements outside the support of an (infinite) atomic FSM object are somehow 'similar', and the related object is characterized by its finite support. FSM is consistent with respect to the finite support requirement stating that every logical construction involving elements from a previously fixed ZF set of basic elements A (called the set of atoms by analogy with the ZFA approach) should be finitely supported under the hierarchically constructed canonical group action of the group of all permutations of A. Actually, FSM represents ZF set theory reformulated by informally replacing '(non-atomic) ZF object' with '(atomic) finitely supported object'. FSM can be adequately reformulated over ZFA if the fixed ZF set A is replaced by the set of atoms in ZFA; this is possible because we did not require an internal structure of the elements of A. Ordinary ZF sets (i.e. sets defined without involving elements of A) are trivial invariant sets. An FSM set is either an invariant set (left unchanged under the effect of each permutation of atoms), or a finitely supported subset of an invariant set (left unchanged under the effect of each permutation of atoms fixing its finite support pointwise). This definition corresponds to the FM approach where sets are hereditary finitely supported subsets of the invariant set $FM(A)$ (in this case A denoting the set of atoms from ZFA). Generally, an algebraic structure defined involving atoms is invariant (or finitely supported) if it can be represented as an invariant set (or as a finitely supported subset of an invariant set) endowed with an equivariant algebraic law/relation (or with an algebraic law/relation which is finitely supported as a subset of an invariant set).

The motivation for studying FSM comes from the idea of modelling infinite algebraic structures that are hierarchically constructed from atoms in a finitary manner,

A. Alexandru, G. Ciobanu, *Foundations of Finitely Supported Structures*,
https://doi.org/10.1007/978-3-030-52962-8_14

by analyzing the finite supports of these structures. FSM also has roots in computer science, where FM sets and nominal sets are used in various areas such as semantics foundation, software verification and proof theory. Moreover, a relaxed notion of 'finiteness' (obtained from the study of generalized nominal sets) called 'orbit-finiteness' (i.e. finiteness modulo the orbits of a symmetric group action) is used in order to study automata, languages or Turing machines that operate over infinite alphabets. The study of FSM algebraic structures (that are FSM sets with consistent FSM internal operations), particularly FSM groups, FSM multisets, FSM lattices and FSM Galois connections have already triggered significant applications in classical algebra, in domain theory, in the theory of abstract interpretation, or in topology (see [7]). Significant applications of nominal sets (that are invariant sets defined over countable families of atoms) are also presented in [44].

A natural question is whether the properties of classical ZF sets (which are trivial invariant sets) remains valid in the framework of atomic FSM sets (which are finitely supported subsets of non-trivial invariant sets). More exactly, our goal was to study if a ZF result remains valid when replacing '(trivial invariant) non-atomic ZF structure' with '(possibly non-trivial) atomic finitely supported structure (under the canonical hierarchically constructed action of the group of all one-to-one transformations of A onto itself)'. In this book, our goal is to present the foundations of FSM algebraic structures (as well as the foundations of FM sets and nominal sets) by employing a set theoretical viewpoint, accessible even to graduate students. We proved that there exist ZF results that can be naturally translated into FSM (i.e. from the 'non-atomic' ZF framework into the 'atomic' framework of objects with finite support). However, not all the ZF results have a correspondent in FSM, and not all the FSM results are natural extensions of related ZF results.

Details regarding the construction of finitely supported structures, together with a meta-theoretical presentation and several limitations regarding the transferability of the results from ZF to FSM are presented in Chapter 1. Basic (old and new) results regarding FSM sets and FSM constructions (i.e. FSM subsets, FSM unions, FSM Cartesian products, FSM relations, and FSM functions) are described in Chapter 2. The Fraenkel-Mostowski axiomatic set theory is integrated in the general framework FSM in the same chapter. According to Theorem 3.1 from Chapter 3, the choice principles **AC** (axiom of choice), **HP** (Hausdorff maximal principle) **ZL** (Zorn lemma), **DC** (principle of dependent choice), **CC** (principle of countable choice), **PCC** (principle of partial countable choice), **AC(fin)** (axiom of choice for finite sets), **Fin** (principle of Dedekind finiteness), **PIT** (prime ideal theorem), **UFT** (ultrafilter theorem), **OP** (total ordering principle), **KW** (Kinna-Wagner selection principle), **OEP** (order extension principle), **SIP** (principle of existence of right inverses for surjective mappings), **FPE** (finite powerset equipollence principle) are inconsistent with FSM. **GCH** (generalized continuum hypothesis) is also inconsistent with FSM. Proving these results was not an easy task since the ZF relationship results between choice principles cannot be involved directly in order to prove the inconsistency of these principles in FSM. In FSM all relationship results between various choice principles should be independently reproved according to the finite support requirement in order to remain valid also when finitely supported atomic structures are involved. Thus, we analyzed each of the previously mentioned choice principles separately, and proved their inconsistency with the finite support requirement in FSM. It follows as a simple consequence (because FSM can be adequately reformulated over ZFA) that these principles are also inconsistent with the axioms of FM set theory. Moreover, the inconsistency results presented in this book are also valid in the framework of nominal sets developed in [44]. The logicality of the FSM approach was proved in Chapter 4 by establishing that invariant sets are logical notions in Tarski sense, while the FSM sets also satisfy a weaker form of logicality (i.e. they are invariant under those permutations of atoms that fix their support pointwise).

An FSM theory of partially ordered sets was developed in Chapter 5 and Chapter 6. The preorder relation $\leq$ on FSM cardinalities defined by involving finitely supported injective mappings is antisymmetric, but not total. The preorder relation $\leq^*$ on FSM cardinalities defined by involving finitely supported surjective mappings is not antisymmetric, nor total. Thus, the Cantor-Schröder-Bernstein theorem (in which cardinalities are ordered by involving finitely supported injective mappings) is consistent with the finite support requirement of FSM. However, the dual of Cantor-Schröder-Bernstein theorem (in which cardinalities are ordered by involving finitely supported surjective mappings) is not valid for finitely supported structures. Furthermore, neither the First Trichotomy Principle for cardinalities, nor the Second Trichotomy Principle for cardinalities is valid in FSM. Other specific FSM order properties for cardinalities were presented in Chapter 5, while we proved that some arithmetic properties of cardinalities are preserved from ZF.

In Chapter 5 and Chapter 6 we also proved that several classical fixed points theorems for partially ordered sets and lattices, namely the Strong Tarski theorem (Theorem 6.2), the Bourbaki-Witt theorem (Theorem 5.6) and the Tarski-Kantorovitch theorem (Theorem 5.8) are preserved in FSM. These theorems can be consistently reformulated according to the finite support requirement and provide new properties of invariant partially ordered sets. We also proved specific fixed points properties (regarding existence and calculability) for self-mappings defined on invariant sets that do not contain infinite uniformly supported subsets. In this sense, we were able to prove that finitely supported monotone (order preserving) self-mappings that are defined on $\wp_{fin}(A)$, on $\wp_{fin}(\wp_{fs}(A))$ or even on $\wp_{fin}(\wp_{fs}(A)^A)$ have least fixed points (that can be iteratively calculated from the value of the related mappings on the empty set), although $\wp_{fin}(A)$, $\wp_{fin}(\wp_{fs}(A))$ and $\wp_{fin}(\wp_{fs}(A)^A)$ are not invariant complete lattices (and so Tarski theorem cannot be directly applied).

Various examples of invariant complete lattices were also emphasized, and their properties were deeply studied in Chapter 7. We particularly mention the finitely supported subsets of an invariant set, the finitely supported functions from an invariant set to an invariant complete lattice, the finitely supported subgroups of an invariant group, and the finitely supported fuzzy subgroups of an invariant group. We also studied FSM groups, and we presented several embedding, correspondence and universality theorems.

The finitely supported Galois connections between invariant partially ordered sets were described in Chapter 8 by generalizing some results from [7]. As an application, by using Pawlak approximation functions, we were able to define the approximations of some finitely supported subsets of a possibly infinite invariant set.

Various definitions for 'FSM infinity' were introduced in Chapter 9, including FSM classical infinity, FSM Kuratowski infinity, FSM Tarski infinity, FSM Mostowski infinity, FSM Dedekind infinity, FSM ascending infinity. We were able to compare them, presenting examples of FSM sets satisfying certain specific infinity properties while not satisfying other infinity properties. We also provided connections with FSM (non-)amorphous sets. The notion of countability was described in FSM in Section 9.2, and we presented connections between countable choice principles and countable union theorem within finitely supported sets.

The purpose of Chapter 10 was to characterize the set A of atoms in FSM (and also its powerset and its finite powerset). Several properties of A are obtained translating classical ZF properties of sets into FSM, by replacing 'ZF object' with 'finitely supported object' (Theorem 10.2). However, the set of atoms (and its powerset) also have some specific FSM properties, such as finiteness properties, choice properties, cardinality properties or fixed point properties (Theorem 10.1). Furthermore, the powerset of atoms satisfies some choice principles such as the prime ideal theorem and the ultrafilter theorem, although these principles are generally not valid in FSM (as proved in Chapter 3). Ramsey theorem for atoms (and so infinite pigeonhole principle for atoms) and Kurepa antichain principle for the powerset of atoms also hold; moreover, they admit constructive proofs (Theorem 10.3). Freshness and abstraction defined in the framework of nominal sets are adapted to our context in Chapter 11 and Chapter 12, respectively.

In Chapter 13 we defined a 'finite-cofinite' set theory (RFM) obtained by relaxing the finite support axiom in the FM set theory (i.e. by requiring only an amorphous structure for the set of atoms). Although in RFM we do not require that any construction is finitely supported, we are able to prove that the set of all permutations of the universe of discourse is locally finite as in the FM approach. The set of atoms in RFM also has similar properties as in the FM case. The related properties of RFM sets were also proved in the framework of P_A-sets that is a ZF alternative to RFM set theory in the same way as FSM is a ZF alternative to FM set theory.

Finally, we note that we proved the non-validity of several classical ZF results when we translated them into FSM by constructing counterexamples represented by finitely supported atomic sets not satisfying the related translated results. We also mention that the non-atomic ZF sets are part of FSM (as they are trivial invariant sets). Thus, the non-validity of the FSM atomic forms of certain classical ZF results does not affect the consistency of their non-atomic forms in ZF (since we constructed relevant counterexamples by involving only structures that use atoms in their construction). This means *no non-atomic ZF result is weakened under the FSM approach*, but there exist classical ZF results (i.e. results obtained for trivial non-atomic sets) that no longer remain valid when extending them from the framework of trivial invariant (non-atomic) structures to the framework of finitely supported atomic structures equipped with canonical permutation actions. We also proved that many ZF results which are originally proved for non-atomic structures remain valid when atomic structures with finite supports are involved. Actually the goal of this book was to study which results in the foundations of set theory remain valid when translating them from the non-atomic ZF framework into the atomic FSM, by replacing 'non-atomic set' with 'atomic finitely supported subset of an invariant set (i.e. atomic FSM set)'.

Below is a table indicating the main results in both non-atomic and atomic FSM.

Result	**Non-Atomic FSM**	**Atomic FSM**
Choice principles	Independent	Not valid
GCH	Independent	Not valid
Construction of reals	Valid	Reals are not atomic
Banach-Tarski paradox	As in ZF	Does not hold
Algebraic Structures - without choice	As in ZF	Can be translated
Definitions of Infinity	Independent	Independent

As a straightforward remark, the results presented in this book are valid in the axiomatic framework of FM sets defined by adding to ZFA an additional axiom of finite support (which claims that for each element x in an arbitrary set we can find a finite set supporting x according to hierarchically constructed group action of the group of all permutation of atoms), as well as in the framework of nominal sets categorically presented in ZF (by considering a fixed countable ZF set A and by equipping the ZF sets with canonical group actions of the group of all permutations of A satisfying a certain finiteness requirement).

References

1. Ajmal, N., Thomas, K.V.: The lattices of fuzzy subgroups and fuzzy normal subgroups. Information Sciences **76**, 1–11 (1994)
2. Ajmal, N., Thomas, K.V.: The lattice of fuzzy normal subgroups is modular. Information Sciences **83**, 199–209 (1995)
3. Alexandru, A.: The Theory of Finitely Supported Structures and Choice Forms. Scientific Annals of Computer Science **28**(1), 1–38 (2018)
4. Alexandru, A., Ciobanu, G.: Nominal fusion calculus. 14th International Symposium on Symbolic and Numeric Algorithms for Scientific Computing, pp. 376–384. IEEE Computer Society Press (2013)
5. Alexandru, A., Ciobanu, G.: Nominal groups and their homomorphism theorems. Fundamenta Informaticae **131**(3-4), 279–298 (2014)
6. Alexandru, A., Ciobanu, G.: Mathematics of multisets in the Fraenkel-Mostowski framework. Bulletin Mathématique de la Société des Sciences Mathématiques de Roumanie **58/106**(1), 3–18 (2015)
7. Alexandru, A., Ciobanu, G.: Finitely Supported Mathematics: An Introduction. Springer (2016)
8. Alexandru, A., Ciobanu, G.: Abstract interpretations in the framework of invariant sets, Fundamenta Informaticae **144**(1), 1–22 (2016)
9. Alexandru, A., Ciobanu, G.: On logical notions in the Fraenkel-Mostowski cumulative universe. Bulletin Mathématique de la Société des Sciences Mathématiques de Roumanie **60/108**(2), 113–125 (2017)
10. Alexandru, A., Ciobanu, G.: Fuzzy sets within Finitely Supported Mathematics. Fuzzy Sets and Systems **339**, 119–133 (2018)
11. Alexandru, A., Ciobanu, G.: On the foundations of finitely supported sets. Journal of Multiple-Valued Logic and Soft Computing **32**(5-6), 541–564 (2019)
12. Alexandru, A., Ciobanu, G.: Infinities within Finitely Supported Structures. arXiv:1902.09570, 1–35 (2019)
13. Alexandru, A. Ciobanu, G.: Finitely supported sets containing infinite uniformly supported subsets. Electronic Proceedings in Theoretical Computer Science **303**, 120–134 (2019)
14. Alexandru, A., Ciobanu, G.: Fixed Point Results for Finitely Supported Algebraic Structures. Fuzzy Sets and Systems, in press, online https://doi.org/10.1016/j.fss.2019.09.014 (2019)
15. Alexandru, A., Ciobanu, G.: Properties of the atoms in finitely supported structures. Archive for Mathematical Logic **59**(1-2), 229–256 (2020)
16. Alexandru, A., Ciobanu, G.: Relaxing the Fraenkel-Mostowski Set Theory, Journal of Multiple-Valued Logic and Soft Computing, in press (2020)
17. Banach, S., Tarski, A.: Sur la décomposition des ensembles de points en parties respectivement congruentes. Fundamenta Mathematicae **6**, 244–277 (1924)
18. Barendregt, H.P.: The Lambda Calculus: Its Syntax and Semantics. North-Holland (1984)

A. Alexandru, G. Ciobanu, *Foundations of Finitely Supported Structures*,
https://doi.org/10.1007/978-3-030-52962-8

19. Barwise, J.: Admissible Sets and Structures: An Approach to Definability Theory. Perspectives in Mathematical Logic vol.7, Springer (1975)
20. Blanchette, J.C., Gheri, L., Popescu, A., Traytel, D.: Bindings as bounded natural functors. In: ACM Symposium on Principles of Programming Languages, pp. 22:1–22:34 (2019)
21. Bojanczyk, M.: Fraenkel-Mostowski sets with non-homogeneous atoms. Lecture Notes in Computer Science vol.7550, 1–5 (2012)
22. Bojanczyk, M., Braud L., Klin, B., Lasota, S.: Towards nominal computation. In: ACM Symposium on Principles of Programming Languages, pp. 401–412 (2012)
23. Bojanczyk, M., Klin, B., Lasota, S.: Automata with group actions. In: 26th Symposium on Logic in Computer Science, pp. 355–364 (2011)
24. Bojanczyk, M., Lasota, S.: A Machine-independent characterization of timed languages. Lecture Notes in Computer Science vol.7392, 92–103 (2012)
25. Brunner, N.: Amorphe Potenzen kompakter Räume. Archiv für Mathematische Logik und Grundlagenforschung **24**, 119–135 (1984)
26. Fraenkel, A.: Zu den Grundlagen der Cantor-Zermeloschen Mengenlehre. Mathematische Annalen **86**, 230–237 (1922)
27. Gabbay, M.J.: A theory of inductive definitions with alpha-equivalence. Ph.D. Thesis, Cambridge University (2001)
28. Gabbay, M.J.: A general mathematics of names. Information and Computation **205**, 982–1011 (2007)
29. Gabbay, M.J., Pitts, A.M.: A new approach to abstract syntax with variable binding. Formal Aspects of Computing **13**(3-5), 341–363 (2001)
30. Gandy, R.: Church's thesis and principles for mechanisms. In: Barwise, J., Keisler, H.J., Kunen, K. (eds.) The Kleene Symposium, pp. 123–148. North-Holland (1980)
31. Gödel, K.: The Consistency of the Axiom of Choice and of the Generalized Continuum-Hypothesis with the Axioms of Set Theory. Annals of Mathematics Studies vol.3. Princeton University Press (1940)
32. Halbeisen, L.: Combinatorial Set Theory, with a Gentle Introduction to Forcing. Springer (2011)
33. Hall, M.: The Theory of Groups. Macmillan (1959)
34. Herrlich, H.: Axiom of Choice. Lecture Notes in Mathematics. Springer (2006)
35. Howard, P., Rubin, J.E.: Consequences of the Axiom of Choice. Mathematical Surveys and Monographs vol.59. American Mathematical Society (1998)
36. Jarvinen, J.: Lattice theory for rough sets. Lecture Notes in Computer Science vol.4374, 400–498 (2007)
37. Jech, T.J.: The Axiom of Choice. Studies in Logic and the Foundations of Mathematics. North-Holland (1973)
38. Klein, F.: Vergleichende Betrachtungen über neuere geometrische Forschungen (A comparative review of recent researches in geometry). Mathematische Annalen **43**, 63–100 (1893)
39. Levy, A.: The independence of various definitions of finite. Fundamenta Mathematicae **46**, 1–13 (1958)
40. Levy, A.: The Fraenkel-Mostowski method for independence proofs in set theory. In: Addison, J.W., Henkin, L., Tarski, A. (eds.) The Theory of Models, pp. 221–228. North-Holland (1965)
41. Lindenbaum, A., Mostowski, A.: Über die Unäbhangigkeit des Auswahlsaxioms und Einiger seiner Folgerungen. Comptes Rendus des Séances de la Société des Sciences et des Lettres de Varsovie **31**, 27–32 (1938)
42. Parrow, P., Victor, B.: The fusion calculus: expressiveness and symmetry in mobile processes. In: 13th Symposium on Logic in Computer Science, pp. 176–185 (1998)
43. Petrisan, D.: Investigations into algebra and topology over nominal sets. Ph.D. Thesis, University of Leicester (2011)
44. Pitts, A.M.: Nominal Sets Names and Symmetry in Computer Science. Cambridge University Press (2013)
45. Rose, J.S.: A Course on Group Theory. Dover Publications (1994)
46. Rotmann, J.: An Introduction to Homological Algebra. Springer (2009)

47. Sher, G.: The Bounds of Logic. MIT Press (1991)
48. Shinwell, M.R.: The fresh approach: functional programming with names and binders. Ph.D. Thesis, University of Cambridge (2005)
49. Tarski, A.: What are logical notions? History and Philosophy of Logic **7**(2), 143–154 (1986)
50. Tarzi, S.: Group actions on amorphous sets and reducts of coloured random graphs. Ph.D. Thesis, University of London (2002).
51. Turner, D.: Nominal domain theory for concurrency. Technical Report no.751, University of Cambridge (2009)
52. Urban, C.: Nominal techniques in Isabelle/HOL. Journal of Automated Reasoning **40**(4), 327–356 (2008)

Index

A. Alexandru, G. Ciobanu, *Foundations of Finitely Supported Structures*,
https://doi.org/10.1007/978-3-030-52962-8

Printed in the United States
by Baker & Taylor Publisher Services